T0189773

Leonard Bolc Piotr Borowik

Many-Valued Logics

1 Theoretical Foundations

Springer-Verlag
Berlin Heidelberg New York
London Paris Tokyo
Hong Kong Barcelona
Budapest

Leonard Bolc
Institute of Computer Science
Polish Academy of Sciences
ul. Ordona 21
PL-01-237 Warszawa

Piotr Borowik
Institute of Mathematics
University of Czestochowa
Al. Armii Krajowej 13/15
PL-42-200 Czestochowa

ISBN 978-3-642-08145-3

45/3140/5 4 3 2 1 0 – Printed on acid-free paper

Preface

In recent years, interest in nonclassical logics has increased considerably, particularly concerning the application of such logics in computational techniques. Topics from the area of many-valued logics have been dealt with in so many papers and articles that it would now be unrealistic to attempt to write a complete monograph covering all of the most important results in this subject. What we present in this volume is therefore a certain selection of the existing theories, reflecting present needs rather than providing an exhaustive exposition. Some classes of many-valued logics are discussed in more detail, others are treated with less concern; some are just mentioned. The same can be said about the methodology of presentation, different for various classes of logics. The way we have chosen to treat the subject material in this first volume has its objective in the intended continuation of our work, the second volume, in which we shall display the methods of automated reasoning using many-valued logics, as well as examples of their application in practice. Just now we wish to encourage those who are interested to read the second part of our book when it appears.

In the present volume we are concerned with finite-valued logics, special attention being paid to three-valued logical calculi. The last three chapters are devoted to fuzzy logics (in a rather specific setting), approximation logics and probability logics. One chapter deals with a formalization of the intuitionistic calculus; apparently, the origins of that calculus have to be sought in the idea of trivalence.

The book is addressed mainly to logicians, mathematicians and computer scientists; it may be of some interest also to specialists in domains other than purely scientific.

We would like to express our special thanks to H. Rasiowa and A. Skowron, whose valuable comments helped us greatly in writing the book. We also owe thanks to A. Szałas as well as to our colleagues from the Institute of Mathematics and Institute of Informatics of the Warsaw University and the Institute of Computer Science of the Polish Academy of Sciences in Warsaw, who have all given us their remarks concerning this text. We also thank Springer-Verlag for the great assistance in editing our book, in particular J.A. Ross who was a critical reviewer of the manuscript and greatly contributed to proper presentation of the mathematical contents of the book. Moreover, we would like to thank M. Wójcik for his help in editing the final version of this book, and A. Lopuch for typing it. The assistance of all these persons greatly contributed to the fact that this book, one of very few publications discussing the theoretical basis of many-valued logics, could be issued in such a comprehensive form.

Warsaw
May 1992

Leonard Bolc
Piotr Borowik

Contents

1	**Preliminaries**	1
1.1	Set Operations	1
1.2	Relations	2
1.3	Partial Functions and Functions	4
1.4	Indexed Families of Sets and Generalized Set Operations	5
1.5	Natural Numbers, Countable Sets	5
1.6	Equivalence Relations, Congruences	6
1.7	Orderings	7
1.8	Trees	9
1.9	Inductive Definitions	10
1.10	Abstract Algebras	12
1.11	Logical Matrices	20
2	**Many-Valued Propositional Calculi**	23
2.1	Remarks on History	23
2.2	The Definition of a Propositional Calculus	25
2.3	Many-Valued Calculi of Łukasiewicz	27
2.4	Finitely Valued Calculi of Łukasiewicz	30
2.4.1	The Formalized Language of Propositional Calculi	30
2.5	Algebraic Characterization of the n-valued Calculi of Łukasiewicz	32
2.5.1	Lattices	32
2.5.2	Quasi-Boolean Algebras and Heyting Algebra	33
2.5.3	Proper Łukasiewicz Algebras	37
2.5.4	The Łukasiewicz Implication	39
2.5.5	Stone Filters in Proper n-valued Łukasiewicz Algebras	41
2.5.6	The Axiom System for the n-valued Propositional Calculus of Łukasiewicz	42
2.6	Many-Valued Calculi of Post	46
2.6.1	Bibliographical Remarks	46
2.6.2	Post Algebras	46
2.6.3	Post Algebra Filters	49
2.6.4	The Axiom System for the n-valued Post Calculus	51
2.6.5	Many-Valued Post Calculi with Several Designated Truth Values	54
2.6.6	Definability of Functors in the n-valued Post Logic	57

3 **Survey of Three-Valued Propositional Calculi** 63

3.1 The Three-Valued Calculus of Łukasiewicz ($Ł_3$) 63
3.2 The Three-Valued Calculus of Bochvar 65
3.3 The Three-Valued Calculus of Finn 66
3.4 The Three-Valued Calculus of Hallden 68
3.5 The Three-Valued Calculus of Åqvist 69
3.6 The Three-Valued Calculi of Segerberg 70
3.7 The Three-Valued Calculus of Piróg-Rzepecka 71
3.8 The Three-Valued Calculus of Heyting 73
3.9 The Three-Valued Calculus of Kleene 74
3.10 The Three-Valued Calculus of Reichenbach 75
3.11 The Three-Valued Calculus of Słupecki 76
3.12 The Three-Valued Calculus of Sobociński 77

4 **Some n-valued Propositional Calculi: A Selection** 79

4.1 The Many-Valued Calculus of Słupecki 79
4.2 The Many-Valued Calculus of Sobociński 82
4.3 The Many-Valued Calculi of Gödel 84
4.4 The Many-Valued Calculus *Cnr* 85

5 **Intuitionistic Propositional Calculus** 95

5.1 The Intuitionistic Propositional Logic in an Axiomatic Setting ... 95
5.2 The Natural-Deduction Method for the Intuitionistic Propositional Logic 98
5.3 Characterization of the Intuitionistic Propositional Logic in Terms of the Consequence Operator Cn_I 100
5.4 Algebraic Characterization of the Intuitionistic Propositional Logic 101
5.5 Kripke's Semantics for the Intuitionistic Propositional Calculus .. 102

6 **First-Order Predicate Calculus for Many-Valued Logics** 105

6.1 The Language of the First-Order Predicate Calculus 105
6.2 Free Variables and Bound Variables 107
6.3 The Rule of Substitution for Individual Variables 108
6.4 Fundamental Semantic Notions 109
6.5 The Many-Valued First-Order Predicate Calculus of Post 113

7 **The Method of Finitely Generated Trees in n-valued Logical Calculi** . . . 123

7.1 Introductory Remarks . . . 123
7.2 Finitely Generated Trees for n-valued Propositional Calculi . . . 123
7.3 The Existence of Models for the Propositional Calculus . . . 130
7.4 Finitely Generated Trees for n-valued First-Order Predicate Calculi . . . 133
7.5 Finitely Generated Trees for n-valued Quantifiers . . . 137

8 **Fuzzy Propositional Calculi** . . . 143

8.1 Introductory Remarks . . . 143
8.2 Fuzzy Sets . . . 143
8.3 Syntactic Introduction . . . 144
8.4 Semantic Basis for Fuzzy Propositional Logics . . . 154
8.5 Remarks on the Incompleteness of Fuzzy Propositional Calculi . . 171
8.6 First-Order Predicate Calculus for Fuzzy Logics . . . 192
8.6.1 Introductory Remarks . . . 192
8.6.2 Generalized Residual Lattices . . . 192
8.6.3 The Language of the Fuzzy First-Order Predicate Calculus . . . 195
8.6.4 Semantic Consequence Operation . . . 199
8.6.5 Syntax of the Fuzzy First-Order Predicate Calculus . . . 202
8.6.6 Syntactic Consequence Operation . . . 203
8.6.7 An Axiom System for the Fuzzy First-Order Predicate Calculus . . 204
8.6.8 Fuzzy First-Order Theories . . . 206

9 **Approximation Logics** . . . 209

9.1 Introduction . . . 209
9.2 Rough Sets . . . 209
9.3 Rough Logics with a Chain of Indistinguishability Relations . . . 212
9.3.1 Basic Concepts . . . 212
9.3.2 Approximate Logical Systems . . . 214
9.3.3 Approximation Theories . . . 219
9.4 Approximation Logics with Partially Ordered Sets of Indiscernibility Relations . . . 221
9.4.1 Plain Semi-Post Algebras . . . 221
9.4.2 Approximation Logic of Type T . . . 225
9.4.3 Approximation Logics of Type T with Many Indiscernibility Relations . . . 228

10 Probability Logics 231

10.1 Introduction 231
10.2 Lukasiewicz' Idea of Logical Probability 232
10.3 An Algebraic Description of Probability Logic 233
10.3.1 Syntax 233
10.3.2 Semantics 234
10.3.3 Constructions 237
10.3.4 Probabilistic Consequence 239
10.4 Axiomatic Approach to Probability Logic 243
10.4.1 Syntax 243
10.4.2 Probability and Probabilistic Consequence 245
10.4.3 Completeness of Probability Logic 247
10.4.4 Applications 252
10.4.5 Unreasonable Inference 253

References 255

Index of Symbols 285

Author Index 287

Subject Index 289

Introduction

The origins of many-valued logics can be traced back to antiquity. In studying ancient Greek philosophy we encounter disputes concerning the problem of the satisfiability of logical propositions. Already in those ancient times it was questioned whether a statement must necessarily bear one of the two features, truth or falsity, and whether there is no other truth status possible. We recognize this as the problem of acceptance of the Law of the Excluded Middle, that $\alpha \vee \sim \alpha$ holds for an arbitrary logical statement α. This problem has to be considered as undecided, up to the present. To accept or reject it is a matter of philosophical standpoint. We are confronted with it each time we meet trouble in setting a line of demarcation or taking a decision. The need for rigid decisions concerning facts or phenomena is imposed by psychological motives; however, this demand can seldom be fulfilled. Therefore, it seems right to accept a logical status other than pure truth or falsity.

This line of thought led J. Łukasiewicz to the construction of his three-valued logic; historically, this was the first many-valued logic devised as a formal system. The third truth value occurring in it can symbolize feasibility, neutrality, indefiniteness or just some intermediate truth status. The idea of trivalence has naturally developed into that of multivalence, and a general many-valued system was also created by Łukasiewicz.

Soon after that, working independently, E. Post presented his many-valued logical system. Post's way of treating the problem was pure formalism. He seems to have paid little attention, if any, to the "logical" interpretation of particular logical values. He spoke of arbitrary elements and functions rather than propositions and functors; his analysis of expressions was purely formal. Apparently, philosophical aspects had no relevance to his considerations. It has to be noted that Post's n-valued system is a direct generalization of the classical two-valued calculus; setting $n = 2$ we obtain classical logic.

Post's systems are functionally complete; Łukasiewiczian systems do not share this property. We will see later that noncomplete logical systems involve certain philosophical subtleties, due to the existence of lexical inexpressibilities. Some three-valued logics are distinguished in this respect.

The works of Łukasiewicz and Post were sources of inspiration for the development of more involved systems, matrix constructions, axiomatizations of propositional calculi, and their methodology. The technique of many-valued logical matrices provides a convenient tool for inspecting the independence of an axiom system. Along with Łukasiewicz and Post, there are several other mathematicians who must be mentioned for their fundamental work on many-valued logics: M. Wajsberg, A. Tarski, J. Słupecki, B. Sobociński, for the early period, and for the more recent time: J. Rosser and A. Turquette, H. Rasiowa, A. Rose and C.C. Chang. Numerous authors have devised various three-valued logics for the solution of very specific theoretical or philosophical problems. Papers by S. Kleene, A. Heyting, K. Gödel and D. Bochvar have to be mentioned in this connection. Other papers of great value, also today, are those of J. Słupecki, N. Martin, A. Rose, E. Foxley and T. Evans, and L. Hardy.

It is not possible to mention all authors and papers of significance for the domain of many-valued logics. The theory is developing fast. Many-valued propositional calculi have become a basis for the creation of systems which should be considered as extensions, generalizations or modifications of those calculi. It is the aim of this book to give a survey of such systems. We begin (Chapter 5) with the intuitionistic calculus, which is now regarded as one of the most important non-classical systems. It has arisen as a result of attempts to axiomatize the three-valued propositional calculus of A. Heyting, which was itself an attempt to formalize the intuitions of L.E.J. Brouwer.

Then (Chapter 8) we pass to a very specific generalization of many-valued logics, the so-called fuzzy logics. Speaking imprecisely and without much going into details, a typical n-valued logic can be viewed as a particular case of fuzzy logic in which the spectrum of values is not spread over a whole real-line interval but is concentrated on a discrete finite set.

New ideas and needs have influenced the development of yet other logical systems, namely, approximation logics and probability logics. Approximation logics (Chapter 9) are adjusted to operating with rough quantities that can only be approached with better or worse accuracy. Probability deduction systems (Chapter 10) enable us to analyze propositional functions from the viewpoint of their likelihood.

Our interest in all these systems, apparently very abstract, is due to the fact that they find application in practice, in automatic inference processes. This practical usefulness has been decisive for the intensive development of these logics.

The present book is intended to be the first volume of a larger work. We wish here to acquaint the reader with theoretical fundamentals of many-valued logics and some of their various modifications. The wide variety of practical applications, together with the methods of automated reasoning using many-valued logics, will be the object of the second volume.

1 Preliminaries

To facilitate clear comprehension of the text and avoid misunderstandings, we begin with a brief survey of the symbolic notation employed in the book. We hope this will make the text easier to read and intelligible to readers having little experience in the use of mathematical symbolism.

1.1 Set Operations

The concept of a set is fundamental for all domains of mathematical science. By sets we shall mean classes of objects, excluding the situations where the notion of a set or the accompanying ideas might result in paradoxes.

Objects that constitute a class will be called its elements. To indicate that an object a is an element of a set A we write: $a \in A$, saying also that a belongs to A.

The converse statement (the negation of $a \in A$) is written as $a \notin A$ or $\sim (a \in A)$.

Suppose A and B are sets. Notation $A \subseteq B$ means that A is contained in B, or A is a subset of B; that is, we write

$$A \subseteq B \quad \text{iff} \quad \forall x(x \in A \Rightarrow x \in B).$$

Sets A and B are considered to be identical, in symbols $A = B$, when they have exactly the same elements:

$$A = B \quad \text{iff} \quad \forall x(x \in A \Leftrightarrow x \in B).$$

Thus, equivalently,

$$A = B \quad \text{iff} \quad A \subseteq B \quad \text{and} \quad B \subseteq A.$$

The symbol $\{x : P(x)\}$ denotes the set of those objects x that satisfy the propositional function $P(x)$. If we wish to restrict attention to objects from a certain given set B, we write $\{x \in B : P(x)\}$. This is the set of those elements of B which satisfy $P(x)$.

The symbols $\{x, y\}, \{x, y, z, \ldots, v\}$ denote, respectively, the unordered pair of elements x, y, and the finite set of elements $x, y, z, \ldots, v$, provided that $x, y, z, \ldots, v$ are pairwise distinct.

A set $\{x\}$ consisting of just one element is called a singleton.

The ordered pair (x, y) whose first element is x and whose second element is y can be defined as follows:

$$(x, y) = \{\{x\}, \{x, y\}\}.$$

Two ordered pairs (x, y) and (u, v) are considered to be identical if and only if $x = u$ and $y = v$. Ordered triples and, in general, n-tuples are defined by

$$(x, y, z) = ((x, y), z),\ (x, y, z, \ldots, u, v) = (((\ldots((x, y)\ldots), u), v).$$

Let A and B be any sets. The set

$$A \cup B = \{x : x \in A \quad \text{or} \quad x \in B\}$$

is called the union (or the join) of A and B; the set

$$A \cap B = \{x : x \in A \quad \text{and} \quad x \in B\}$$

is called the intersection (or the meet) of A and B; the set

$$A \setminus B = \{x : x \in A \quad \text{and} \quad x \notin B\}$$

is the difference of A and B, or the complement of B relative to A.

The empty set is denoted by $\emptyset$.

Let A by any set. The set

$$P(A) = \{Y : Y \subseteq A\}$$

is named the power set of A.

Thus $P(A)$ is the family of all subsets of A, including the empty set.

For any two sets A, B, the class

$$A \times B = \{(x, y) : x \in A \quad \text{and} \quad y \in B\}$$

is called the Cartesian product (or just the product) of A and B.

The product of more factors is defined similarly:

$$A \times B \times \ldots \times C = \{(x, y, \ldots, z) : x \in A, y \in B, \ldots, z \in C\}.$$

The set

$$A^n = A \times A \times \ldots \times A = \{(x_1, x_2, \ldots, x_n) : x_i \in A \quad \text{for} \quad 1 \leq i \leq n\}$$

is called the n-th Cartesian power of A.

1.2 Relations

A two-argument (or binary) relation with domain in a set A and range in a set B is defined as any subset of the product $A \times B$.

Suppose $r \subseteq A \times B$ is a binary relation. The sets

$$D(r) = \{x : \exists y (y \in B,\ (x, y) \in r)\}$$

and

$$D^{-1}(r) = \{y : \exists x (x \in A,\ (x, y) \in r)\}$$

are called the domain and range of r, respectively.

Thus, in rigorous terms, a (binary) relation is a triple (r, A, B), in which A, B are some sets and $r \subseteq A \times B$.

Of course, we do not exclude the case where $A = B$; in fact, this is the most frequently encountered situation. We then say that r is defined in A.

Let $r \subseteq A \times B$ be a relation. Then

$$r^{-1} = \{(x, y) : (y, x) \in r\} \subseteq B \times A$$

is called the inverse relation to r, or simply the inverse (or the converse) of r. Clearly,

$$D(r^{-1}) = D^{-1}(r),\ D^{-1}(r^{-1}) = D(r).$$

Suppose $r \subseteq A \times B$ and $s \subseteq B \times C$ are relations. The composition of r and s is defined as

$$s \circ r = \{(x,y) : \exists z(z \in B,\ (x,z) \in r \quad \text{and} \quad (z,y) \in s)\}.$$

This operation is associative,

$$(r \circ s) \circ p = r \circ (s \circ p),$$

yet it is not commutative.

Let A be any set. The identity relation on A is the set

$$i_A = \{(x,x) : x \in A\}.$$

The full relation in A is the whole product

$$A^2 = A \times A.$$

In most cases, by relation one means a binary relation. However, one can also consider n-argument relations, for an arbitrary n. Formally, an n-argument relation is defined as an $(n+1)$-tuple $(r, A_1, A_2, \ldots, A_n)$ where $A_1, A_2, \ldots, A_n$ are some sets and

$$r \subseteq A_1 \times A_2 \times \ldots \times A_n.$$

Again, the case of $A_1 = A_2 = \ldots = A_n = A$ is not excluded. Then simply $r \subseteq A^n$.

The following terminology is used with regard to binary relations. We say that a relation

$r \subseteq A \times A$ is reflexive in A iff $i_A \subseteq r$;
$r \subseteq A \times A$ is symmetric in A iff $r^{-1} \subseteq r$;
$r \subseteq A \times A$ is transitive in A iff $r \circ r \subseteq r$;
$r \subseteq A \times A$ is antisymmetric in A iff $r \cap r^{-1} \subseteq i_A$;
$r \subseteq A \times A$ is connected in A iff $r \cup r^{-1} \cup i_A = A^2$;
$r \subseteq A \times B$ is univalent on A iff $r^{-1} \circ r \subseteq i_A$.

Accordingly:

$r \subseteq A \times A$ is reflexive in A iff
$(x,x) \in r$ for every $x \in D(r) \cup D(r^{-1})$;
$r \subseteq A \times A$ is symmetric in A iff
$(x,y) \in r$ implies $(y,x) \in r$, for all x, y;
$r \subseteq A \times A$ is transitive in A iff
$(x,y) \in r$ and $(y,z) \in r$ imply
$(x,z) \in r$, for all x, y, z;
$r \subseteq A \times A$ is antisymmetric in A iff
$(x,y) \in r, (y,x) \in r$ do not hold
simultaneously unless $x = y$;
$r \subseteq A \times A$ is connected in A iff
for any x, y, either $(x,y) \in r$ or
$(y,x) \in r$, or $x = y$ holds;
$r \subseteq A \times B$ is univalent on A iff, for any x, y_1, y_2, the conditions
$(x,y_1) \in r$ and $(x,y_2) \in r$ imply $y_1 = y_2$.

Moreover, the following equalities hold for any relations (r, A, B), (s, C, D), (p, E, F):

$$s \circ r = \emptyset \quad \text{wherever} \quad B \cap C = \emptyset,$$
$$r \circ (s \circ p) = (r \circ s) \circ p,$$
$$(r \circ s)^{-1} = s^{-1} \circ r^{-1},$$
$$(s^{-1})^{-1} = s,$$
$$r \circ i_B = i_A \circ r = r.$$

1.3 Partial Functions and Functions

A univalent relation is often called a partial function. Thus, formally, a partial function is a triple (f, A, B), with f satisfying the condition of univalence (see above); we also write $f: A -\!\circ\!\rightarrow B$ to indicate that f is a partial function from A to B. If the domain $D(f)$ is the whole A, we call f a function.

A function $f: A \rightarrow B$ is otherwise called a mapping (or map, or transformation) of A into B.

A mapping $f: A \rightarrow B$ is injective when $f^{-1} \circ f = i_A$; surjective when $f \circ f^{-1} = i_B$; bijective if it is both injective and surjective. A bijective (or injective or surjective) map is also called a bijection (or injection or surjection, respectively). A surjective mapping $f: A \rightarrow B$ is said to map the set A onto B. A bijection is otherwise called a one-to-one map.

It is not hard to show that a function $f: A \rightarrow B$ is injective if, for any pair of functions $g, h: C \rightarrow A$, the equality $f \circ g = f \circ h$ implies $g = h$. Similarly, f is surjective if, for any $g, h: B \rightarrow C$, the equality $g \circ f = h \circ f$ forces $g = h$.

Now suppose that f is a function (or a partial function) from A to B and let $X \subseteq A, Y \subseteq B$. The set

$$f(X) = \{y \in B : \exists x(x \in X,\, y = f(x))\}$$

is called the image of X under f, and the set

$$f^{-1}(Y) = \{x \in A : \exists y(y \in Y,\, y = f(x))\}$$

is the inverse image (or preimage) of Y under f. This notation should not cause ambiguity. The symbol f^{-1} alone stands for the inverse relation to f; when it is a function, f is said to be invertible and f^{-1} is called the inverse function (map, mapping, transformation) to f. Thus $f: A \rightarrow B$ is invertible if and only if it is a bijection. The set of all bijections of a set A onto itself constitutes a group, with composition of maps as group operation.

By an n-argument function (or partial function) defined in a set A, with values in a set B, we mean any function (partial function) from A^n to B.

1.4 Indexed Families of Sets and Generalized Set Operations

Let I and X be any sets. By an indexed family of sets (with I as the index set) we understand any function from I into $P(X)$; the usual notation is

$$(A_i)_{i \in I} \quad \text{or} \quad \{A_i : i \in I\},$$

with A_i (a subset of X) being the value of this function at i, for $i \in I$.

The set

$$\bigcup (A_i)_{i \in I} = \bigcup_{i \in I} A_i = \{x : \exists i (i \in I,\ x \in A_i)\}$$

is called the (generalized) union (join) of the family $(A_i)_{i \in I}$. The set

$$\bigcap (A_i)_{i \in I} = \bigcap_{i \in I} A_i = \{x : \forall i (i \in I \ \text{ implies } \ x \in A_i)\}$$

is called the (generalized) intersection (meet) of the family $(A_i)_{i \in I}$.

The product of the family $(A_i)_{i \in I}$ is defined as the set

$$\prod_{i \in I} A_i = \{f : (f : I \to \bigcup_{i \in I} A_i \ \text{ such that } \ \forall i \in I\, f(i) \in A_i)\}$$

When $A_i = A$ for all $i \in I$, we write A^I for the product $\prod_{i \in I} A_i$ and call it the (generalized) Cartesian power of the set A.

When I is a two-element set, the "generalized" concepts just introduced coincide with the usual operations of join, meet and product (of two sets).

1.5 Natural Numbers, Countable Sets

Positive integers are called natural numbers. It is often convenient to consider 0 as a natural number too. In this book, the symbol N will occur in two meanings: sometimes it will stand for the set of nonnegative integers and sometimes for positive integers only. It will be always clear from the context which is the actual meaning. (In most cases, however, this will be quite irrelevant.)

Initial segments of N will be denoted thus:

$$(n) = \{1, 2, \ldots, n\},$$
$$\bar{n} = \{0, 1, \ldots, n-1\}.$$

A nonempty set A is called finite if there exists a surjection h from (n) onto A, for some positive integer n. When h is a bijection, the number n is called the cardinality of A.

A set A is countable if there exists a surjection h from N onto A. When h is bijective, A is said to be countably infinite.

1.6 Equivalence Relations, Congruences

A relation $r \subseteq A \times A$ is said to be an equivalence relation if and only if it is reflexive, symmetric and transitive:

$$i_A \subseteq r,$$
$$r^{-1} \subseteq r,$$
$$r \circ r \subseteq r.$$

If r is an equivalence relation in a set A and x is any element of A, then the set

$$[x]_r = \{y : (x, y) \in r\}$$

is called the equivalence class, or the coset, of x modulo r. We also say that this class is represented by x, or x is a representative of the class. Clearly, any other element related to x by r is also a representative of that class. If it is clear from the context what relation is being under consideration, the subscript r in $[x]_r$ can be omitted.

The set of all cosets modulo r,

$$A/r = \{[x]_r : x \in A\},$$

is termed the quotient (set) of A modulo r. Thus an equivalence relation induces a partition of A into pairwise disjoint sets, its cosets (equivalence classes). With this partition we associate the natural surjection $h_r : A \to A/r$, called the canonical map of A to A/r and defined by

$$h_r(x) = [x]_r.$$

Let $f : A \to B$ and let $r = f \circ f^{-1}$. Then r is an equivalence relation on A. Moreover, the function f admits decomposition of the form

$$f = m \circ f' \circ h$$

such that

$h : A \to A/r$ is the canonical surjection,
$f' : A/r \to f(A)$ is a bijection,
$m : f(A) \to B$ is the inclusion map.

Let $r \subseteq A \times A$ be any relation. We define r^n for $n \in N$ as follows:

$$r^0 = i_A,$$
$$r^1 = r,$$
$$r^{n+1} = r^n \circ r.$$

The transitive closure of r,

$$r^+ = \bigcup_{n \in N_+} r^n,$$

(where $N_+ = N - \{0\}$) is the least (in the sense of inclusion) transitive relation on A containing r. The union

$$r^* = \bigcup_{n \in N} r^n,$$

called the transitive reflexive closure of r, is the least transitive and reflexive relation on A containing r.

Obviously,

$$r^+ = r \circ r^* = r^* \circ r,$$
$$r^* = i_A \cup r^+.$$

Consequently, if $r \subseteq A \times A$ is any relation, then

$$(r \cup r^{-1})^*$$

is the least equivalence relation on A containing r.

Suppose $r \subseteq A \times A$ is an equivalence relation and let $f : A^n \to A$. We call f a congruence modulo f if and only if, for any elements $x_1, x_2, \ldots, x_n, y_1, y_2, \ldots, y_n$, the conditions $(x_i, y_i) \in r$ for $i = 1, 2, \ldots, n$ entail

$$(f(x_1, x_2, \ldots, x_n), f(y_1, y_2, \ldots, y_n)) \in r.$$

We then also say that r preserves the function f, or r is f-preserving.

1.7 Orderings

Let A be a nonempty set. By an ordering of A (or an order in A) we mean a relation $r \subseteq A \times A$ which is reflexive, antisymmetric and transitive; this means that

(1) $i_A \subseteq r$,

(2) $r \cap r^{-1} \subseteq i_A$,

(3) $r \circ r \subseteq r$.

A set A with a given order relation is called an ordered set. (Thus, rigorously, an ordered set should be rather defined as a pair (A, r), with r an ordering of A.) It is customary to denote order relations by the symbol $\leq$ and to write "$x \leq y$" rather than "$(x, y) \in \leq$".

We say that A is linearly ordered by $\leq$ (or $\leq$ is a linear order in A) if

(4) $r \cup r^{-1} = A^2$,

i.e., for any two elements $x, y \in A$, either $x \leq y$ or $y \leq x$ holds. A linearly ordered set is also called a chain.

Suppose A is an ordered set, with an order relation $\leq$. Let B be any subset of A and let a be an element of A.

We say that a is an upper (or lower) bound for B if every element $x \in B$ satisfies $x \leq a$ (or $a \leq x$, respectively). When a belongs to B and is an upper (or lower) bound for B, we call a the greatest (or least, respectively) element of B.

When a belongs to B and there exists no element $x \in B$ other than a such that $a \leq x$ (or $x \leq a$), we say that a is a maximal (or minimal, respectively) element of B.

If the set of all elements in A which are upper (or lower) bounds for B has a least (or greatest) element, we call it the least upper bound (or greatest lower bound, respectively) of A.

The following is a very important tool in many proofs.

Theorem 1.1 (Kuratowski–Zorn Lemma)

Suppose that A is an ordered set in which every chain has an upper bound. Then there exists a maximal element in A. ∎

(Dually, a valid statement results on replacing "upper" by "lower" and "maximal" by "minimal".)

Let A be an ordered set with an order relation $\leq$. Define strict ordering by

$$x < y \quad \text{iff} \quad x \leq y \quad \text{and} \quad x \neq y.$$

We say that A is well-founded (or well-ordered) if there is no infinite sequence $x_1, x_2, x_3, \ldots$ such that $x_{i+1} < x_i$ for all $i \in N$. Equivalently, an ordered set A is well-founded if and only if every nonempty subset of A has a minimal element.

Now suppose A is a well-founded set and let p be a property characterizing the elements of a certain subset of A. We identify p with the characteristic function of that subset, $p : A \to \{0, 1\}$, so that $p(x) = 1$ if and only if the element x has the property in question (i.e., iff $p(x)$ holds).

Induction Principle

Suppose $p(x)$ holds for every x which is a minimal element of a subset of A (a well-founded set). Further, suppose that if $z \in A$ is such that $p(y)$ holds for all $y < z$ then $p(z)$ holds too. Then $p(x)$ holds for all x in A.

If A is a set ordered by $\leq$, we can define an ordering $\leq^*$ in $A \times A$ as follows:

$$(x, y) \leq^* (z, w) \quad \text{iff either} \quad (x = z \quad \text{and} \quad y \leq w) \quad \text{or} \quad (x < z).$$

We call $\leq^*$ the lexicographic order in $A \times A$. If $(A, \leq)$ is well-founded, then so is $(A \times A, \leq^*)$.

Let now A, B be any sets and let U be the set of all partial functions with domains in A and ranges in B. Let $f, g \in U$. The relation defined by

$$f \subseteq^* g \quad \text{iff} \quad gr(f) \subseteq gr(g),$$

where $gr(f) = \{(x, y) : f(x, y)\}$, is an ordering in U. If $f \subseteq^* g$ holds, we say that g is an extension of f (or g extends f), and f is a restriction of g.

Let $f_1, f_2, \ldots$ be a sequence of partial functions $f_n : A -\!\circ\!\to B$ such that $f_n \subseteq f_{n+1}$ for all n. Writing each f_n formally as a triple (f_n, A, B), we can define a partial function g to be the triple $(\bigcup(f_n)_{n \in N}, A, B)$. This function is the least upper bound of the sequence (f_n), in the sense of the ordering just introduced.

Let A be any set and let n be any positive integer. A sequence of length n (or an n-string, or an n-tuple) of elements of A is defined as a function $u : (n) \to A$.

If we allow n to be 0, we are led to considering a map from the empty set to A. We agree that there is exactly one such map; it is convenient to regard it as the "empty" sequence.

The set of all finite sequences over A (i.e., with terms in A) will be denoted by A^*. The empty sequence over A will be written as e_A. The elements (or terms, or entries) of a sequence u are often (but not always) written as $u_1, u_2, \ldots, u_n$ rather than $u(1), u(2), \ldots, u(n)$.

If $u : (m) \to A, v : (n) \to A$ are two finite sequences ($m, n \geq 0$), we define their concatenation, denoted by uv, to be the sequence of length $m + n$,

$$w : (m+n) \to A$$

whose entries are

$$w(i) = \begin{cases} u(i) & \text{for } \ 1 \leq i \leq m, \\ v(i-m) & \text{for } \ m+1 \leq i \leq m+n. \end{cases}$$

In particular, when either m or n is zero, we have $ue_A = e_Au = u$ for every $u \in A^*$.

Concatenation is an associative operation, not commutative in general. With this operation, the set A^* becomes a semigroup.

Let $u \in A^*$. By a prefix of u we mean any sequence $v \in A^*$ such that $u = vw$ for some $w \in A^*$; by a suffix of u we mean any $v \in A^*$ such that $u = wv$ for some $w \in A^*$. A subsequence (substring) of u is any $v \in A^*$ such that $u = x \vee y$ for some $x, y \in A^*$.

A prefix (suffix, subsequence) v of a sequence u is proper if $v \neq u$.

1.8 Trees

Let N be the set of positive integers. According to the notation of the preceding section, N^* is the set of all finite sequences over N. A nonempty set D in N^* can be taken for a tree domain if:

(a) together with any sequence $u \in D$, every prefix of u belongs to D;

(b) for every $u \in D$ and each $i \in N$, if $ui \in D$ and $1 \leq j \leq i$ then $uj \in D$;

where, of course, ui denotes the concatenation of u and the sequence of length 1 whose unique element is i.

Suppose D is a tree domain and let E be any set, called in the following the set of labels (the label set). By a tree we mean an arbitrary function $t : D \to E$. Then D, the domain of t, is denoted by $D(t)$. Any sequence $u \in D(t)$ is named an address, a node, or a point of the tree.

Let $u \in D(t)$. The degree of u is defined by

$$d(u) = \text{card } \{i : ui \in D(t)\}.$$

Nodes of degree 0 are called leaves. The point defined by the empty address is called the root of the tree. A tree is finite when its domain is a finite set.

Let u be any point of the domain of a tree t. Every point of the form $ui \in D(t)$, $i \in N$, is called an immediate successor of u.

Consider the relation $\leq$ in $D(t)$ defined as follows:

$$u \leq v \quad \text{iff} \quad \begin{array}{l} \text{either } u \text{ is a prefix of } v \\ \text{or there exists sequences } x, y, z \in N^* \\ \text{and integers } i, j \in N \text{ with } i < j \\ \text{such that } u = xiy, v = xjz. \end{array}$$

This is a linear ordering of $D(t)$. (As it generalizes the notion of lexicographic ordering in a Cartesian product $A \times A$, it is also called lexicographic.)

When $u \leq v$ and u is a prefix of v, we say that u precedes v, or v dominates u; when the second condition of the definition of "$u \leq v$" is satisfied, we say that u is left of v.

A finite path with origin u and endpoint v (path from u to v) is any finite sequence of points $u_0, u_1, \ldots, u_n$ such that $u_0 = u, u_n = v$ and, for each j with $1 \leq j \leq n$, there is an $i_j \in N$ such that $u_j = u_{j-1}i_j$. Then n is called the length of the path. A path from the root to a leaf is called a branch or chain.

An infinite path with origin u is defined as an infinite sequence $u_0, u_1, u_2, \ldots$ of points of $D(t)$ such that $u_0 = u$ and, for each j with $j \geq 1$, there exists $i_j \in N$ such that $u_j = u_{j-1}i_j$.

For a finite tree t and any point $u \in D(t)$, by the height of u we mean the number

$$w(u) = \max\{l(p) : p \text{ is a path from } u \text{ to a leaf }\},$$

$l(p)$ denoting the length of p. The height (or depth) of a finite tree t is defined to be the height of its root, i.e., the maximum length of a path from the root to a leaf.

Let $u \in D(t)$, where t is any tree. The subtree of t with root u is defined as the tree t/u with domain $\{v : uv \in D(t)\}$,

$$t/u(v) = t(uv) \quad \text{for} \quad v \in D(t/u).$$

Suppose t_1, t_2 are trees and let $u \in t_1$. The substitution of t_2 to address u in t_1 is the function whose graph is the set

$$\{(v, t_1(v)): \ u \text{ is not a prefix of } v\} \cup \{(uv, t_2(v))\}.$$

It is denoted by $t_1[u \leftarrow t_2]$.

1.9 Inductive Definitions

Induction is an important tool, for proving theorems as well as for defining notions. Numerous definitions in logic, and most of mathematical definitions in computer science, follow the inductive pattern.

An inductive definition consists of two sets of conditions: the initial data and the induction step, often accompanied by additional minimization conditions.

Let A be a nonempty set and let X and Y be subset of A. Further, let F be any set of functions $f : A^n \to A$, n arbitrary natural. We say that Y is inductive over X with respect to F if $X \subseteq Y$ and Y is closed under each function $f \in F$. The set

$$X^+ = \bigcup\{Y : Y \text{ is inductive over } X \text{ with respect to } F\}$$

is called the inductive closure of X under F (with respect to F, relative to F). Clearly, X^+ is the smallest inductive set over X. If $X = \emptyset$ then also $X^+ = \emptyset$.

Let X and F be as above. Define

$$X_0 = X;$$

$$X_{i+1} = X_i \cup \{f(x_1, \ldots, x_n) : n \in N, f : A^n \to A, f \in F; x_1, x_2, \ldots, x_n \in X_i\}$$

for $i = 0, 1, 2, \ldots$, and consider the set

$$X_+ = \bigcup_{i \geq 0} X_i.$$

Evidently, $X_+ = X^+$. This yields the following

Induction Principle

Let Y be a set closed under the operations of F and such that $X \subseteq Y \subseteq X_+$. Then $Y = X_+$.

We say that X_+ is freely generated by the set X and the function family F if:

(a) the restriction of every function $f \in F, f : A^m \to A$, to X_+^m is an injection;

(b) for any two distinct functions $f, g \in F, f : A^m \to A, g : A^n \to A$, we have $f(X_+^m) \cap g(X_+^n) = \emptyset$;

(c) for every $f \in F, f : A^m \to A$, and every m-tuple $(x_1, x_2, \ldots, x_m) \in X_+^m$ we have $f(x_1, x_2, \ldots, x_m) \notin X$.

Setting additionally $X_{-1} = \emptyset$ we see that $X_{i-1} \neq X_i$ for each $i \geq 0$. Moreover, for every $f \in F, f : A^n \to A$, and every $(x_1, x_2, \ldots, x_n) \in X_i^n - X_{i-1}^n$ we have $f(x_1, x_2, \ldots, x_n) \notin X_i$.

Now, let A, X, F, X_+ have the same meaning as above, let B be another nonempty set and let G be a family of finite-argument functions $g : B^n \to B$, n arbitrary natural. Suppose there exists a mapping $d : F \to G$ such that, for any n-argument function $f \in F$, its transform $g = d(f) \in G$ is also an n-argument function. Further, assume that X_+ is freely generated by [illegible] and F. Under these assumptions, every function $h : X \to B$ extends uniquely to a function $h^* : X_+ \to B$ so that the following condition is satisfied: for any $f \in F, f : A^n \to A$, and any $x_1, x_2, \ldots, x_n \in X_+^n$ we have the equality

$$h^*(f(x_1, x_2, \ldots, x_n)) = g(h^*(x_1), h^*(x_2), \ldots, h^*(x_n)),$$

where $g = d(f)$.

In other words, the diagrams

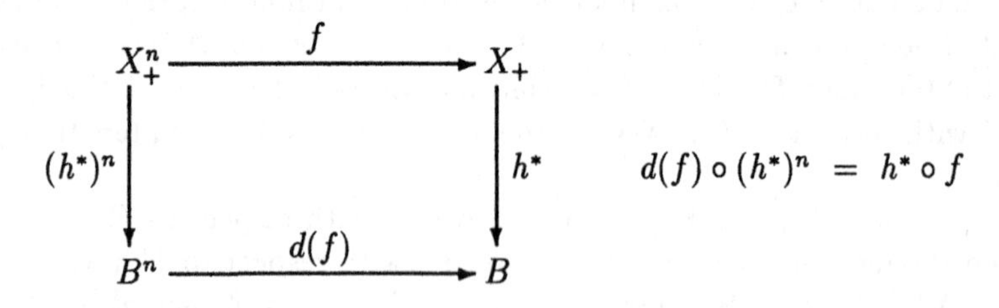

and

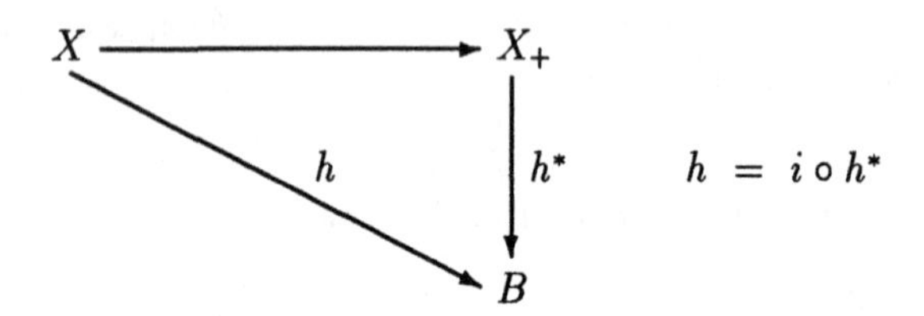

are commuting; here i denotes the inclusion map and $(h^*)^n$ is given by

$$(h^*)^n(x_1, x_2, \ldots, x_n) = (h^*(x_1), h^*(x_2), \ldots, h^*(x_n)).$$

1.10 Abstract Algebras

Let U be any nonempty set. A mapping

$$s : U^m \to U,$$

defined on the Cartesian product of m copies of U, with values in U, is called an m-argument operation in U, or an m-ary operation (for $m = 1, 2, 3, \ldots$: unary, binary, ternary,...); the number m itself is called the arity of s. The case of $m = 0$ is also admitted: each element of U is identified with a constant operation, depending on no arguments at all, hence of arity 0.

A set $W \subseteq U$ is said to be closed under an m-ary operation s in U if

$$s(u_1, u_2, \ldots, u_m) \in W \quad \text{wherever} \quad u_1, u_2, \ldots, u_m \in W.$$

By an abstract algebra (or an algebra) we mean a pair

$$\mathfrak{A} = (U, \{s_i : i \in I\})$$

where U is a nonempty set, called the universe or the underlying space of $\mathfrak{A}$; each s_i is an operation in U and I is some index set. When I is finite, the algebra is denoted by

$$\mathfrak{A} = (U, \{s_1, \ldots, s_k\}) \text{ or just } \mathfrak{A} = (U, s_1, \ldots, s_k).$$

When U is a singleton, $U = \{u\}$, the algebra is degenerate. Obviously, each operation then has the form $s(u, \ldots, u) = u$.

If $\mathfrak{A} = (U, \{s_i : i \in I\})$ is an algebra and if $K \subset I$ is any proper subset of the index set, then we call the algebra $(U, \{s_i : i \in K\})$ a reduct of $\mathfrak{A}$.

Let $\mathfrak{A} = (U, \{s_i : i \in I\})$ be any algebra. By a subalgebra of $\mathfrak{A}$ we mean an algebra

$$\mathfrak{B} = (W, \{s_i : i \in I\})$$

whose universe W is a subset of U, closed with respect to all operations of $\mathfrak{A}$, and the operations in $\mathfrak{B}$ are defined as the restrictions to W of the respective operations in $\mathfrak{A}$.

It is often convenient to identify a subalgebra with the set $W \subseteq U$ which is its universe.

Theorem 1.2

The intersection of any class of subalgebras of a given algebra $\mathfrak{A}$ is a subalgebra of $\mathfrak{A}$.

Proof

Obvious. ∎

Let $\mathfrak{A} = (U, \{s_i : i \in I\})$ be any algebra and let $W \subseteq U$ be a nonempty set. There exists a smallest subalgebra $\mathfrak{B} = (V, \{s_i : i \in I\})$ such that $W \subseteq V$, namely, the intersection of all subalgebras whose universes contain W. We say that $\mathfrak{B}$ is generated by W and call W the set of generators of $\mathfrak{B}$. In particular, if no subalgebra of $\mathfrak{A}$, other than $\mathfrak{A}$ itself, contains the given set W, then W is a set of generators of $\mathfrak{A}$.

Two algebras

$$\mathfrak{A} = (U, \{s_i : i \in I\}) \quad \text{and} \quad \mathfrak{B} = (W, \{p_k : k \in K\})$$

are called similar if $I = K$ and $\arg(s_i) = \arg(p_i)$ holds for each $i \in I$ ($\arg(s)$ denoting the arity of s). We will often denote the corresponding operations by the same symbols.

Suppose $\mathfrak{A} = (U, \{s_i : i \in I\}), \mathfrak{B} = (W, \{p_i : i \in I\})$ are similar. A homomorphism of $\mathfrak{A}$ into $\mathfrak{B}$ is a map

$$h : U \to W$$

such that

$$h(s_i(u_1, u_2, \ldots, u_{\arg(s_i)})) = p_i(h(u_1), h(u_2), \ldots, h(u_{\arg(s_i)})),$$

for each $i \in I$ and any $u_1, u_2, \ldots, u_{\arg(s_i)} \in U$.

A homomorphism of $\mathfrak{A}$ into the same algebra $\mathfrak{A}$ is called an endomorphism.

A homomorphism defined by an "onto" map ($h(U) = W$) is called an epimorphism.

A homomorphism defined by a one-to-one map ($h(u_1) \neq h(u_2)$ wherever $u_1 \neq u_2$) is called a monomorphism.

A homomorphism which is both a monomorphism and an epimorphism is called an isomorphism.

Speaking of homomorphism between algebras, we shall often write $h : \mathfrak{A} \to \mathfrak{B}$ rather than $h : U \to W$, thus avoiding the need to exhibit the universe sets.

Theorem 1.3

If $\mathfrak{A}, \mathfrak{B}, \mathfrak{C}$ are similar algebras and $h : \mathfrak{A} \to \mathfrak{B}, g : \mathfrak{B} \to \mathfrak{C}$ are homomorphisms, then the composition $gh : \mathfrak{A} \to \mathfrak{C}$ is a homomorphism.

Proof

Denoting the operations in $\mathfrak{A}, \mathfrak{B}, \mathfrak{C}$ by s_i, p_i, r_i, respectively, we have for each i (writing $m = \arg(s_i)$):

$$\begin{aligned}(gh)(s_i(u_1, u_2, \ldots, u_m)) = {} & g(h(s_i(u_1, u_2, \ldots, u_m)) \\ & g(p_i(h(u_1), h(u_2), \ldots, h(u_m)) \\ & r_i(g(h(u_1)), g(h(u_2)), \ldots, g(h(u_m)) \\ & r_i((gh)(u_1), (gh)(u_2), \ldots, (gh)(u_m)),\end{aligned}$$

whence the claim. ∎

Theorem 1.4

Let h be a homomorphism of an algebra $\mathfrak{A}$ into a similar algebra $\mathfrak{B}$. If h carries a set of generators of $\mathfrak{A}$ onto a set of generators of $\mathfrak{B}$, then h is an epimorphism.

Proof

Let U be the universe of $\mathfrak{A}$, W the universe of $\mathfrak{B}$. The image $h(U)$ is a subalgebra of $\mathfrak{B}$ and contains the set of generators of $\mathfrak{B}$. Hence $h(U) = W$. ∎

Theorem 1.5

Let G be a set of generators of an algebra $\mathfrak{A} = (U, \{s_i : i \in I\})$ and let h and g be homomorphisms of $\mathfrak{A}$ into a similar algebra $\mathfrak{B} = (W, \{p_i : i \in I\})$. If $h(u) = g(u)$ holds for every $u \in G$, then $h = g$, and $h(u) = g(u)$ for all $u \in U$.

Proof

Consider the set

$$F = \{u \in U : h(u) = g(u)\}.$$

It contains G, by assumption. Further, we have for each $i \in I$ and every $u_1, u_2, \ldots, u_m \in F$ (where $m = \arg(s_i)$)

$$\begin{aligned}& h(s_i(u_1, u_2, \ldots, u_m)) = p_i(h(u_1), h(u_2), \ldots, h(u_m)) = \\ & p_i(g(u_1), g(u_2), \ldots, g(u_m)) = g(s_i(u_1, u_2, \ldots, u_m)),\end{aligned}$$

showing that $s_i(u_1, u_2, \ldots, u_m) \in F$. Thus the set F is closed under each operation s_i and hence defines a subalgebra of $\mathfrak{A}$. And since F contains the generators of $\mathfrak{A}$, it must be the entire algebra. So $f = U$, and we are done. ∎

Theorem 1.6

Let f be a one-to-one mapping from a set G of generators of an algebra $\mathfrak{A}$ onto a set of generators of a similar algebra $\mathfrak{B}$. Suppose there exist homomorphisms $h : \mathfrak{A} \to \mathfrak{B}$ and $g : \mathfrak{B} \to \mathfrak{A}$ such that h extends f and g extends f^{-1}. Then h is an isomorphism of $\mathfrak{A}$ onto $\mathfrak{B}$, and $h^{-1} = g$.

Proof

For $u \in G$ we have

$$(gh)(u) = g(h(u)) = f^{-1}(f(u)) = u.$$

The assertion follows immediately from Theorems 1.3 and 1.5. ∎

Now, let R be a class of pairwise similar algebras. An algebra $\mathfrak{A} \in \mathrm{R}$ is said to be R-free, or free in class R, if it has the following property: there exists a set G of generators of $\mathfrak{A}$ such that every mapping f from G into the universe of any algebra $\mathfrak{B} \in \mathrm{R}$ can be extended to a homomorphism of $\mathfrak{A}$ to $\mathfrak{B}$. The elements of G are then called free generators of $\mathfrak{A}$ (R-free, to be precise), and $\mathfrak{A}$ is said to be freely generated by G.

Theorem 1.7

Suppose $\mathfrak{A}$ and $\mathfrak{B}$ are R-free algebras, freely generated by the respective sets G and F of equal cardinalities. Then $\mathfrak{A}$ and $\mathfrak{B}$ are isomorphic, and, moreover, every one-to-one map f from G onto F extends uniquely to an isomorphism between $\mathfrak{A}$ and $\mathfrak{B}$.

Proof

Immediate, in view of Theorem 1.6. ∎

Let $\mathfrak{A} = (U, \{s_i : i \in I\})$ be any algebra and let $\approx$ be an equivalence relation in U. We call $\approx$ a congruence in $\mathfrak{A}$ if the following condition is satisfied for each $i \in I$:

$$\text{if} \quad u_1, \approx w_1, u_2 \approx w_2, \ldots, u_m \approx w_m, \quad \text{where} \quad m = \arg(s_i),$$
$$\text{then} \quad s_i(u_1, u_2, \ldots, u_m) \approx s_i(w_1, w_2, \ldots, w_m).$$

In other words, a congruence is defined as an equivalence relation which preserves the algebra operations.

Given a congruence $\approx$, we form the quotient set

$$U/\approx = \{[x] : x \in U\}$$

whose elements are the cosets $[x] = \{y \in U : x \approx y\}$. Each operation s_i of the algebra $\mathfrak{A}$ induces an operation s_i^* in $U/\approx$ by

$$s_i^*([u_1], [u_2], \ldots, [u_m]) = [s_i(u_1, u_2, \ldots, u_m)],$$

where $m = \arg(s_i)$; when $\arg(s_i) = 0$, we set $s_i^* = [s_i]$. This is correct (does not depend on the choice of representatives) because $\approx$ is a congruence.

The structure

$$\mathfrak{A}/\approx = (U/\approx, \{s_i^* : i \in I\})$$

is an algebra, similar to $\mathfrak{A}$. We call it the quotient algebra of $\mathfrak{A}$ modulo relation $\approx$.

The mapping

$$h : U \to U/\approx \quad \text{defined by} \quad h(u) = [u] \quad \text{for} \quad u \in U$$

is an epimorphism of $\mathfrak{A}$ onto the quotient algebra, called the natural epimorphism.

Theorem 1.8

If h is any epimorphism of an algebra $\mathfrak{A} = (U, \{s_i : i \in I\})$ onto a similar algebra $\mathfrak{B} = (W, \{p_i : i \in I\})$, then the condition

$$u \approx w \quad \text{iff} \quad h(u) = h(w)$$

defines a congruence in $\mathfrak{A}$. The quotient algebra $\mathfrak{A}/\approx$ is then isomorphic to $\mathfrak{B}$; the isomorphism is established by the map

$$f : U/\approx \to W \quad \text{defined by} \quad f([u]) = h(u) \quad \text{for} \quad u \in U.$$

Proof

Evidently, $\approx$ is an equivalence relation. It remains to verify that it preserves the operations. Thus choose an operation s_i, write $m = \arg(s_i)$, and suppose

$$u_1 \approx w_1, u_2 \approx w_2, \ldots, u_m \approx w_m.$$

Then

$$h(u_1) = h(w_1), h(u_2) = h(w_2), \ldots, h(u_m) = h(w_m).$$

Hence

$$h(s_i(u_1, u_2, \ldots, u_m)) = p_i(h(u_1), h(u_2), \ldots, h(u_m))$$
$$p_i(h(w_1), h(w_2), \ldots, h(w_m)) = h(s_i(w_1, w_2, \ldots, w_m)),$$

showing that

$$s_i(u_1, u_2, \ldots, u_m) \approx s_i(w_1, w_2, \ldots, w_m).$$

Thus $\approx$ is a congruence.

It follows from the definition of f that f is a one-to-one map of $U/\approx$ onto W. To claim that f is an isomorphism, it will be enough to show that f is a homomorphism. Now,

$$f(s_i^*([u_1], [u_2], \ldots, [u_m]) = f([s_i(u_1, u_2, \ldots, u_m)])$$
$$h(s_i(u_1, u_2, \ldots, u_m)) = p_i(h(u_1), h(u_2), \ldots, h(u_m))$$
$$p_i(f([u_1], [u_2], \ldots, [u_m])),$$

ending the proof. ∎

Suppose that $\{\mathfrak{A}_t : t \in T\}$ is an arbitrary class of similar algebras, indexed by t ranging over some index set T. Let $\mathfrak{A}_t = (U_t, \{s_i^t : i \in I\})$. Consider the product

$$U = \prod_{t \in T} U_t.$$

By definition, U consists of all mappings

$$u : T \to \bigcup_{t \in T} u_t \quad \text{with} \quad u_t \in U_t \quad \text{for} \quad t \in T.$$

Clearly, u_t stands for the value of u at t. It is customary to write $u = (u_t)_{t \in T}$.

The algebra structure is naturally introduced in U. Fix $i \in I$, and let $m = m_i$ be the common arity of all operations $s_i^t, t \in T$:

$$m = m_i = \arg(s_i^t) \quad \text{for} \quad t \in T.$$

We define an operation s_i in U (of the same arity m) coordinatewise, as follows:

for $u_1, u_2, \ldots, u_m \in U$,

$u_1 = (U_{1t})_{t\in T}, u_2 = (u_{2t})_{t\in T}, \ldots, u_m = (u_{mt})_{t\in T}$:

$s_i(u_1, u_2, \ldots, u_m) = (s_i^t(u_{1t}, u_{2t}, \ldots, u_{mt}))_{t\in T}$.

The structure that arises, i.e., the algebra

$$\mathfrak{A} = (U, \{s_i : i \in I\}),$$

is called the Cartesian product of the family $\{\mathfrak{A} : t \in T\}$.

Theorem 1.9

Let $\mathfrak{A}$ be the Cartesian product of a family of algebras $\{\mathfrak{A} : t \in T\}$ and let $\mathfrak{B}$ be another algebra, similar to each $\mathfrak{A}_t$. Let W be the universe of $\mathfrak{B}$. Suppose that for each $t \in T$ there is defined a homomorphism of $\mathfrak{B}$ into $\mathfrak{A}_t$. Then the formula

$$h(u) = (h_t(u))_{t\in T} \quad \text{for} \quad t \in W$$

defines a homomorphism of $\mathfrak{B}$ into $\mathfrak{A}$.

Proof

Immediate from the definition of a homomorphism and the definition of operations in the product algebra. ∎

Now we pass to the concept of a generalized algebra.

Let U be any nonempty set and let R be a subset of $P(U)$, the power set of U. It is assumed that the empty set does not belong in R. Any mapping

$$Q : \mathrm{R} \to U$$

will be considered as a generalized operation in U; the class R is its domain. Sets $X \in \mathrm{R}$ will be referred to as admissible for Q. When there is at least one infinite set admissible for Q, we say that Q is an infinite operation. The element of U obtained as a result of applying operation Q to a set $X \in \mathrm{R}$ will be denoted by $Q(X)$.

When $X = \{u_t : t \in T\}$, with T being some index set, we write $Q_{t\in T}u_t$ rather than $Q(X)$.

By a generalized abstract algebra (a generalized algebra), we mean a system

$$\mathfrak{A} = (U, \{s_i : i \in I\}, \{Q_j : j \in J\})$$

in which U is a nonempty set (the universe of $\mathfrak{A}$), each s_i is an operation in U and each Q_j is a generalized operation in U; the index sets I and J can be quite arbitrary.

If the domains of generalized operations Q_j exhaust the whole $P(U) - \emptyset$ as j ranges over J, then the generalized algebra $\mathfrak{A}$ is called complete.

A set $X \subseteq U$ is said to be closed under a generalized operation Q with domain R if

$$Q(Y) \in X \quad \text{wherever} \quad Y \subseteq X, Y \in \mathrm{R}.$$

Every nonempty subset W of the universe of a generalized algebra

$$\mathfrak{A} = (U, \{s_i : i \in I\}, \{Q_j : j \in J\})$$

determines the generalized algebra

$$\mathfrak{B} = (W, \{s_i : i \in I\}, \{Q_j : j \in J\})$$

with operations and generalized operations defined as restrictions of the respective s_i and Q_j to W. Every generalized algebra obtained in this way is called a subalgebra of $\mathfrak{A}$.

A generalized algebra is generated by a set $G \subseteq U$ (or G is a set of generators of $\mathfrak{A}$) of $\mathfrak{A}$ is the smallest subalgebra of $\mathfrak{A}$ containing G.

Two generalized algebras

$$\mathfrak{A} = (U, \{s_i : i \in I\}, \{Q_j : j \in J\})$$

and

$$\mathfrak{B} = (W, \{p_k : k \in K\}, \{S_l : l \in L\})$$

are called similar when $I = K$, $J = L$ and $\arg(s_i) = \arg(p_i)$ for $i \in I$.

Let $\mathfrak{A}$ and $\mathfrak{B}$ be similar and let $h : U \to W$ be a mapping satisfying the conditions:

(a) $h(s_i(u_1, u_2, \ldots, u_{\mathbf{arg}(s_i)})) = p_i(h(u_1), h(u_2), \ldots, h(u_{\mathbf{arg}(p_i)}))$

for each $i \in I$ and every $u_1, u_2, \ldots, u_{\mathbf{arg}(s_i)} \in U$;

(b) $h(Q_j(Y)) = S_j(h(Y))$

for each $j \in J$ and every set $Y \subseteq U$ admissible for Q_j. Then h is called a homomorphism of $\mathfrak{A}$ into $\mathfrak{B}$. As before, we write also $h : \mathfrak{A} \to \mathfrak{B}$ instead of the rigorously correct $h : U \to W$, to indicate that h is a homomorphism between the algebras in question.

It is easily proved that the composition of two homomorphisms (between generalized algebras) is a homomorphism, too.

Let $G \subseteq U$ be a set of generators of a generalized algebra $\mathfrak{A}$ and let g be a mapping from G into the universe of a generalized algebra $\mathfrak{B}$, similar to $\mathfrak{A}$. If g admits extension to a homomorphism of $\mathfrak{A}$ into $\mathfrak{B}$, then this extension is unique.

Given a generalized algebra $\mathfrak{A}$ with universe U, consider the class K of complete generalized algebras similar to $\mathfrak{A}$. We say that $\mathfrak{A}$ is K-free, or free in class K, if $\mathfrak{A}$ is generated by a set $G \subseteq U$ such that every mapping g from G into the universe of any algebra $\mathfrak{B} \in$ K can be extended to a homomorphism of $\mathfrak{A}$ into $\mathfrak{B}$. Then $\mathfrak{A}$ is said to be freely generated by G, and G is a set of free generators of $\mathfrak{A}$.

A generalized algebra

$$\mathfrak{B} = (W, \{p_i : i \in I\}, \{S_j : j \in J\})$$

is an extension of a generalized algebra

$$\mathfrak{A} = (U, \{s_i : i \in I\}, \{Q_j : j \in J\})$$

if:

(a) $U \subseteq W$;

(b) for each $i \in I$ we have $\arg(s_i) = \arg(p_i)$

and the equality

$s_i(u_1, u_2, \ldots, u_{\arg(s_i)}) = p_i(u_1, u_2, \ldots, u_{\arg(p_i)})$

holds for any $u_1, u_2, \ldots, u_{\arg(s_i)} \in U$;

(c) for each $j \in J$, if $X \subseteq U$ is a set admissible for Q_j then X is admissible for S_j and

$$Q_j(X) = S_j(X).$$

When $J = \emptyset$, it is obvious that $\mathfrak{B}$ is an extension of $\mathfrak{A}$ if and only if $\mathfrak{A}$ is a subalgebra of $\mathfrak{B}$.

Using the Kuratowski–Zorn lemma (Theorem 1.1) it is not hard to prove the following

Theorem 1.10

Every generalized algebra can be extended to a complete generalized algebra.

■

It is often convenient to identify the structure of an algebra by means of the set of its function symbols. This set is usually denoted by Ω and is called the alphabet, signature or type. An algebra

$$\mathfrak{A} = (U, \{s_k : k \in K\})$$

can be regarded as the pair

$$\mathfrak{A} = (U, i)$$

where $i : \Omega \to \{s_k : k \in K\}$ is a function that determines the set $\{s_k : k \in K\}$.

Let Ω be any signature. By an Ω-algebra we mean a pair (U, i) in which U is a nonempty set and i is a function as above, and such that

(a) for every symbol $f \in \Omega$, if $\arg(f) = n > 0$ then $i(f) : U^n \to U$;

(b) if c is a symbol representing a constant in Ω then $i(c) \in U$.

Here $\arg(f)$ denotes the arity of the function symbol f. Thus, formally, arg is a function from Ω into N. The type of Ω is often described by just listing the values of arg.

1.11 Logical Matrices

Let $\mathfrak{A} = (A, \{s_i : i \in I\})$ be an algebra. Every pair of the form $(\mathfrak{A}, A^*)$, with A^* a nonempty subset of A, is called a matrix. If $\mathfrak{M} = (\mathfrak{A}, A^*)$ is a matrix, then $\mathfrak{A}$ is called the algebra of $\mathfrak{M}$, and we write $A = \mid \mathfrak{M} \mid, A^* = \mid \mathfrak{M} \mid^*$.

We say that $\mathfrak{N} = (\mathfrak{B}, B^*)$ is a submatrix of $\mathfrak{M}$, in symbols $\mathfrak{N} \subseteq \mathfrak{M}$, if $\mathfrak{B}$ is a subalgebra of $\mathfrak{A}$ and $\mid \mathfrak{N} \mid^* = \mid \mathfrak{M} \mid^* \cap \mid \mathfrak{N} \mid$.

Suppose we consider the language of an arbitrary logic. Let S be the set of all formulas of that language. Then S can be regarded as an algebra with operations induced by logical connectives. By a logical matrix for S we mean any matrix $\mathfrak{M} = (\mathfrak{A}, A^*)$ with $\mathfrak{A}$ an algebra similar to the algebra of S and such that S embeds injectively in $\mathfrak{A}$, tautologies of S being carried to elements of A^*. The embedding itself is called an interpretation of S in $\mathfrak{A}$.

Algebra S is freely generated by V, the set of propositional variables. Thus every function $h : V \rightarrow \mid \mathfrak{M} \mid$ uniquely extends to a homomorphism $h^* : S \rightarrow \mid \mathfrak{M} \mid$, so that:

(a) $h^*(p) = h(p)$ for $p \in V$;

(b) $h^*(s_i(\alpha_1, \alpha_2, \ldots, \alpha_n)) = s_i(h^*(\alpha_1), h^*(\alpha_2), \ldots, h^*(\alpha_n))$,
for every operation (connective) s_i and
for all $\alpha_1, \alpha_2, \ldots, \alpha_n \in S$, where $n = \arg(s_i)$;

(c) $h^*(c_S) = c_{\mid \mathfrak{M} \mid}$ for symbols representing constants.

Every such function is called a valuation of propositional variables in $\mathfrak{M}$ (or over $\mathfrak{M}$).

Let $\alpha \in S$. We say that α is satisfiable in $\mathfrak{M}$ if there exists a valuation h such that $h^*(x) \in \mid \mathfrak{M} \mid^*$. When $h^*(x) \in \mid \mathfrak{M} \mid^*$ for every valuation h, we say that α is tautologous over $\mathfrak{M}$, or α is a tautology (of the logic in question). The set of all tautologies over $\mathfrak{M}$ will be denoted by $E(\mathfrak{M})$. If α is not satisfiable in $\mathfrak{M}$, we say that α is a contradiction over $\mathfrak{M}$. An important fact to notice is that $\mathfrak{N} \subseteq \mathfrak{M}$ implies $E(\mathfrak{M}) \subseteq E(\mathfrak{N})$.

By a rule over S we mean a nonvoid relation $r \subseteq P(S) \times S$ (with $P(S)$ denoting the power set of S) satisfying for a certain positive integer n the following condition:

$$\text{card } X = n \quad \text{wherever} \quad (X, \alpha) \in r.$$

(There exist logics in which rules are defined in a different way; however, in any case, rules are relations of some type.)

Let r be a rule over S. It is called infallible in $\mathfrak{M}$ if the conditions $X \subseteq E(\mathfrak{M})$, $\alpha \in S$, $(X, \alpha) \in r$ imply $\alpha \in E(\mathfrak{M})$. The set of all rules infallible with respect to a given matrix $\mathfrak{M}$ will be denoted by $V(\mathfrak{M})$.

A rule r is called normal in $\mathfrak{M}$ if, for every valuation $h : V \to \mid \mathfrak{M} \mid$ (extending to a homomorphism $h^* : S \to \mid \mathfrak{M} \mid$), the conditions $X \subseteq S, \alpha \in S, (X, \alpha) \in r$ and $h^*(X) \subseteq \mid \mathfrak{M} \mid^*$ imply $h^*(\alpha) \in \mid \mathfrak{M} \mid^*$. The set of all rules normal in $\mathfrak{M}$ will be denoted by $N(\mathfrak{M})$.

It follows from this definition that if $\mathfrak{N} \subseteq \mathfrak{M}$ then $N(\mathfrak{M}) \subseteq N(\mathfrak{N})$; moreover, the set $N(\mathfrak{M})$ is contained in the intersection of the sets $V(\mathfrak{N})$ over all $\mathfrak{N} \subseteq \mathfrak{M}$.

Now suppose that R is a certain set of rules over S and let $X \subseteq S$ be a set closed under each rule $r \in R$. Every such pair (X, R) is called a propositional calculus over S. Assume, as above, $S = \mid \mathfrak{M} \mid$. Matrix $\mathfrak{M}$ is said to be adequate for the calculus (X, R) if $X = E(\mathfrak{M})$.

Theorem 1.11 (Lindenbaum)

Let (X, R) be a propositional calculus such that the rule of substitution belongs to R. Then there exists a finite or countable matrix adequate for (X, R).

■

A propositional calculus can admit adequate matrices of different cardinalities (clearly, by the cardinality of $\mathfrak{M}$ we mean the cardinality of $\mid \mathfrak{M} \mid$). Given a calculus (X, R), a natural problem is to characterize its adequate matrix of minimum cardinality, called the minimal adequate matrix of (X, R).

Let now J be any index set. For each $j \in J$, let $\mathfrak{M}_j = (\mathfrak{A}_j, A_j^*)$ be a certain matrix for the language S. We can form the product algebra $\mathfrak{A} = \prod_{i \in J} \mathfrak{A}_j$ and its subset $A^* = \prod_{j \in J} A_j^*$. The resulting matrix $\mathfrak{M} = (\mathfrak{A}, A^*)$ is called the product of matrices $\mathfrak{M}_j$ and is denoted by $\prod_{j \in J} \mathfrak{M}_j$.

Theorem 1.12 (Jaśkowski)

If $\mathfrak{M} = \prod_{j \in J} \mathfrak{M}_j$, then $E(\mathfrak{M}) = \prod_{j \in J} E(\mathfrak{M}_j)$.

Moreover,

$$\prod_{j \in J} V(\mathfrak{M}_j) \subseteq V(\prod_{j \in J} \mathfrak{M}_j),$$

$$\prod_{j \in J} N(\mathfrak{M}_j) \subseteq N(\prod_{j \in J} \mathfrak{M}_j). \quad ■$$

Assume that $\mathfrak{N}$ and $\mathfrak{M}$ are two matrices for a language S. Suppose there exists a surjective map $F : \mid \mathfrak{N} \mid \to \mid \mathfrak{M} \mid$ which is a homomorphism between the corresponding algebras, preserves the symbols representing logical constants and carries $\mid \mathfrak{N} \mid^*$ onto $\mid \mathfrak{M} \mid^*$ and $\mid \mathfrak{N} \mid - \mid \mathfrak{N} \mid^*$ onto $\mid \mathfrak{M} \mid - \mid \mathfrak{M} \mid^*$. Every such mapping is called a homomorphism of $\mathfrak{N}$ onto $\mathfrak{M}$; we say that $\mathfrak{M}$ is a homomorphic image of $\mathfrak{N}$.

In view of this definition, if $\mathfrak{M}$ is a homomorphic image of $\mathfrak{N}$ then $V(\mathfrak{N}) = V(\mathfrak{M})$ and $N(\mathfrak{M}) \subseteq N(\mathfrak{N})$. Moreover, we then have $E(\mathfrak{N}) = E(\mathfrak{M})$. The converse implication does not hold in general; matrices inducing the same sets of tautologies need not be homomorphic.

2 Many-Valued Propositional Calculi

2.1 Remarks on History

The origins of multivalued logics can be traced as deep as the treatises of Aristotle. He was the first to object to rigid bivalence of statements. His doubts concerned the so-called Law of the Excluded Middle ($p \vee \sim p$), considered as undeniable truth in classical logic. The idea of accepting sentences which (in a given instance) fail to be either absolutely true or absolutely false aroused contention between the Epicureans, on the one side, and the Stoics (including Chrysippus) on the other. The latter represented the standpoint of extreme determinism, with its orthodox bivalence in logic. The former, rejecting absolute determinism, admitted the possibility that neither of two statements, one of which negated the other, must necessarily be true; in particular, when the statements involved events that were to come. This is the reason why many-valued logics (or, more generally, non-classical logics) are sometimes referred to as non-Chrysippean.

In more recent times, G. Boole, C. S. Peirce [Peirce 1885] and N. A. Vasil'ev [Vasil'ev 1924] have to be considered as pioneers of many-valued logic. Yet, the actual founders of mature multivalued logical systems were (independently) J. Łukasiewicz [Łukasiewicz 1920] and E. Post [Post 1920]. In his investigation on the role of modality in ontological activities of the human being, Łukasiewicz made an attempt to modify the propositional calculus so as to achieve a kind of conceptual synthesis of determinism and indeterminism. This gave rise to three-valued logic, presented for the first time in the publication of 1920. It has become a starting point for constructions of finitely-valued logics, as well as for systems with an infinite number of logical values.

Working independently, Post devised a functionally complete n-valued logic, with n different logical values, representing distinct truth statuses. Unlike Łukasiewicz, Post was not led by any philosophical motivations. In his view, the essence of the problem appeared as purely theoretical, purely formal. Post's system can be regarded as formalizing some logic; in this system, or rather systems, the set of assertions (tautologies) shrinks as the number of logical values grows.

While n-valued logics with n greater than three have come into being as a product of more or less formal speculations, certain three-valued logical systems (not necessarily complete) have been created for actual, quite concrete purposes.

A specific position is taken by those three-valued systems which attempt to resolve certain philosophic-logical problems arising in connection with quantum mechanics. These problems were the object of interest of G. Birkhoff and J. von Neumann [Birkhoff, von Neumann 1936], P. Destouches-Février [Destouches-Février 1949, 1951, 1952], and in particular H. Reichenbach [Reichenbach 1951, 1962, 1963, 1964], in the 1940s and 1950s. The three-valued logic devised by S. C. Kleene [Kleene 1952] was regarded as a tool for settling several questions concerning recursive partial functions. Further, B. Kirkerud [Kirkerud 1982] and H. Rasiowa [Rasiowa 1974] have pursued the applicability of three-valued logic in investigations on the correctness of computational programs. The works of K. Piróg-Rzepecka [Piróg-Rzepecka 1977], provide a method for analyzing, with the aid of a three-valued logic, some mathematical problems involving expressions that can lack sense. To close this list, let us mention the works of J. Słupecki [Słupecki 1936], B. Sobociński [Sobociński 1936] and D. L. Webb [Webb 1936a], in which three-valued calculi were used for analyzing and deciding certain questions of purely logical nature.

Logics admitting more than three truth values constitute a further refinement of those three-valued calculi. As mentioned above, the variety of n-valued systems is a consequence of their functional incompleteness. Let us recall that a propositional calculus is said to be complete if every propositional connective is definable in terms of connectives of this calculus. Functionally incomplete many-valued calculi differ in their sets of tautologies, as well as in the forms which their tautologies can take. Functionally complete n-valued propositional calculi with identical sets of tautologies admit further distinction with regard to the basic set of connectives (the primitives), applied to formation of formulas, and also with respect to philosophical background influencing the choice of connectives.

The construction of propositional calculi can follow various methodological lines. The so-called algebraic method is, in our opinion, the leading one. As regards others, let us list here the axiomatic method, natural deduction, Gentzen's method of sequents (together with its modifications) and the method of finitely generated trees. Each of them requires some form of completeness theorem; and this in fact provides an algebraic interpretation of the calculus in question. The algebraic construction of propositional calculi has to be viewed as fundamental, and also prior to other ones (also in the case of the classical two-valued logic).

In each logical system there is always specified (sometimes implicitly) a set of values that can be assumed by formulas of the system. Assigning values to formulas (e.g., truth and falsity in classical logic) is nothing but a mapping of one algebraic structure into another – a similar one, though not written in formal terms. These remarks seem not to concern the intuitionistic calculus or modal calculi, which were not constructed on the basis of algebra. For instance, the intuitionistic calculus results from the axiomatized classical calculus by rejecting the law of excluded middle ($p \vee \sim p$). Yet, it is precisely the rejection of this law that indicates the non-bivalence of the system and of the algebra that corresponds to it. Of course, algebraic descriptions of intuitionism and modal calculi have been discovered in the course of further devolopment.

For each new method of creating propositional calculi, one requires a theorem on the completeness of the system. Completeness alone does guarantee the correctness of construction. Roughly, the completeness theorem (for a given method of constructing a propositional calculus) asserts that any statement which is a tautology in that calculus can be proved in it. The converse assertion is called the consistency theorem. The two theorems taken jointly establish the equivalence of the particular method of construction to the algebraic method; or, in some rare occasions, to some other construction in which the completeness theorem does hold.

2.2 The Definition of a Propositional Calculus

The statement that algebra provides the right framework for the construction of propositional calculi may be disputable. However, it can be justified by two arguments. The first is the fact that the set of well formed formulas of a given language constitutes an algebra. Moreover, its quotient algebra modulo the relation of deductive equivalence (based on the given calculus) is homomorphic to its adequate matrix. In most cases, for finitely valued logics, this turns out to be an isomorphism. The other argument is the fact that, in a given propositional calculus, a formula possesses a proof if and only if it is a tautology over any algebra isomorphic to the minimal adequate matrix of the calculus.

Thus, if the starting point of a definition of a propositional calculus is not provided by its minimal adequate matrix, the problem arises of finding such a matrix. This can be extremely difficult sometimes.

Let M be an arbirary Ω-algebra with a finite universe. Without loss of generality it can be assumed that

$$M = ((n), (n) - (m), i)$$

with $m < n$; i.e., $(n) = \mid M \mid$, $(n) - (m) = \mid M \mid^*$, and i is an interpretation function defined on Ω.

By a propositional calculus we mean a quadruple

$$\mathcal{X} = (M,\ L,\ v,\ Pr)$$

where

(a) $M = ((n), (n) - (m), i)$ is a minimal adequate matrix for $\mathcal{X}$;

(b) $L = (V, S)$, with V an infinite set of propositional variables and S a set of Ω-terms called well formed formulas, or simply formulas;

(c) $v : V \longrightarrow (n)$ is a valuation of propositional variables;

(d) $Pr : P(S) \longrightarrow P(S)$ is a function subject to some further conditions, specified below.

Pr has to be considered as an operation of consequence, providing a mechanism for acceptance or refutation of (sets of) formulas.

According to this definition, the language L is in fact an Ω-algebra with universe V. It can be proved that $F = (V, S)$ is the Ω-algebra freely generated by V. Evidently, every valuation $v : V \longrightarrow (n)$ uniquely extends to a homomorphism h of the entire set S into M.

A formula $\alpha \in S$ is called a tautology of $\mathcal{X}$ if and only if $h(\alpha) \in (n) - (m)$ holds for every valuation $v : V \longrightarrow (n)$, with h denoting the extension of v to a homomorphism on the whole of S.

The consequence operation Pr is subject to the following conditions (to be satisfied for any $X \subseteq S$, $Y \subseteq S$, $\alpha \in S$, $\beta \in S$):

(a) $X \subseteq Pr(X) \subseteq S$;

(b) if $X \subseteq Pr(Y)$ then $Pr(X) \subseteq Pr(Y)$;

(c) if $\alpha \in Pr(X)$ then there exists a finite set Y such that $Y \subseteq X$ and $\alpha \in Pr(Y)$;

(d) $Pr(Pr(X)) \subseteq Pr(X)$.

Speaking informally, a propositional calculus is axiomatically defined if and only if

(a) its language algebra S is specified;

(b) there is isolated a finite or recursively countable set of axioms, a subset of F; it is then referred to as the axiom system of the calculus;

(c) there is specified a set of rules $r_1, r_2, \ldots, r_m$ for the acceptance of formulas from among previously accepted ones (and the axioms).

In that situation, the operation Pr is fully determined by the axiom system and the set of rules $r_1, r_2, \ldots, r_m$.

Suppose $\mathcal{X} = (M, L, v, Pr)$ is a propositional calculus. By a proof in $\mathcal{X}$ we understand any finite chain of formulas $\alpha_1, \alpha_2, \ldots, \alpha_s$ such that, for each i with $1 \leq i \leq s$, either α_i belongs to the axiom system of $\mathcal{X}$ or there exist indices $i_1, i_2, \ldots, i_k$ smaller than i and such that

$$\langle \alpha_{i_1}, \alpha_{i_2}, \ldots, \alpha_{i_k}, \alpha_i \rangle \in r_j$$

for a certain r_j in $r_1, r_2, \ldots, r_m$.

A formula $\alpha \in F$ is called a theorem of $\mathcal{X}$ (or an assertion of $\mathcal{X}$) if and only if there exists a proof $\alpha_1, \alpha_2, \ldots, \alpha_s$ in $\mathcal{X}$ such that $\alpha_s = \alpha$. It is then called a proof of α.

Now, let Y be any set contained in S. A formula α is in $Pr(Y)$ if and only if it admits a proof (in $\mathcal{X}$) from Y; that is, iff there exists a chain of formulas $\alpha_1, \alpha_2, \ldots, \alpha_s$ such that $\alpha_s = \alpha$ and, for each i, $1 \leq i \leq s$, either α_i is an assertion of Y or there exist $i_1, i_2, \ldots, i_k$ smaller than i and such that

$$\langle \alpha_{i_1}, \alpha_{i_2}, \ldots, \alpha_{i_k}, \alpha_i \rangle \in r_j$$

for an r_j in $r_1, r_2, \ldots, r_m$.

Thus, according to this definition, the set of all assertions of a propositional calculus coincides with $Pr(Y)$.

The operation Pr thus defined satisfies conditions corresponding to (a)–(d) above.

2.3 Many-Valued Calculi of Łukasiewicz

Let n be an integer greater than 2 and let $\Omega = \Omega_n$ be a signature such that

$$i(\Omega_0) = (n) = 1, 2, 3, \ldots, n$$
$$i(\Omega_1) = -$$
$$i(\Omega_2) = \max, \min, \longrightarrow, \longleftrightarrow, \ldots$$

where

$$-x = n - x + 1$$
$$x \longrightarrow y = \begin{cases} n & \text{for } x \leq y \\ y & \text{for } x > y \end{cases}$$
$$x \longleftrightarrow y = \min(x \longrightarrow y, y \longrightarrow x).$$

Further, let

$$j(\Omega_1) = \sim$$
$$j(\Omega_2) = \vee, \wedge, \Longrightarrow, \Longleftrightarrow, \ldots,$$

the symbols standing respectively for negation, disjunction, conjunction, implication and equivalence in the n-valued calculus of Łukasiewicz.

Suppose $V = p_i : i \in N$ is the set of propositional variables. Then S, the set of correctly defined formulas, is the least among all sets S' that fulfill the conditions:

(a) $V \subseteq S'$,

(b) if $a \in S'$ then $\sim a \in S'$,

(c) if $\alpha, \beta \in S'$ then $\alpha \vee \beta \in S'$, $\alpha \wedge \beta \in S'$, $\alpha \Longrightarrow \beta \in S'$ and $\alpha \Longleftrightarrow \beta \in S'$.

The pair $L = (V, S)$ will be called the language of the Łukasiewicz calculus. In fact, S is the algebra freely generated by V. Therefore the language L can be also defined as the algebra $S = (V, j)$ with j satisfying the conditions imposed on an interpretation function (see p.27).

Now let $v : V \longrightarrow (n)$ be a valuation. It extends uniquely to a homomorphism

$$|\cdot|: S \longrightarrow M$$

as follows:

$$\begin{aligned}
| p_i | &= v(p_i) \text{ for } p_i \in V, i \in N, \\
|\sim \alpha | &= - | \alpha |, \\
| \alpha \vee \beta | &= \max\{| \alpha |, | \beta |\}, \\
| \alpha \wedge \beta | &= \min\{| \alpha |, | \beta |\}, \\
| \alpha \Longrightarrow \beta | &= | \alpha | \longrightarrow | \beta |, \\
| \alpha \Longleftrightarrow \beta | &= | \alpha | \longleftrightarrow | \beta |.
\end{aligned}$$

The following axiom system is assumed (for $\alpha, \beta, \gamma \in F$):

(L1) $\alpha \Longrightarrow (\beta \Longrightarrow \alpha)$,

(L2) $(\alpha \Longrightarrow \beta) \Longrightarrow ((\beta \Longrightarrow \gamma) \Longrightarrow (\alpha \Longrightarrow \gamma))$,

(L3) $(\sim \beta \Longrightarrow \sim \alpha) \Longrightarrow (\alpha \Longrightarrow \beta)$,

(L4) $((\alpha \Longrightarrow \beta) \Longrightarrow \alpha) \Longrightarrow ((\beta \Longrightarrow \alpha) \Longrightarrow \alpha)$,

(L5n) $n\alpha \Longrightarrow (n-1)\alpha$,

(L6n) $(n-1)((\sim \alpha)^j \vee (\alpha \wedge (j-1)\alpha))$,

where $n > 3,\ 1 < j < n-1$,

$$\begin{aligned}
n\alpha &= \alpha \vee \alpha \vee \ldots \vee \alpha \quad (n \text{ times}), \\
\alpha^j &= \alpha \wedge \alpha \wedge \ldots \wedge \alpha \quad (j \text{ times}),
\end{aligned}$$

provided j does not divide $n-1$.

The only primitive rule is the rule of detachment (modus ponens), abbreviated (mp), defined by the pattern

(mp) $$\frac{\alpha, \alpha \Rightarrow \beta}{\beta}$$

The n-valued calculus of Łukasiewicz is defined as the quadruple

$$Ł_n = (M, L, | \cdot |, Pr)$$

where $M, L, |\cdot|$ are as above and the operation Pr is determined by axioms $(L1) - (L6n)$ and the rule of detachment.

In n-valued propositional calculi ($n > 2$), the following tautologies of the classical two-valued logic are not tautologies any more:

(1) $(\sim \alpha \Longrightarrow \alpha) \Longrightarrow \alpha$,

(2) $\sim (\alpha \wedge \sim \alpha)$,

(3) $\alpha \vee \sim \alpha$,

(4) $(\alpha \Longrightarrow \sim \alpha) \Longrightarrow \sim \alpha$,

(5) $(\alpha \Longrightarrow \beta \wedge \sim \beta) \Longrightarrow \sim \alpha$.

A proof of this fact can be found, e.g., in A. Zinov'ev [Zinov'ev 1960]. The formulas $\sim \alpha \Longrightarrow \beta$ and $(\alpha \Longrightarrow \beta) \Longrightarrow \beta$ are equivalent in two-valued logic, but they are not equivalent already in $Ł_3$.

Finally, we introduce functions J_k for $k \in (n) = \{1, 2, \ldots, n\}$:

$$J_k(x) = \begin{cases} 1 & \text{if } x = k \\ n & \text{if } x \neq k. \end{cases}$$

Lemma 2.1

For each $k \in (n)$, J_k can be defined in terms of $\longrightarrow$ and $-$.

Proof

Let

$$h_1(x) = -x,$$
$$h_{m+1}(x) = x \longrightarrow h_m(x).$$

Obviously,

$$h_m(x) = \begin{cases} n - mx & \text{if } n \geq mx, \\ 0 & \text{if } n < mx \end{cases}$$

holds for each $m \in (n)$.

Define

$$I_k(x) = \begin{cases} n & \text{for } x \leq k, \\ 0 & \text{for } x > k. \end{cases}$$

Note that $I_1(x) = h_n(x)$. Suppose we have already defined I_l for $l \leq k$ and let r be the greatest integer such that $h_r(k+1) > 0$. Write $p = h_r(k+1) - 1$. Then $p \leq k$ and we have

$$I_{k+1}(x) = -I_p(h_r(x)).$$

Hence

$$J_1(x) = -I_1(x),$$
$$J_k(x) = (-I_k(x)) \wedge I_{k-1}(x),$$

ending the proof. ■

2.4 Finitely Valued Calculi of Łukasiewicz

In Section 2.2 we have defined a propositional calculus as a quadruple

$$\mathcal{X} = (M, L, |\cdot|, Pr),$$

in which

(a) M is a minimal Ω-algebra of a given type Ω,

(b) L is the language of $\mathcal{X}$,

(c) $|\cdot|$ is the valuation function,

(d) Pr is the operation of consequence.

We are new going to specify these concepts in some detail. General algebra will be our starting point for constructing many-valued logics. The language L shall itself constitute an algebra, similar to M, at least. It is therefore vital, in the investigation of concrete propositional calculi, to have the alphabet Ω exactly specified. The formalization which we present below is a slight modification of that given by H. Rasiowa in [Rasiowa 1974].

2.4.1 The Formalized Language of Propositional Calculi

Let $\Omega = \{\Omega_0, \Omega_1, \Omega_2, \Omega_3, \ldots\}$ be any alphabet, Ω_j denoting the set of j-argument function symbols, and let $V = \{p_i : i \in N\}$ be a countable set of propositional variables.

The formalized language of a propositional calculus $\mathcal{X}$ is defined as the pair $L_{\mathcal{X}} = (V, i_{\mathcal{X}})$ in which

(a) $i_{\mathcal{X}}(\Omega_0)$ is the set of propositional constants,

(b) $i_{\mathcal{X}}(\Omega_1)$ is the set of unary connectives,

(c) $i_{\mathcal{X}}(\Omega_2)$ is the set of binary connectives,

and, in general,

(d) $i_{\mathcal{X}}(\Omega_j)$ is the set of j-argument propositional connectives, for each $j \in N$.

Subscript $\mathcal{X}$ can be omitted if there can arise no confusion as to the underlying calculus.

The sets $i_{\mathcal{X}}(\Omega_j)$ in general do not contain all possible j-argument connectives. In the sequel we will write $\Omega_j^{\mathcal{X}}$, or just Ω_j, rather than $i_{\mathcal{X}}(\Omega_j)$. Of course, the alphabet includes auxiliary symbols common for all alphabets, namely the parentheses and comma.

It is sometimes convenient to regard the language as the pair $L_{\mathcal{X}} = (S, \Omega^{\mathcal{X}})$ where $\Omega^{\mathcal{X}}$ is the alphabet for $\mathcal{X}$ and S is the smallest of all sets $S'_{\mathcal{X}}$ that fulfill the following conditions:

(a) $V \subseteq S'_{\mathcal{X}}$,

(b) $\Omega_0^{\mathcal{X}} \subseteq S'_{\mathcal{X}}$,

(c) if $\alpha \in S'_{\mathcal{X}}$ and $m \in \Omega_1^{\mathcal{X}}$ then $m\alpha \in S'_{\mathcal{X}}$,

(d) if α, $\beta \in S'_{\mathcal{X}}$ and $s \in \Omega_2^{\mathcal{X}}$ then $\alpha s \beta \in S'_{\mathcal{X}}$ and, in general, if $\alpha_i \in S'_{\mathcal{X}}$ for $i = 1, 2, \ldots, k$ and $r \in \Omega_k^{\mathcal{X}}$ then $r(\alpha_1, \alpha_2, \ldots, \alpha_k) \in S'_{\mathcal{X}}$, for all $k = 3, 4, \ldots$.

We will call $S_{\mathcal{X}}$ the set of correctly defined formulas, or just the set of formulas. In concrete situations the subscript (or superscript) $\mathcal{X}$ will be omitted.

It follows that $L_{\mathcal{X}}$ is the algebra freely generated by the set of variables V. Hence, if $\mathfrak{A} = (A, \Omega^{\mathcal{X}})$ is any $\Omega^{\mathcal{X}}$-algebra, then every map $v : V \longrightarrow A$ can be extended to a homomorphism $|\cdot| : S \longrightarrow A$ by setting

(a) $| p_i | = v(p_i)$ for $p_i \in V$,

(b) $| \alpha_S | = \alpha_{\mathfrak{A}}$ for $\alpha \in \Omega_0$,

(c) $| r_S(\alpha_1, \alpha_2, \ldots, \alpha_k) | = r_{\mathfrak{A}}(| \alpha_1 |, | \alpha_2 |, \ldots, | \alpha_k |)$
for $\alpha_1, \alpha_2, \ldots, \alpha_k \in S$, $k = 1, 2, \ldots$ and $r \in \Omega_k$.

By a valuation of the Ω-algebra $L_{\mathcal{X}}$ in an Ω-algebra $\mathfrak{A} = (A, \Omega^{\mathcal{X}})$ we mean any map

$$v : V \longrightarrow A,$$

i.e., an arbitrary point of the Cartesian product A^V. With each formula $\alpha \in S$ we associate exactly one mapping

$$| \alpha |_{\mathfrak{A}} : A^V \longrightarrow A$$

defined as follows:

(a) $| p_i |_{\mathfrak{A}} (v) = v(p_i)$ for $p_i \in V$,

(b) $| \alpha |_{\mathfrak{A}} (v) = \alpha_{\mathfrak{A}}$ for $\alpha \in \Omega_0$,

(c) $| r_s(\alpha_1, \alpha_2, \ldots, \alpha_k) |_{\mathfrak{A}} (v) = r_{\mathfrak{A}}(| \alpha_1 |_{\mathfrak{A}} (v), | \alpha_2 |_{\mathfrak{A}} (v), \ldots, | \alpha_k |_{\mathfrak{A}} (v))$
for $r \in \Omega_k$, $k = 1, 2, \ldots$.

This mapping will be called the interpretation of formula α in the Ω-algebra $\mathfrak{A}$. Obviously, the value $| \alpha | (v)$ depends only on the values of $p_1, p_2, \ldots, p_n$, all propositional variables that actually occur in α.

2.5 Algebraic Characterization of the n-valued Calculi of Łukasiewicz

We now introduce the concept of Łukasiewicz algebras. They will be defined as symmetric Heyting algebras fulfilling certain additional axioms. We shall see later on that Łukasiewicz algebras turn out to be reducts of Post algebras. Our approach is similar (in many points identical) to those of H. Rasiowa [Rasiowa 1974] and R. Cignoli [Cignoli 1982].

2.5.1 Lattices

It will be convenient to consider the type of an algebra as the ordered k-tuple $(n_1, n_2, \dots, n_k)$ where $n_i = r(f_i)$ for $f_i \in \Omega$, $i = 1, 2, \dots, k$. Instead of f_j we may of course use any other symbols to denote the algebra operations.

A lattice is defined as an Ω-algebra of type $(2, 2)$ with the following identities satisfied:

(L1) $a \vee b = b \vee a, \ a \wedge b = b \wedge a,$

(L2) $a \vee (b \vee c) = (a \vee b) \vee c, \ a \wedge (b \wedge c) = (a \wedge b) \wedge c,$

(L3) $(a \wedge b) \vee b = b, \ a \wedge (a \vee b) = a.$

The elements $a \vee b$ and $a \wedge b$ will be called the join and meet (or the sum and product) of a and b; the two-argument functions $\vee$ and $\wedge$ are called the lattice addition and multiplication, respectively.

Lemma 2.2

For any lattice $(A, \vee, \wedge)$ and any elements $a, b \in A$ we have

(a) $a \vee b = b$ iff $a \wedge b = a$.

This allows us to define:

(b) $a \leq b$ iff $a \vee b = b$ or $a \wedge b = a$. ∎

The relation $\leq$ defined by (b) is reflexive, antisymmetric and transitive. We call it the lattice order. Obviously,

$$a \vee b = \sup\{a, b\}, \quad a \wedge b = \inf\{a, b\},$$

the supremum and infimum taken with respect to $\leq$. Hence we have

Lemma 2.3

(c) $a \vee a = a, \quad a \wedge a = a;$

(d) $a \leq a \vee b, \quad a \wedge b \leq a;$

(e) $b \leq a \vee b, \quad a \wedge b \leq b;$

(f) if $a \leq c$ and $b \leq c$ then $a \vee b \leq c$;
if $c \leq a$ and $c \leq b$ then $c \leq a \wedge b$;

(g) if $a \leq c$ and $b \leq d$ then $a \vee b \leq c \vee d$ and $a \wedge b \leq c \wedge d$. ∎

Lemma 2.4

Suppose $(A, \leq)$ is a partially ordered set in which every two elements $a, b \in A$ have a least upper bound $\sup\{a, b\}$ and a greatest lower bound $\inf\{a, b\}$. Define

$$a \vee b = \sup\{a, b\}, \quad a \wedge b = \inf\{a, b\}.$$

Then $(A, \vee, \wedge)$ is a lattice and the relation $\leq$ coincides with the induced lattice order. ∎

Let $\mathfrak{A} = (A, \vee, \wedge)$ be a lattice and let $\leq$ be its lattice order. If A has a greatest element, we call it the lattice unit and denote by 1. Similarly, if A has a least element, it is called the lattice zero and denoted by 0. Evidently,

(h) $a \leq 1, \ 0 \leq a,$

(i) $a \vee 1 = 1, \ a \wedge 1 = a,$

(j) $a \vee 0 = a, \ a \wedge 0 = 0,$

for every element $a \in A$.

A lattice $\mathfrak{A} = (A, \vee, \wedge)$ is said to be distributive if the equalities

$$\text{(L4)} \quad a \wedge (b \vee c) = (a \wedge b) \vee (a \wedge c),$$
$$a \vee (b \wedge c) = (a \vee b) \wedge (a \vee c)$$

hold for any $a, b, c \in A$.

2.5.2 Quasi-Boolean Algebras and Heyting Algebra

An Ω-algebra $\mathfrak{A} = (A, \vee, \wedge, -, 1)$ of type $(2, 2, 1, 0)$ is called quasi-Boolean when the reduced algebra $(A, \vee, \wedge)$ is a distributive lattice with unit 1 and the unary operation $-$ satisfies the axioms

(Q1) $-(-a) = a$,

(Q2) $-(a \vee b) = -a \wedge -b$

for all $a, b \in A$.

This definition immediately implies

(a) $0 \in A, \quad -0 = 1, \quad -1 = 0$,

(b) $-(a \wedge b) = -a \vee -b$.

An algebra $\mathfrak{A} = (A, \vee, \wedge, \Rightarrow, -, 0, 1)$ of type $(2,2,2,1,0,0)$ is called a symmetric Heyting algebra when $(A, \vee, \wedge, -, 0, 1)$ is quasi-Boolean, 0 and 1 being its zero and unit, and operation $\Rightarrow$ is defined as follows:

$a \Rightarrow b =$ the greatest element in A such that $a \wedge (a \Rightarrow b) \leq b$.

The following properties are an easy consequence of this definition:

(H1) $a \Rightarrow -b = b \Rightarrow -a$,

(H2) $(a \Rightarrow -b) \Rightarrow -a = b$,

(H3) $(a \Rightarrow b) \Rightarrow c = -((-a \Rightarrow c) \Rightarrow -(b \Rightarrow c))$.

We are now in position to introduce Łukasiewicz algebras. Let $n \geq 2$ be an integer.

An algebra $(A, \vee, \wedge, \Rightarrow, \sim, s_1^n, \ldots, s_n^n - 1, 0, 1)$ of type $(2,2,2,1,1,\ldots,1,0,0)$ is an n-valued Łukasiewicz algebra if $(A, \vee, \wedge, \sim, 0, 1)$ is a symmetric Heyting algebra and, for each i, $1 \leq i \leq n-1$, the unary operations s_i^n satisfy the following axioms:

(L1) $s_i^n(a \vee b) = s_i^n a \vee s_i^n b$,

(L2) $s_i^n(a \Rightarrow b) = \bigwedge_{j=i}^{n-1}(s_j^n a \Rightarrow s_j^n b)$,

(L3) $s_i^n s_j^n = a = s_j^n a \quad \text{for} \quad 1 \leq i \leq n-1, 1 \leq j \leq n-1$,

(L4) $s_1^n a \vee a = a$,

(L5) $s_i^n \sim a = \sim s_{n-1}^n a$,

(L6) $s_1^n a \vee \sim s_1^n a = 1$,

where

$$\bigwedge_{j=i}^{n-1} b_j$$

stands for

$$b_i \wedge b_{i+1} \wedge \ldots \wedge b_{n-1}.$$

By a Łukasiewicz algebra we shall always mean an n-valued Łukasiewicz algebra for a certain $n \geq 2$.

Example

Fix $n \geq 2$. Consider the set

$$A = \left\{ \frac{j}{n-1} : 0 \leq j \leq n-1 \right\}.$$

Define

$$\begin{aligned} x \vee y &= \max\{x, y\}, \\ x \wedge y &= \min\{x, y\}, \\ x \Rightarrow y &= \begin{cases} 1 & \text{if } x \leq y, \\ y & \text{if } x > y, \end{cases} \\ \sim x &= 1 - x, \\ s_i^n\left(\frac{j}{n-1}\right) &= \begin{cases} 1 & \text{if } i + j \geq n, \\ 0 & \text{if } i + j < n, \end{cases} \\ (1 \leq i &\leq n-1) \\ 0 &= \frac{0}{n-1}, \\ 1 &= \frac{n-1}{n-1}. \end{aligned}$$

Then

$$L = (A, \vee, \wedge, \Rightarrow, \sim, s_1^n, \ldots, s_{n-1}^n, 0, 1)$$

is an n-valued Łukasiewicz algebra.

Later on we will show that this algebra is minimal.

Let again A be an arbitrary Łukasiewicz algebra.[1]

It is convenient to set

$$s_0^n a = 0, \quad s_n^n a = 1$$

for $a \in A$. Consider the functions

$$J_i^n : A \longrightarrow A, \quad i = 0, 1, \ldots, n-1,$$

defined by

(a) $J_i^n(a) = s_{n-i}^n a \vee \sim s_{n-i-1}^n a.$

In the algebra of the example above, the functions J_i^n take the form

$$J_i^n\left(\frac{j}{n-1}\right) = \begin{cases} 1 & \text{if } j = i, \\ 0 & \text{if } j \neq i. \end{cases}$$

[1] As usual in mathematics, we write "A is an algebra" meaning that there is an underlying structure in the set A making it into an algebra; formally, we ought to write: "let $\mathfrak{A} = (A, \ldots)$ be an algebra".

Moreover,

(b) $s_i^n a = \bigvee_{j=1}^{i} J_{n-j}^n(a) \quad$ for $1 \leq i \leq n$.

A nonempty subset F of a Łukasiewicz algebra A is called a Stone filter if the following conditions are fulfilled:

(c) if $a \in F$ and $a \leq b$ then $b \in F$;

(d) $1 \in F$;

(e) if $a \in F$ then $s_1^n a \in F$.

Suppose F is a Stone filter in a Łukasiewicz algebra A. Then the relation

$$r(F) = \{(a, b) : a \wedge c = b \wedge c \text{ for some } c \in F\}$$

is a congruence in A. And conversely, if r is a congruence in a Łukasiewicz algebra A, then

$$F = \{a \in A : (a, 1) \in r\}$$

is a Stone filter and $r = r(F)$.

Let A be any distributive lattice. The symbol B_A will denote the Boolean algebra consisting of the complements to all elements of A. When A is a Łukasiewicz algebra, an element a belongs to B_A if and only if there exists i with $1 \leq i \leq n-1$ such that $a = s_i^n a$. Moreover, $s_1^n a$ is the greatest element in B_A which satisfies $s_1^n a \leq a$. Thus:

(f) $s_1^n(a \Rightarrow b)$ is the greatest element of B_A such that $a \wedge s_1^n(a \Rightarrow b) \leq b$;

and hence

(g) $a \Rightarrow b = s_1^n(a \Rightarrow b) \vee b$.

Lemma 2.5

(a) If A is a Heyting algebra and if $a \in B_A$, then

$$a \Rightarrow b = -a \vee b$$

where $-a$ denotes the Boolean complement to a.

(b) If A is a Łukasiewicz algebra and if $b \in B_A$, then

$$a \Rightarrow b = \sim s_{n-1}^n a \vee b.$$

Proof

(a) Obviously, $a \wedge (-a \vee b) = (a \wedge -a) \vee (-a \wedge b) = -a \wedge b \leq b$. Supposing $a \wedge c \leq b$ we get $c \leq c \vee -a = (c \wedge a) \vee -a \leq b \vee -a$, and hence $a \Rightarrow b = -a \vee b$.

(b) Property (f) implies in view of (L2)

$$a \Rightarrow b = s_1^n(a \Rightarrow b) \vee b = b \vee \bigwedge_{i=1}^{n-1} (s_i^n a \Rightarrow s_i^n b).$$

Since $b \in B_A$, we have $S_i^n b = b$ for $i = 1, 2, \ldots, n-1$. Thus if $s_i^n a \in B_A$ for $i = 1, 2, \ldots, n-1$, then $s_i^n a \Rightarrow s_i^n b = s_i^n a \vee b$, implying

$$a \Rightarrow b = b \vee ((\bigwedge_{i=1}^{n-1} -s_i^n a) \vee b) = -s_{n-1}^n a \vee b. \blacksquare$$

2.5.3 Proper Łukasiewicz Algebras

Let N denote the set of nonnegative integers. Define for $n \in N$:
$Q_n = \{(i,j) \in N \times N:\ 3 \leq i \leq n-2, 1 \leq j \leq n-4, j < i\}$ if $n \geq 5$,
$Q_n = \emptyset$ if $n < 5$,
$T_n = \{(i,j) \in N \times N:\ 2 \leq i \leq n-2, 1 \leq j \leq n-3, j < i\}$ if $n \geq 4$,
$T_n = \emptyset$ if $n < 4$.

It is easily checked that for $n \geq 5$,

$$\text{card } Q_n = \frac{n(n-5)+2}{2}, \qquad \text{card } T_n = \text{card } Q_n + 2,$$

while card $T_4 = 1$.

Let $n \geq 2$. A proper n-valued Łukasiewicz algebra (or just a proper Łukasiewicz algebra) is defined as a system

$$L = (A, \vee, \wedge, \Rightarrow, -, \{s_i^n\}_{1 \leq i \leq n-1}, \{F_{ij}^n\}_{(i,j) \in Q_n}, 0, 1)$$

such that

$$(A, \vee, \wedge, \Rightarrow, -, \{s_i^n\}_{1 \leq i \leq n-1}, 0, 1)$$

is an n-valued Łukasiewicz algebra and the functors

$$F_{ij}^n : A^2 \longrightarrow A$$

are connected with its structure through the identities

(p) $$s_k^n F_{ij}^n(a,b) = \begin{cases} 0 & \text{if } k \leq i-j \\ J_i^n(a) \wedge J_i^n(b) & \text{if } k > i-j \end{cases}$$

for $1 \leq k \leq n-1$.

The class of all proper n-valued Łukasiewicz algebras will be denoted by P_n. Evidently, for $2 \leq n \leq 4$, all Łukasiewicz algebras are proper, since $Q_n = \emptyset$.

Example

In the preceding subsection we constructed an example of an n-valued Łukasiewicz algebra. Defining the functors F^n_{ij} for $(i,j) \in Q_n$ by

$$F^n_{ij}\left(\frac{r}{n-1}, \frac{s}{n-1}\right) = \begin{cases} \frac{n-1-i+j}{n-1} & \text{if } (r,s) = (i,j), \\ 0 & \text{otherwise,} \end{cases}$$

we get an example of a proper Łukasiewicz algebra.

Lemma 2.6

Let A be a proper n-valued Łukasiewicz algebra and let $(i,j) \in T_n$. Then we have for any $x, y \in A$ and $a, b \in B_A$.

(a) $F^n_{ij}(-y, -x) = F^n_{(n-1-j)(n-1-i)}(x, y)$,

(b) $F^n_{ij}(x \wedge a,\ y \wedge b) = F^n_{ij}(x, y) \wedge a \wedge b$,

(c) $F^n_{ij}(x \vee a,\ y \vee b) = F^n_{ij}(x, y) \wedge -a \wedge -b$,

(d) $F^n_{ij}(x, b) = F^n_{ij}(a, y) = 0$.

Proof

To obtain (a) and (b) it suffices to apply equality (p) and the fact known as the Moisil principle [Moisil 1972] asserting that in every Łukasiewicz algebra the equality $x = y$ holds if and only if $s^n_i x = s^n_i y$. Claim (c) is an immediate consequence of (a) and (b). Equality (d) follows immediately from (b) and (c); consider for instance the first expression in (d):

$$F^n_{ij}(x, b) = F^n_{ij}(x \vee 0,\ b \vee 0) = F^n_{ij}(x, 0) \wedge -b,$$

while

$$F^n_{ij}(x, 0) = F^n_{ij}(x \wedge 1,\ 0 \wedge 0) = F^n_{ij}(x, 0) \wedge 1 \wedge 0 = 0. \quad ■$$

Suppose F is a Stone filter in a proper Łukasiewicz algebra A and let $r(F)$ be the induced algebra congruence. Suppose that pairs (x, x') and (y, y') belong in $r(F)$. Then there exist elements $a, b \in F \cap B_A$ such that

$$x \wedge a = x' \wedge a, \quad y \wedge b = y' \wedge b$$

and, according to Lemma 2.6 (b),

$$F^n_{ij}(x, y) \wedge a \wedge b = F^n_{ij}(x \wedge a,\ y \wedge b) = F^n_{ij}(x' \wedge a,\ y' \wedge b) = F^n_{ij}(x', y') \wedge a \wedge b.$$

Since $a \wedge b \in F$, we have

$$(F_{ij}^n(x', y'),\ F_{ij}^n(x, y)) \in r(F)$$

for all $(i, j) \in T_n$, showing that $r(F)$ is a proper-algebra congruence. This justifies the following lemma.

Lemma 2.7

Let A be a proper Łukasiewicz algebra. Assigning to each Stone filter F the relation $r(F)$ defines an isomorphism between the lattice of Stone filters of A (with inclusion as lattice order) and the lattice of congruences in A. ■

The class of all n-valued Łukasiewicz algebras together with their homomorphisms constitutes a category. It is not hard to see that the class of all proper n-valued Łukasiewicz algebras with their homomorphisms is a complete subcategory of the former.

We will denote by L_n the class of n-valued Łukasiewicz algebras and by $\mathfrak{L}_n$ the corresponding category. The class and category of proper n-valued algebras will be denoted by P_n and $\mathfrak{P}_n$.

2.5.4 The Łukasiewicz Implication

Let $A \in P_n$ and let $x, y \in A$. The Łukasiewicz implication "$\rightarrow$" is defined by

(a) $x \rightarrow y = (x \Rightarrow y) \vee -x \vee \bigvee_{(i,j) \in T_n} F_{ij}^n(x, y).$

Since $T_3 = \emptyset$ and $T_4 = \{(2,1)\}$, we conclude:

(b) if $A \in P_3 = L_3$ then $x \rightarrow y = (x \Rightarrow y) \vee -x$;

(c) if $A \in P_4 = L_4$ then
$x \rightarrow y = (x \Rightarrow y) \vee -x \vee F_{21}(x, y) = -x \vee (J_2^n(x) \wedge J_1^n(y) \wedge -y).$

Theorem 2.1

In every proper Łukasiewicz algebra $A \in P_n$:

(a) $s_1^n(x \rightarrow y) = s_1^n(x \Rightarrow y)$,

(b) $x \Rightarrow y = s_1^n(x \rightarrow y) \vee y$,

(c) if $a \in B_A$ then $x \rightarrow a = -x \vee a$,

(d) if $b \in B_A$ then $b \rightarrow x = -b \vee x$,

(e) if $a, b \in B_A$ then $a \rightarrow b = -a \vee b$,

(f) $1 \to x = x$,

(g) $x \to y = 1$ if and only if $x \leq y$ (with respect to the lattice order $\leq$ in A).

Proof

(a) Since $1 \leq i - j$ for all $(i,j) \in T_n$, we see that $s_i^n F_{ij}^n(x,y) = 0$ and $x \wedge s_i^n(-x) = x \wedge -s_{n-1}^n x = 0 \leq y$. Moreover, $s_1^n(-x) \leq s_1^n(x \Rightarrow y)$. Hence $s_1^n(x \to y) = s_1^n(x \Rightarrow y) \vee s_1^n(-x) = s_1^n(x \Rightarrow y)$.

(b) Since $x \Rightarrow y = s_1^n(x \Rightarrow y) \vee y$, we conclude (b) from (a).

(c) This follows from (d), Lemma 2.6 and equality (b).

(d) Immediate from Lemma 2.6.

(e) A direct consequence of (c) and (d).

(f) Immediate from (d).

(g) In the following chain of equalities, each one is equivalent to the preceding one:

$$x \to y = 1$$
$$s_1^n(x \to y) = 1$$
$$s_1^n(x \Rightarrow y) = 1$$
$$x \Rightarrow y = 1$$
$$x \leq y.$$

This completes the proof of the theorem. ■

We conclude this subsection by stating several equalities that hold in proper Łukasiewicz algebras. The proofs can be found in R. Cignoli [Cignoli 1982] and J. B. Rosser and A. Turquette [Rosser, Turquette 1952].

(a) $x \vee y = \max(x,y) = (x \to y) \to y$,

(b) $x \wedge y = \min(x,y) = -(-x \vee -y)$.

Denoting by H_k^n the unary operation in $A \in P_n$ given by

(i) $H_i^n(x) = -x$,

(ii) $H_{k+1}^n(x) = -H_k^n(x)$,

we have the following equalities:

(c) $J_{n-1}^n(x) = -H_{n-1}^n(x)$ and $J_0^n(x) = J_{n-1}^n(-x)$.

For any integer k with $1 \leq k \leq n-2$ let $u(k)$ be the greatest integer less than $\dfrac{n-1}{n-1-k}$ and let

$$r(k) = (n-1) \cdot H^n_{u(k)}\left(\tfrac{k}{n-1}\right).$$

If $n - i = r(n-i)$, then

(d) $J^n_{n-i}(x) = J^n_{n-1}((H^n_{u(n-i)}(x) \vee x) \to (H^n_{u(n-i)}(x) \wedge x)$;

and if $n - i < r(n-i)$, then

(e) $J^n_{n-i}(x) = J^n_{r(n-i)}(H^n_{u(n-i)}(x))$.

(f) $s^n_i(x) \;=\; \bigvee\limits_{j=1}^{i} J^n_{n-1}(x)$;

(g) $F^n_{ij}(x,y) \;=\; (x > y) \wedge J^n_i(x) \wedge J^n_i(y)$;

(h) $x \Rightarrow y = s^n_1(x \to y) \vee y = J^n_{n-1}(x \to y) \vee y = (-H^n_{n-1}(x \to y) \to y) \to y$.

The functions s^n_i can be of course defined in proper Łukasiewicz algebras in terms of $-$ and $\to$, without introducing the J^n_i; see Suchoń [Suchoń 1974]. Also, $\vee$, $\wedge$ and F^n_{ij} can be defined in terms of $-$ and $\to$, according to a theorem of McNaughton.

2.5.5 Stone Filters in Proper n-valued Łukasiewicz Algebras

In a proper Łukasiewicz algebra $A \in P_n$, a Stone filter is defined as a subset $F \subseteq A$ satisfying the conditions

(a) $1 \in F$,

(b) if $x \in F$ and $x \to y \in F$ then $y \in F$,

(c) if $x \in F$ then $s^n_i x \in F$.

Let $A \in P_n$ and let $D \subseteq A$. We call D a deductive system when

(d_1) $1 \in D$,

(d_2) if $x \in D$ and $x \to y \in D$ then $y \in D$.

Theorem 2.2

Suppose $A \in P_n$, $D \subseteq A$. The set D is a Stone filter in A if and only if D is a deductive system.

Proof

Assume that D is a Stone filter in A. Since $1 \in D$, condition (d_1) is fulfilled.

Suppose $x \in D$ and $x \to y \in D$. Then $s_1^n(x \to y) \in D$ and $x \Rightarrow y = s_1^n(x \to y) \vee y \in D$, implying $y \in D$. Thus D is a deductive system.

Conversely, assume D is a deductive system and let $x \in D$, $x \Rightarrow y \in D$. Condition (d_2) entails (b). Condition (d_2) is equivalent to (a). Finally, it is not hard to see that $x \in D$ yields $s_1^n x \in D$. Thus (c) is also fulfilled, and we are done. ∎

Theorem 2.3

Let $A \in P_n$ and let $D \subseteq A$ be a deductive system. For every x, y in the universe of A the following statements are equivalent:

(a) $(x, y) \in r(D)$;

(b) $x \to y \in D$ and $y \to x \in D$;

(c) $x \Rightarrow y \in D$ and $y \Rightarrow x \in D$.

Proof

Assume (a); i.e., let $(x, y) \in r(D)$. Since $x \to y$ is a polynomial in algebra A, we have $(x \to y, x \to x) \in r(D)$ and $(y \to x, y \to y) \in r(D)$. And since $x \to x = y \to y = 1$, we get $x \to y \in D$ and $y \to x \in D$, so that (b) holds.

Now assume (b). Then clearly $s_1^n(x \to y) \in D$ and $s_1^n(y \to x) \in D$, whence $s_1^n(x \to y) \vee y \in D$ and $s_1^n(y \to x) \vee x \in D$. Consequently $x \Rightarrow y \in D$ and $y \Rightarrow x \in D$, proving (c).

Finally, assume (c). Thus $s_1^n(x \Rightarrow y) \wedge s_1^n(y \Rightarrow x) \in D$. Now, $x \wedge s_1^n(x \Rightarrow y) \wedge s_1^n(y \Rightarrow x) \leq y \wedge s_1^n(x \Rightarrow y) \wedge s_1^n(y \Rightarrow x) \leq x \wedge s_1^n(x \Rightarrow y) \wedge s_1^n(y \Rightarrow x)$. Hence $(x, y) \in r(D)$, and so condition (a) is satisfied. ∎

2.5.6 The Axiom System for the n-valued Propositional Calculus of Łukasiewicz

The axiomatization of the n-valued Łukasiewicz calculus, which we present below, has been adopted from R. Cignoli [Cignoli 1982].

The calculus will be denoted by $Ł_n$. The basic connectives are: $\wedge, \vee, \Rightarrow, F_{ij}^n$ for $(ij) \in T_n$ (two-argument connectives) and $-, s_i^n$ for $i = 1, 2, \ldots, n-1$ (one-argument ones). The use of identical symbols to denote logical connectives and functions in a Łukasiewicz algebra will cause no confusion and emphasizes the algebraic nature of our approach.

Let S_n be the set of well-formed formulas over the alphabet containing the set V of propositional variables and the connectives just listed. We will denote by A_n the following system of axioms for the calculus $Ł_n$ (assumed to hold for any $\alpha, \beta, \gamma \in S_n$); we write $\alpha \Leftrightarrow \beta$ for $(\alpha \Rightarrow \beta) \wedge (\beta \Rightarrow \alpha)$:

(a$_1$) $\alpha \Rightarrow (\beta \Rightarrow \alpha)$,

(a$_2$) $(\alpha \Rightarrow (\beta \Rightarrow \gamma)) \Rightarrow ((\alpha \Rightarrow \beta) \Rightarrow (\alpha \Rightarrow \gamma))$,

(a$_3$) $\alpha \Rightarrow (\alpha \vee \beta)$

(a$_4$) $\beta \Rightarrow (\alpha \vee \beta)$

(a$_5$) $(\alpha \Rightarrow \gamma) \Rightarrow ((\beta \Rightarrow \gamma) \Rightarrow (\alpha \vee \beta \Rightarrow \gamma))$

(a$_6$) $\alpha \wedge \beta \Rightarrow \alpha$

(a$_7$) $\alpha \wedge \beta \Rightarrow \beta$

(a$_8$) $(\alpha \Rightarrow \beta) \Rightarrow ((\alpha \Rightarrow \gamma) \Rightarrow (\alpha \Rightarrow \beta \wedge \gamma))$

(a$_9$) $\alpha \Leftrightarrow - - \alpha$,

(a$_{10}$) $s_1^n(\alpha \Rightarrow \beta) \Leftrightarrow s_1^n(-\beta \Rightarrow -\alpha)$

(a$_{11}$) $s_i^n(\alpha \vee \beta) \Leftrightarrow s_i^n\alpha \vee s_i^n\beta$ for $i = 1, 2, \ldots, n-1$,

(a$_{12}$) $s_i^n(\alpha \Rightarrow \beta) \Leftrightarrow \bigwedge_{j=1}^{n-1}(s_j^n\alpha \Rightarrow s_j^n\beta)$ for $i = 1, 2, \ldots, n-1$.

(a$_{13}$) $s_j^n s_j^n\alpha \Leftrightarrow s_j^n\alpha$ for $i, j = 1, 2, \ldots, n-1$,

(a$_{14}$) $s_1^n\alpha \Rightarrow \alpha$

(a$_{15}$) $s_i^n\alpha \Leftrightarrow -s_{n-i}^n - \alpha$

(a$_{16}$) $s_1^n\alpha \vee -s_1^n\alpha$

(a$_{17}$) $-s_k^n F_{ij}^n(\alpha, \beta)$ for $k = 1, 2, \ldots, i-j$, $(i, j) \in T_n$,

(a$_{18}$) $s_k^n F_i^n j(\alpha, \beta) \Leftrightarrow s_{n-i}^n\alpha \wedge -s_{n-i-1}^n\alpha \wedge s_{n-i}^n\beta \wedge s_{n-i-1}^n\beta$.

We assume the following rules of inference:

(rd) $\dfrac{\alpha, \alpha \Rightarrow \beta}{\beta}$,

(rn) $\dfrac{\alpha}{s_1^n\alpha}$.

They determine the consequence operation Pr (see Section 2.2).

Theorem 2.4

Let X be any set contained in S_n (the set of all admissible formulas). Then

(a) if $\alpha \Leftrightarrow \beta \in Pr(X)$ then $-\alpha \Leftrightarrow -\beta \in Pr(X)$;

(b) if $\alpha \Leftrightarrow \beta \in Pr(X)$ then $s_i^n\alpha \Leftrightarrow s_i^n\beta \in Pr(X)$ for $i = 1, 2, \ldots, n-1$;

(c) if $\alpha \Leftrightarrow \gamma$ and $\beta \Leftrightarrow \delta$ belong in $Pr(X)$ then $F_{ij}^n(\alpha, \beta) \Leftrightarrow F_{ij}^n(\gamma, \delta) \in Pr(X)$, for every $(i, j) \in T_n$;

(d) if $s_i^n\alpha \Rightarrow s_i^n\beta \in Pr(X)$ for all i with $1 \leq i \leq n-1$ then $\alpha \Rightarrow \beta \in Pr(X)$;

(e) $\alpha \Leftrightarrow \beta \in Pr(X)$ if and only if $J_i^n(\alpha) \Leftrightarrow J_i^n(\beta) \in Pr(X)$
for $i = 0, 1, \ldots, n-1$;

(f) if $-\alpha \in Pr(X)$ then $\alpha \Rightarrow \beta \in Pr(X)$ for every $\beta \in S_n$.

Proof

(a) Immediate from axioms (a_{10}) and (a_{14}), in view of (rn).

(b) If $\alpha \Rightarrow \beta \in Pr(X)$, then by (rn) and ($a_{12}$) we have $\bigwedge_{i=1}^{n-1}(s_i^n\alpha \Rightarrow s_i^n\beta) \in Pr(X)$, and hence $s_i^n\alpha \Rightarrow s_i^n\beta \in Pr(X)$ for $i = 1, 2, \ldots, n-1$.
The proof of (c) is preceded by (d), (e) and (f).

(d) Suppose $s_i^n\alpha \Rightarrow s_i^n\beta \in Pr(X)$ for $i = 1, 2, \ldots, n-1$. Then $s_1^n(\alpha \Rightarrow \beta) \in Pr(X)$, and thus axiom ($a_{14}$) yields by (mp): $\alpha \Rightarrow \beta \in Pr(X)$.

(e) Since $J_i^n(\alpha) \Leftrightarrow (s_{n-1}^n\alpha \wedge -s_{n-i-1}^n\alpha)$, the statement of (e) follows from (a), (b) and (d).

(f) It suffices to notice that if $-\alpha \in Pr(X)$ then $-\beta \Rightarrow -\alpha \in Pr(X)$, for each $\beta \in S_n$. Using axioms (a_{10}) and (a_{14}) we infer $\alpha \Rightarrow \beta \in Pr(X)$.

(c) Now assume that $\alpha \Leftrightarrow \gamma$ and $\beta \Leftrightarrow \delta$ are in $Pr(X)$. Choose k with $1 \leq k \leq n-1$. Let $(i, j) \in T_n$. If $1 \leq k \leq i-j$, then by (a_{17}) and (f) the equivalence $s_k^nF_{ij}^n(\alpha, \beta) \Leftrightarrow s_k^nF_{ij}^n(\gamma, \delta)$ belongs to $Pr(X)$.

And if $i - j < k \leq n-1$, then by (a_{18}) and (e) we get $s_k^nF_{ij}^n(\alpha, \beta) \Leftrightarrow J_i^n(\alpha) \wedge J_i^n(\beta) \Leftrightarrow J_i^n(\gamma) \wedge J_i^n(\delta) \Leftrightarrow s_k^nF_{ij}^n(\gamma, \delta)$, a chain of equivalences belonging in $Pr(X)$. So the claim follows and the theorem is proved. ∎

Let $\alpha, \beta \in S_n$. We write $\alpha \equiv \beta$ if and only if $\alpha \Leftrightarrow \beta \in Pr(\emptyset)$. The relation $\equiv$ thus defined is a congruence in S_n. Let $\mathfrak{A}_n$ be the quotient algebra of S_n modulo this relation; i.e., let

$$\mathfrak{A}_n = (S_n / \equiv, \vee, \wedge, \Rightarrow, -, \{s_i^n : i = 1, 2, \ldots, n-1\}, \{F_{ij}^n : (ij) \in Q_n\}, 0, 1),$$

where $1 = Pr(\emptyset)$ and $0 = -1$. The functors induced in $\mathfrak{A}_n$ are denoted by the same symbols as in S_n; this should not lead to misunderstandings.

Lemma 2.8

$\mathfrak{A}_n$ is a proper n-valued Łukasiewicz algebra.

Proof

By virtue of axioms (a_1)–(a_8) and the (rd) rule, $\mathfrak{A}_n$ is a relative-complementary lattice with unit. Further, it follows from axioms (a_9), (a_{10}), (a_{14}) and the (rn) rule that $\mathfrak{A}_n$ is a symmetric Heyting algebra. The demanded properties of functors s_i^n for $i = 1, 2, \ldots, n-1$ and F_{ij}^n for $(i, j) \in Q_n$ are a consequence of Theorem 2.4. ∎

Theorem 2.5

For an $\alpha \in S_n$, each of the following statements implies the other ones:

(a) $\alpha \in Pr(\emptyset)$.

(b) $[\alpha]_\equiv = 1$ in the algebra $\mathfrak{A}_n$.

(c) For every algebra $A \in L_n$ and every valuation $v : S_n \to A$ we have

$$| \alpha | = 1_A,$$

$| \cdot |$ denoting the extension of v to a homomorphism of S_n into A.

(d) For every proper n-valued Łukasiewicz algebra A and every homomorphism $h : S_n \to A$ we have

$$h(\alpha) = 1_A.$$

Proof

It is obvious that (a) and (b) are equivalent, by the definition of the unit in $\mathfrak{A}_n$.

Suppose $A \in L_n$, v and $| \cdot |$ are as in (c). Consider the set

$$B = \{\beta \in S_n : | \beta | = 1_A\}.$$

Evidently, all axioms (a_1)–(a_{18}) belong in B, and moreover, B is closed with respect to rules (rd) and (rn). Thus $Pr(\emptyset) \subseteq B$, so statement (b) implies (c). Clearly, (c) implies (d).

To conclude the proof, it remains to show that (d) forces (a). Assume the contrary: suppose there exists a formula β, equivalent to α and not belonging to $[\alpha]_\equiv$. This is a plain contradiction.

The proof is now complete. ∎

We have already mentioned that the n-valued propositional calculus can be axiomatized in terms of just two primitive connectives, the Łukasiewicz arrow ($\to$) and negation ($-$); see the papers by R. Cignoli [Cignoli 1982] and W. Suchoń [Suchoń 1974]. The resulting calculus in equivalent to that presented above.

2.6 Many-Valued Calculi of Post

2.6.1 Bibliographical Remarks

In 1921, E. Post devised an n-valued propositional calculus; his discovery was made independently of Łukasiewicz. The calculus differs from those of Łukasiewicz in two respects: more than one value can be considered as designated, and Post calculi are functionally complete.

The lack of completeness in Łukasiewicz calculi is caused by philosophical subtleties and should not be taken for a drawback. The discovery of Post logics inspired the creation of so-called Post algebras, which provide semantics for the corresponding calculi. The definition of a Post algebra goes back to Rosenbloom [Rosenbloom 1942]. Ever since, the theory of Post algebras and their generalizations has developed dynamically. Some of the most important papers in this subject have been: [Epstein 1960], [Traczyk 1962, 1964, 1967], [Dwinger 1966, 1977], [Rasiowa 1973, 1977], [Rousseau 1969], [Cat-Hao 1973], [Epstein-Horn 1974], along with many others, as well as many newer ones.

Our presentation of Post's n-valued propositional calculi (for $n > 2$) follows the elaboration work of Rasiowa [Rasiowa 1974] and the papers [Orłowska 1985], [Saloni 1972] and [Rousseau 1970]. However, the algebraic characterization in these papers concerns the particular case of Post calculi with only one designated value. The case of more than one designated value is presented without an algebraic characterization.

2.6.2 Post Algebras

By an n-valued Post algebra (or a Post algebra of order n; $n \geq 2$) we mean a system

$$P_n = (P_n, \vee, \wedge, \Longrightarrow, \sim, 1, d_1, d_2, \ldots, d_{n-1}, e_0, e_1, \ldots, e_{n-1})$$

of type $(2,2,2,1,0,1,1,\ldots,1,0,0,\ldots,0)$, such that $(P_n, \vee, \wedge, \Rightarrow, \sim, 1)$ is a Heyting algebra and the remaining structure is subject to the following axioms, holding for i with $1 \leq i \leq n-1$ and for any $a, b \in P_n$:

(p1) $d_i(a \vee b) = d_i(a) \vee d_i(b)$,

(p2) $d_i(a \wedge b) = d_i(a) \wedge d_i(b)$,

(p3) $d_i(a \Rightarrow b) = \bigwedge_{j=1}^{i} (d_j(a) \Rightarrow d_j(b))$,

(p4) $d_i(\sim a) = \sim d_i(a)$,

(p5) $d_i(d_j(a)) = d_j(a)$ for $j = 1, 2, \ldots, n-1$,

(p$_6$) $d_i(e_j) = \begin{cases} 1 & \text{if } i \leq j \\ 0 & \text{otherwise,} \end{cases}$ for $j = 0, 1, \ldots, n-1$; 0 standing for ~ 1,

(p$_7$) $a = \bigvee_{i=1}^{n-1} (d_i(a) \wedge e_i)$,

(p$_8$) $d_1(a) \vee \sim d_1(a) = 1$.

The elements $e_0, e_1, \ldots, e_{n-1}$ are assumed to be all distinct. It follows that P_n is nondegenerate (has at least two elements).

A two-valued Post algebra

$$P_2 = (P_2, \vee, \wedge, \Rightarrow, \sim, 1, d_1, e_0, e_1)$$

is a Boolean algebra. This is justified as follows: Since $a = d_1(a) \wedge e_1$, we have $a \leq e_1 = 1$ for each a. Hence $1 = e_1, 0 = \sim 1 = e_0$. Similarly, we have $e_0 = d_1(e_0) \wedge e_1 = \sim 1 \wedge e_1 = 0 \wedge e_1 = 0$ and $a \vee \sim a = 1$ for all $a \in P_2$.

Lemma 2.9

Suppose P_n is an n-valued Post algebra. Then:

(a) $(P_n, \vee, \wedge, 1)$ is a distributive lattice with zero and unit;

(b) $0 = e_0 \leq e_1 \leq \ldots \leq e_{n-1} = 1$;

(c) $d_i(a) \leq d_j(a)$ for $j \leq i$; $i, j = 1, 2, \ldots, n-1$;

(d) if $a \leq b$ then $d_i(a) \leq d_i(b)$ for $i = 1, 2, \ldots, n-1$;

(e) $d_{n-1}(a) \leq a \leq d_1(a)$;

(f) $e_0 \Rightarrow a = e_{n-1}$,

$e_1 \Rightarrow a = d_1(a)$,

$e_{n-1} \Rightarrow a = a$;

(g) $e_i \Rightarrow a = \bigvee_{j=1}^{i-1} (d_j(a) \wedge e_j) \vee d_i(a)$ for $1 < i < n-1$;

(h) $a \Rightarrow e_0 = \sim a$,

$a \Rightarrow e_{n-1} = e_{n-1}$;

(i) $a \Rightarrow e_i = e_i \vee \sim d_{i+1}(a)$ for $0 \leq i \leq n-1$;

(j) $e_i \Rightarrow e_j = \begin{cases} e_{n-1} & \text{if } i \leq j, \\ e_j & \text{if } i > j \end{cases}$

for $i, j = 0, 1, \ldots, n-1$;

(k) if $e_{i+1} \leq d_j(a) \vee e_i$ holds for a certain i, $0 \leq i < n+1$, then $d_j(a) = e_{n-1}$.

■

The proofs of all these statements can be found in [Rasiowa 1974], where it is also shown that, in every Post algebra P_n, the set of all elements of the form $d_i(a)$, where $i = 1, 2, \ldots, n-1$ and $a \in P_n$, constitutes a Boolean algebra.

Example

Let $P_n = \{e_0, e_1, \ldots, e_{n-1}\}$, $n \geq 2$, $e_i \neq e_j$ for $i \neq j$. Without loss of generality it may be assumed that $e_i = i$. Thus $P_n = \{0, 1, \ldots, n-1\}$; the elements $n-1$ and 0 are respectively the unit and zero of the algebra. Operations $\vee, \wedge, \Rightarrow, \sim, d_i$ are defined by

$$i \vee j = \max\{i, j\},$$

$$i \wedge j = \min\{i, j\},$$

$$i \Rightarrow j = \begin{cases} n-1 & \text{if } i \leq j, \\ j & \text{if } i > j, \end{cases}$$

$$\sim i = \begin{cases} n-1 & \text{if } i = 0, \\ 0 & \text{if } i > 0, \end{cases}$$

$$d_i(j) = \begin{cases} n-1 & \text{if } i \leq j, \\ 0 & \text{if } i > j. \end{cases}$$

This is certainly a Post algebra. We will call it the n-element Post algebra.

Theorem 2.6

Every n-valued Post algebra is functionally complete.

Proof

We show by induction on $k = 0, 1, 2, \ldots$ that every function $f : P_n^k \to P_n$ can be written as a composition of $\wedge, \vee, \Rightarrow, \sim, d_1, \ldots, d_{n-1}, e_0, \ldots, e_{n-1}$.

The claim is true for $k = 0$, as $e_0, \ldots, e_{n-1}$ are the only constant functions in P_n. Fix m and assume the claim for every k with $0 \leq k \leq m$. That is to say, assume that every k-argument function $g : P_n^k \to P_n$ is represented by means of the algebra operations and constants.

Let $f : P_n^{m+1} \to P_n$. Define

$$J_i(x) = d_i(x) \wedge \sim d_{i+1}(x)$$

for $i = 0, 1, \ldots, n-1$. That is,

$$J_0(x) = \sim d_1(x),$$
$$J_1(x) = d_1(x) \wedge \sim d_2(x),$$
$$\cdots$$
$$J_{n-2}(x) = d_{n-2}(x) \wedge \sim d_{n-1}(x),$$
$$J_{n-1}(x) = d_{n-1}(x).$$

Then clearly

$$J_i(e_j) = \begin{cases} e_{n-1} & \text{if } i = j, \\ e_0 & \text{if } i \neq j. \end{cases}$$

Now it suffices to write

$$f(x_1, x_2, \ldots, x_m, x_{m+1}) = \bigvee_{i=0}^{n-1} (f(x_1, x_2, \ldots, x_m, e_i) \wedge J_i(x_{m+1})).$$

This completes the induction. ∎

2.6.3 Post Algebra Filters

Let $P_n = (P_n, \wedge, \vee, \Rightarrow, \sim, d_1, \ldots, d_{n-1}, e_0, \ldots, e_{n-1})$ be an n-valued Post algebra. A set $\nabla \subseteq P_n$ is said to be a filter when

(a) $1 \in \nabla$;

(b) if $a \in \nabla$ and $a \Rightarrow b \in \nabla$ then $b \in \nabla$.

A filter ∇ is called a D-filter when

(c) if $a \in \nabla$ then $d_i(a) \in \nabla$ for $i = 1, 2, \ldots, n-1$.

A D-filter ∇ is said to be:

— proper, if there exists an $a \in P_n$, $a \notin \nabla$;

— irreducible, if it is proper and is not equal to the intersection of any two D-filters other than itself;

— prime, if $a \vee b \in \nabla$ or $b \in \nabla$, for any $a, b \in P_n$;

— maximal, if it is proper and there exists no other proper D-filter properly containing ∇.

Evidently, each maximal D-filter is irreducible.

Let $a \in P_n$. The set

$$\{x \in P_n : d_{n-1}(a) \leq x\}$$

is the D-filter generated by a. A filter of this form is called a principal D-filter.

Now let A be any subset of P_n. The D-filter generated by A is defined to consist of all elements $a \in P_n$ for which there exist $a_1, a_2, \ldots, a_m$ in A such that $\bigwedge_{i=1}^{m} d_{n-1}(a_i) \leq a$.

Lemma 2.10

For every D-filter ∇ in an n-valued Post algebra P_n the following statements are mutually equivalent:

(a) ∇ is a maximal D-filter;

(b) ∇ is an irreducible D-filter;

(c) ∇ is a prime D-filter;

(d) for each $a \in P_n$, either $d_{n-1}(a) \in \nabla$ or $\sim d_{n-1}(a) \in \nabla$. ∎

For a proof of this lemma, as well as for further information about D-filters in Post algebras, we refer to [Traczyk 1964, 1967] or to [Rasiowa 1974].

A topological space X is called an n-valued Post space if

(a) $X = \bigcup_{i=1}^{n-1} X_i$, $X_i \cap X_j = \emptyset$ for $i \neq j$;

(b) there exists a totally disconnected compact Hausdorff space X_0 and there exist homomorphisms g_i mapping X_i onto X_0, $i = 1, 2, \ldots, n-1$;

(c) the set family

$$B(X) = \left\{ \bigcup_{i=1}^{n-1} g_i^{-1}(U) : U \text{ closed-open in } X_0 \right\}$$

is a base for the topology of X.

It follows directly from this definition that an n-valued Post space is compact.

If $X_i, g_i, B(X)$ are as above, we write

$$X = (\{X_i, g_i\}_{i=1,2,\ldots,n-1}, X_0, B(X)).$$

Let an n-valued Post topological space be given. Define the system

$$\mathfrak{B}(X) = (B(X), \cup, \cap, \Rightarrow, -, X)$$

where

$$U \Rightarrow V = (X - U) \cup V,$$
$$-U = X - U;$$

this is the field of all closed-open subsets of X.

Evidently, set fields of that type are instances of Post algebras.

A detailed investigation of those fields and a representation theorem for Post algebras can be found in the papers [Traczyk 1964, 1967], [Dwinger 1977] or in the book [Rasiowa 1974].

2.6.4 The Axiom System for the n-valued Post Calculus

We commence, just as we did in Section 2.4.6, by defining S_n, the set of correctly defined formulas over the alphabet composed of propositional variables, two-argument connectives $\vee, \wedge, \Rightarrow$, one-argument connectives $\sim, d_1, d_2, \ldots, d_{n-1}$, and zero-argument connectives (logical constants) $e_0, e_1, \ldots, e_{n-1}$.

As before, in the case of Łukasiewicz calculus and algebras, we now denote Post-logical connectives by the same symbols which we have used for operations in Post algebras. This should not involve confusion; the context will always indicate the actual meaning of the symbol.

Thus assume $\alpha, \beta, \gamma \in S_n$ and let $\alpha \Leftrightarrow \beta$ be an abbreviation for $(\alpha \Rightarrow \beta) \wedge (\beta \Rightarrow \alpha)$. We will denote by A_n the following system of axioms for the Post calculus:

(a_1) $\alpha \Rightarrow (\beta \Rightarrow \alpha)$,

(a_2) $(\alpha \Rightarrow (\beta \Rightarrow \gamma)) \Rightarrow ((\alpha \Rightarrow \beta) \Rightarrow (\alpha \Rightarrow \gamma))$

(a_3) $\alpha \Rightarrow \alpha \vee \beta$

(a_4) $\beta \Rightarrow \alpha \vee \beta$

(a_5) $(\alpha \Rightarrow \gamma) \Rightarrow ((\beta \Rightarrow \gamma) \Rightarrow (\alpha \vee \beta \Rightarrow \gamma))$

(a_6) $\alpha \wedge \beta \Rightarrow \alpha$

(a_7) $\alpha \wedge \beta \Rightarrow \beta$

(a_8) $(\alpha \Rightarrow \beta) \Rightarrow ((\alpha \Rightarrow \gamma) \Rightarrow (\alpha \Rightarrow \beta \wedge \gamma))$

(a_9) $(\alpha \Rightarrow \sim \beta) \Rightarrow (\beta \Rightarrow \sim \alpha)$

(a_{10}) $\sim (\alpha \Rightarrow \alpha) \Rightarrow \beta$

(a_{11}) $d_i(\alpha \vee \beta) \Leftrightarrow d_i\alpha \vee d_i\beta$ for $i = 1, 2, \ldots, n-1$,

(a_{12}) $d_i(\alpha \wedge \beta) \Leftrightarrow d_i\alpha \wedge d_i\beta$ for $i = 1, 2, \ldots, n-1$,

(a_{13}) $d_i(\alpha \Rightarrow \beta) \Leftrightarrow \bigwedge_{j=1}^{i} (d_j\alpha \Rightarrow d_j\beta)$ for $i = 1, 2, \ldots, n-1$,

(a_{14}) $d_i \sim \alpha \Leftrightarrow \sim d_i\alpha$ for $i = 1, 2, \ldots, n-1$,

(a_{15}) $d_i d_j\alpha \Leftrightarrow d_j\alpha$ for $i, j = 1, 2, \ldots, n-1$,

(a_{16}) if $i \leq j$ then $d_i e_j$; if $i > j$ then $\sim d_i e_j$, for $i = 1, 2, \ldots, n-1$, $j = 0, 1, \ldots, n-1$,

(a_{17}) $\alpha \Leftrightarrow \bigvee_{j=1}^{n-1} (d_j\alpha \wedge e_j)$,

(a_{18}) $d_1\alpha \vee \sim d_1\alpha$.

Primitive rules of inference are

(rd) $\dfrac{\alpha, \alpha \Rightarrow \beta}{\beta}$

(the rule of detachment) and

(rn) $\dfrac{\alpha}{d_{n-1}\alpha}$

(the validity of the d_{n-1}-connective).

The operation of consequence determined by this axiom system and rules (rd), (rn) will be denoted by Cn, and the n-valued Post calculus by P_n. Thus $P_n = (S_n, Cn)$.

It is not hard to verify that P_n, viewed as an algebra, is a Post algebra of order n (see [Rousseau 1970]). Also the following fact is obvious: if a formula $\alpha \in S_n$ is derivable in P_n, then $|\alpha|(v) = 1$ holds for every valuation v of S_n in an arbitrary Post algebra of order n; clearly, $|\alpha|(v)$ denotes the value of α under valuation v.

Lemma 2.11

Calculus P_n is consistent.

Proof

Take any propositional variable; it certainly does not belong to P_n. Thus $P_n \neq S_n$, proving the claim. ∎

Theorem 2.7

Let $\alpha \in S_n$. The following statements are mutually equivalent:

(a) $\alpha \in P_n$;

(b) α is a tautology in the class of Post algebras of order n;

(c) $\mid \alpha \mid_P (X)(v) = 1$ for every Post set-field $P(X)$ of order n and every valuation $v : S_n \to P(X)$;

(d) $\mid \alpha \mid_P (v) = 1$ for every valuation $v : S_n \to P$, P denoting the n-element Post algebra of order n.

Proof

Evidently, (a) is equivalent to (b). Condition (b) trivially implies (c), and (c) implies (d). It remains to show that (a) follows from (d).

Suppose a formula α does not belong in S_n. Then

$$\mid \alpha \mid_P (v) \neq 1,$$

and since

$$d_{n-1} \mid \alpha \mid_P (v) \leq \mid \alpha \mid_P (v),$$

we see that

$$d_{n-1} \mid \alpha \mid_P (v) \notin \nabla_0.$$

So there exists a prime D-filter ∇ such that $\mid \alpha \mid_P (v) \notin \nabla$.

The quotient algebra P/∇ is an n-element Post algebra of order n. Let $h : P \to P/\nabla$ be the induced epimorphism. Since $\mid \alpha \mid_P (v) \notin \nabla$, we see that $h(\mid \alpha \mid_P (v))$ is different from 1, the unit of P/∇.

Let v_0 be a valuation of S_n in P. Then hv_0 is a valuation of S_n in P/∇ and we have

$$\mid \alpha \mid_{P/\nabla} (hv_0) = h(\mid \alpha \mid_P (v_0)) \neq 1,$$

showing that (d) implies (a). ∎

The equivalence between (a) and (d) is known as the *completeness theorem.* A consequence of it is that every P_n is a decidable calculus.

Over each calculus P_n one can construct theories in language of order zero. A theory of that type is defined as a triple $T_n = (S_n, Cn, X)$, where X is any set of formulas contained in S_n. A thorough description of T_n and a study of its properties can be found in [Rasiowa 1974].

2.6.5 Many-Valued Post Calculi with Several Designated Truth Values

Without digging deep into the philosophical controversion of whether it makes sense to designate intermediate values between "absolute truth" and "absolute falsity," Post has emphasized that the logic-values make a chain. The maximal element of this chain certainly expresses "full truth." Along with systems in which the status of truth is attributed to the maximal element alone, he has devised systems with multiple "truths" and "falsities," also arrayed into chains. The result are two-parameter systems P_{nr} with $n > 2$, $r \leq n$. We have already made it clear that our starting point for the construction of logical systems, with n logical values accepted, is a certain n-element algebra with precisely defined operations which implicitly induce an ordering in its universe.

We now present Post propositional calculi P_{nr}. The enumeration of logical values is reversed, as compared with Post's original definition; an insignificant change.

As before, we employ identical symbols to denote propositional connectives and the corresponding functions in the n-element algebra providing the "minimal" interpretation of the calculus. This algebra will be sometimes referred to as the minimal algebra of P_{nr}.

Let $e_0, e_1, \ldots, e_{n-1}$ be the chain of logical values. Without loss of generality we may assume $e_0 = 0, e_1 = 1, \ldots, e_{n-1} = n - 1$. Logical connectives in P_{nr} are interpreted in terms of functions defined on the set $\bar{n} = \{0, 1, \ldots, n-1\}$. Then $\{r, r+1, \ldots, n-1\}$ is the set of designated values (those considered as "true"). In Post logic P_{nr} there are two negations $\sim$, $-$ and disjunction $\vee$. These connectives are characterized by

$$\sim x = (x-1) \bmod n = \begin{cases} x-1 & \text{if } x \neq 0, \\ n-1 & \text{if } x = 0, \end{cases}$$

$$-x = n - 1 - x,$$

$$x \vee y = \max\{x, y\},$$

or, equivalently, by the tables:

x	$\sim x$
0	$n-1$
1	0
2	1
$\vdots$	$\vdots$
$n-3$	$n-4$
$n-2$	$n-3$
$n-1$	$n-2$

x	$-x$
0	$n-1$
1	$n-2$
2	$n-3$
$\vdots$	$\vdots$
$n-3$	2
$n-2$	1
$n-1$	0

$\vee$	0	1	2	...	$n-3$	$n-2$	$n-1$
0	0	1	2	...	$n-3$	$n-2$	$n-1$
1	1	1	2	...	$n-3$	$n-2$	$n-1$
2	2	2	2	...	$n-3$	$n-2$	$n-1$
$\vdots$	$\vdots$	$\vdots$	$\vdots$		$\vdots$	$\vdots$	$\vdots$
$n-3$	$n-3$	$n-3$	$n-3$	...	$n-3$	$n-2$	$n-1$
$n-2$	$n-2$	$n-2$	$n-2$	...	$n-2$	$n-2$	$n-1$
$n-1$	$n-1$	$n-1$	$n-1$	...	$n-1$	$n-1$	$n-1$

Accordingly, the $\sim$ symbol acts as cyclic shift; it decreases each logical value, except 0, by one. Thus, loosely speaking, it is rather weakly negating; the exception is 0, the "falsest" value, which, negated, produces the maximum designated value $n-1$. The disjunction, as in Łukasiewicz systems, takes the value of the "higher truth status" of its two arguments. The $-$ negation is identical to its Łukasiewicz counterpart and is definable by means of $\sim$ and $\vee$.

Conjunction is defined in P_{nr} by

$$\alpha \wedge \beta = -(-\alpha \vee -\beta) = \min\{\alpha, \beta\},$$

or by the table

$\wedge$	0	1	2	...	$n-3$	$n-2$	$n-1$
0	0	0	0	...	0	0	0
1	0	1	1	...	1	1	1
2	0	1	2	...	2	2	2
$\vdots$	$\vdots$	$\vdots$	$\vdots$		$\vdots$	$\vdots$	$\vdots$
$n-3$	0	1	2	...	$n-3$	$n-3$	$n-3$
$n-2$	0	1	2	...	$n-3$	$n-2$	$n-2$
$n-1$	0	1	2	...	$n-3$	$n-2$	$n-1$

A distinguished position in P_{nr} calculi is occupied by implication. Also this functor can be defined either in terms of $\sim$ and $\vee$, or by a table. It is really close to what could be considered a "natural" mode of inference.

The Post implication $\Rightarrow$ is characterized by the following table:

$\Rightarrow$	0	1	2	...	$r-1$	r	$r+1$	...	$n-3$	$n-2$	$n-1$
0	$n-1$	$n-1$	$n-1$	...	$n-1$	$n-1$	$n-1$	...	$n-3$	$n-2$	$n-1$
1	$n-2$	$n-1$	$n-1$	...	$n-1$	$n-1$	$n-1$	...	$n-1$	$n-1$	$n-1$
2	$n-3$	$n-2$	$n-1$	...	$n-1$	$n-1$	$n-1$	...	$n-1$	$n-1$	$n-1$
$\vdots$	$\vdots$	$\vdots$	$\vdots$		$\vdots$	$\vdots$	$\vdots$		$\vdots$	$\vdots$	$\vdots$
$r-1$	$n-r$	$n-r+1$	$n-r+2$	...	$n-1$	$n-1$	$n-1$	...	$n-1$	$n-1$	$n-1$
r	0	1	2	...	$r-1$	$n-1$	$n-1$	...	$n-1$	$n-1$	$n-1$
$r+1$	0	1	2	...	$r-1$	r	$n-1$	...	$n-1$	$n-1$	$n-1$
$\vdots$	$\vdots$	$\vdots$	$\vdots$		$\vdots$	$\vdots$	$\vdots$		$\vdots$	$\vdots$	$\vdots$
$n-3$	0	1	2	...	$r-1$	r	$r+1$	...	$n-1$	$n-1$	$n-1$
$n-2$	0	1	2	...	$r-1$	r	$r+1$	...	$n-3$	$n-1$	$n-1$
$n-1$	0	1	2	...	$r-1$	r	$r+1$	...	$n-3$	$n-2$	$n-1$

r denoting the least one among the designated values. An equivalent definition is by the equality

$$x \Rightarrow y = \begin{cases} n-1 & \text{when } x \leq y, \\ y & \text{when } x > y \text{ and } x \geq r, \\ n-x+y-1 & \text{when } x > y \text{ and } x < r. \end{cases}$$

In the case where $n-1$ is the unique designated value, Post implication coincides with Łukasiewicz implication.

Examples

(Designated values are marked by an asterisk.)

(a) $n = 5, r = 2$:

$\Rightarrow$	0	1	2*	3*	4*
0	4	4	4	4	4
1	3	4	4	4	4
2*	0	1	4	4	4
3*	0	1	2	4	4
4*	0	1	2	3	4

(b) $n = 5, r = 4$:

$\Rightarrow$	0	1	2	3	4*
0	4	4	4	4	4
1	3	4	4	4	4
2	2	3	4	4	4
3	1	2	3	4	4
4*	0	1	2	3	4

The implication defined by the table in (b) coincides with the implication in the 5-valued Łukasiewicz calculus.

In the axiomatic approach to Post calculus, with Post implication as a primitive functor, the modus ponens pattern

$$\frac{\alpha, \alpha \Rightarrow \beta}{\beta}$$

is a primitive rule of inference.

2.6.6 Definability of Functors in the n-valued Post Logic

As remarked above, the n-valued Post connectives are characterized in terms of functions acting in a certain set of n elements, viewed as logical values. Whether they are denoted by $0, 1, \ldots, n-1$ or by any other symbols is of no significance. The connectives are interpreted in the set $\bar{n} = \{0, 1, 2, \ldots, n-1\}$ in a "minimal" fashion. There are many functions whose interpretation in the language of propositional calculus may seem very unnatural. Yet, they are indispensable, e.g., in the proofs of functional completeness of this or another set of connectives.

We now list those functions considered to be fundamental.

(a) $J_i(x) = \begin{cases} n-1 & \text{if } x = i, \\ 0 & \text{if } x \neq i, \end{cases}$

(b) $j_i(x) = \begin{cases} 1 & \text{if } x = i, \\ 0 & \text{if } x \neq i, \end{cases}$

(c) $\max\{x, y\} = x \vee y$,

(d) $\min\{x, y\} = x \wedge y$,

(e) $(x + y) \bmod n$,

(f) $(x - y) \bmod n$.

Functions J_i (for $i \neq n-1$) imitate some properties of negation. Each j_i is the characteristic function of its index i; also the j_i's can be viewed as reflecting some aspects of the concept of negation.

The functions defined in (e) and (f) can be considered as generalizations of disjunction and conjunction.

Now, let $x \circ y$ denote any one of the functions defined by (c), (d), (e) or (f). Then:

(a) $x \circ y = y \circ x$,

(b) $x \circ (y \circ z) = (x \circ y) \circ z$,

(c) $(x \vee y) \wedge z = (x \wedge z) \vee (y \wedge z)$,

(d) $(x \wedge y) \vee z = (x \vee z) \wedge (x \vee z)$,

(e) $J_k J_i(x) = J_k(J_i(x)) = \begin{cases} \bigvee\limits_{\substack{s=0 \\ s \neq i}}^{n-1} J_s(x) & \text{if } k = 0, \\ 0 & \text{if } 0 < k < n-1, \\ J_i(x) & \text{if } k + n - 1, \end{cases}$

(f) $J_k(x \wedge y) = (J_k(x) \wedge \bigvee_{s=k}^{n-1} J_s(y)) \vee (J_k(y) \wedge \bigvee_{s=k}^{n-1} J_s(x))$,

(g) $J_k(x \vee y) = (J_k(x) \wedge \bigvee_{s=0}^{k} J_s(y)) \vee (J_k(y) \wedge \bigvee_{s=0}^{k} J_s(x))$,

(h) $x = \bigvee_{i=1}^{n-1} (i \wedge J_i(x))$

(elimination of "pure" occurrences of a variable),

(i) $x = x \wedge \bigvee_{i=0}^{n-1} J_i(y)$

(introduction of a variable),

(j) cancellation laws:

$$J_s(x)J_t(x) = \begin{cases} J_s(x) & \text{if } s = t, \\ 0 & \text{if } s \neq t, \end{cases}$$

$(n-1) \wedge x = x$,

$(n-1) \vee x = n-1$,

$0 \wedge x = 0$,

$0 \vee x = x$,

(k) every function $f : (\bar{n})^m \to \bar{n}$ can be represented in the form

$$f(x_1, x_2, \ldots, x_m) = \bigvee_{\substack{(s_1, s_2, \ldots, s_m) \\ s_i \in \bar{n}}} \left(\bigwedge_{i=1}^{m} J_{s_i}(x_i) \wedge f(s_1, s_2, \ldots, s_m) \right),$$

called the disjunctive normal form of f.

To write a given function $f : (\bar{n})^m \to \bar{n}$ in its alternative normal form is not a difficult task. Simple application of the above-listed identities usually does the job.

Examples of functionally complete sets of connectives

(a) The set $\{0, 1, 2, \ldots, n-1, J_0, J_1, \ldots, J_{n-1}, \wedge, \vee\}$ is functionally complete. This follows immediately from property (k) above; each function $f : (\bar{n})^m \to \bar{n}$, written in its disjunctive normal form, proves to be a composition of functions from the set.

(b) Define a one-argument operator – by the table

x	$-x$
0	1
1	2
2	3
$\vdots$	$\vdots$
$n-2$	$n-1$
$n-1$	0

or, equivalently, by the equality

$$-x = (x+1) \bmod n.$$

Repeated application of – produces

$$\begin{aligned} x+1 &= -x, \\ x+2 &= -^2x, \\ x+3 &= -^3x, \\ &\ldots \\ x+(n-1) &= (x+(n-2))+1 = -^{n-1}x \end{aligned}$$

and finally

$$-^0x = x = -^nx.$$

Hence

$$n-1 = \bigvee_{i=0}^{n-1} -^ix$$

and

$$k = -^{k+1}(n-1) \quad \text{for} \quad k = 0, 1, 2, \ldots, n-1.$$

We claim that the set $\{-,\vee\}$ is functionally complete. We have just shown how the constants $0, 1, 2, \ldots, n-2, n-1$ are produced using –.
Now we show how the J_i can be constructed.

Forget the definition of J_i and set

$$J_i(x) = -\left(\bigwedge_{t \neq n-1-i} \{x+t\}\right).$$

If $x = i$ then $J_i(x) = n - 1$.
If $x \neq i$ then

$$J_i(x) = \bigwedge_{t \neq n-1-i} \{x + t\} + 1 = (x + (n-1) - x) + 1 = 0.$$

Hence,

$$J_i(x) = \begin{cases} n-1 & \text{for } x = i, \\ 0 & \text{for } x \neq i, \end{cases}$$

as needed.

Finally, define functions $j_{s,i}$ by

$$j_{s,i}(x) = -s + (J_i(x) \vee n - 1 - s).$$

Suppose $f : \bar{n} \to \bar{n}$ is any function. Then

$$f(x) = \bigvee_{i=0}^{n-1} j_{f(i),i}(x).$$

In particular,

$$\sim x = \bigvee_{i=0}^{n-1} j_{n-1-i,i}(x).$$

Consequently $x \wedge y = \sim x \vee \sim y$ is expressed by $-$ and $\vee$.

We have thus shown that the two symbols $-$ and $\vee$ can produce all the functions of the set considered in the preceding example. Hence the claim.

Now look at the two-argument function $|$ defined by the equality

$$x \mid y = -(x \vee y).$$

Consider the one-element set that consists of this function alone. It is functionally complete.

Here we give the table of values of $|$:

$\mid$	0	1	2	3	...	$n-2$	$n-1$
0	1	2	3	4	...	$n-1$	0
1	2	2	3	4	...	$n-1$	0
2	3	3	3	4	...	$n-1$	0
3	4	4	4	4	...	$n-1$	0
$\vdots$	$\vdots$	$\vdots$	$\vdots$	$\vdots$	...	$\vdots$	$\vdots$
$n-2$	$n-1$	$n-1$	$n-1$	$n-1$	...	$n-1$	0
$n-1$	0	0	0	0	...	0	0

This function has been introduced by Webb and corresponds to Sheffer's stroke function from two-valued logic. The logical connective associated with this function is anything but natural.

We do not rigorously distinguish (in notation or elsewhere) between algebraic operations defined on logical values and the respective logical connectives; in most cases this does not cause any confusion. However, the interpretation of logical values is troublesome. It is hard to tell what sort of entities in fact they are. In two-valued logic one calls them truth and falsity; still there remains the problem, pertaining to philosophy, of what is truth and what is falsity. A satisfactory definition has never been given. So it is perhaps safer to take the theoretical standpoint and accept these values just as algebraic objects. The symbols used to denote them are of no importance at all.

The motivation for accepting logical values other than truth and falsity is rather empirical. They serve to provide a working analysis of linguistic phenomena involving such metalanguage expressions as "it's hardly possible" and the like.

In some situations we shall make a precise distinction between a language and its algebraic characterization. As mentioned in the preceding chapter, a language can admit many algebraic interpretations. The minimal interpretation is that decisive for the number of logical values to accept.

3 Survey of Three-Valued Propositional Calculi

What we present below, is an algebraically oriented survey of logical systems with a fixed set of truth-values, of cardinality 3. Calculi will be characterized in terms of their adequate minimal matrices. The general lines of classification and notation have been adopted, with minor changes and additions, from N. Rescher's book [Rescher 1969] and K. Piróg-Rzepecka's paper [Piróg-Rzepecka 1977]. Owing to its historical position and significance, the three-valued calculus of Łukasiewicz (thus a particular case of what has been presented in the preceding chapter) is here again included.

3.1 The Three-Valued Calculus of Łukasiewicz ($Ł_3$)

The primitives of this calculus are: negation $\sim$, implication $\Rightarrow$, and three logical values $0, 1, 2$, viewed respectively as falsity, indefiniteness (or neutrality) and truth. "Truth" (value 2) is the only designated value.

The negation and implication functors are characterized by the tables

x	$\sim x$
0	2
1	1
2	0

$\Rightarrow$	0	1	2
0	2	2	2
1	1	2	2
2	0	1	2

Three further functors, namely conjunction, disjunction and equivalence, are defined by means of $\sim$ and $\Rightarrow$ as follows:

(a) $\alpha \vee \beta$ stands for $(\alpha \Rightarrow \beta) \Rightarrow \beta$;

(b) $\alpha \wedge \beta$ stands for $\sim ((\sim \alpha \Rightarrow \sim \beta)$;

(c) $\alpha \Leftrightarrow \beta$ stands for $(\alpha \Rightarrow \beta) \wedge (\beta \Rightarrow \alpha)$.

Accordingly, these three functors are characterized by the following tables:

$\vee$	0	1	2
0	0	1	2
1	1	1	2
2	2	2	2

$\wedge$	0	1	2
0	0	0	0
1	0	1	1
2	0	1	2

$\Leftrightarrow$	0	1	2
0	2	1	0
1	1	2	1
2	0	1	2

Thus the three-valued Łukasiewicz calculus is given by the minimal matrix

$$\mathfrak{M}\text{Ł}_3 = \langle \{0,1,2\}, \{2\}, \sim, \Rightarrow, \vee, \wedge, \Leftrightarrow \rangle$$

with functions $\sim, \Rightarrow, \vee, \wedge, \Leftrightarrow$ acting in the set of logical values and subject to the tables above.

This calculus, originally defined by matrix $\mathfrak{M}\text{Ł}_3$, was axiomatized in 1931 by M. Wajsberg. His axiom system was the following:

(a_1) $\alpha \Rightarrow (\beta \Rightarrow \alpha)$,

(a_2) $(\alpha \Rightarrow \beta) \Rightarrow ((\beta \Rightarrow \gamma) \Rightarrow (\alpha \Rightarrow \gamma))$,

(a_3) $(\sim \beta \Rightarrow \sim \alpha) \Rightarrow (\alpha \Rightarrow \beta)$,

(a_4) $((\alpha \Rightarrow \sim \alpha) \Rightarrow \alpha) \Rightarrow \alpha$,

with the rule of detachment (modus ponens)

(rd) $$\frac{\alpha, \alpha \Rightarrow \beta}{\beta}$$

For this axiom system, the completeness theorem does hold.

However, this calculus fails to be functionally complete (as has been already remarked). In 1936 J. Słupecki extended Ł$_3$ adding a one-argument propositional functor $T(\cdot)$ characterized by the table

x	Tx
0	1
1	1
2	1

and governed by two further axioms

(a_5) $T\alpha \Rightarrow \sim T\alpha$,

(a_6) $\sim T\alpha \Rightarrow T\alpha$.

The resulting system, called the three-valued calculus of Łukasiewicz-Słupecki, Ł$_3$S, is functionally complete.

3.2 The Three-Valued Calculus of Bochvar

The first system of so-called nonsense-logic was devised by D. A. Bochvar (Bočvar) [Bochvar 1939], intended as a tool for overcoming problems that arise from logical antinomies. Bochvar makes a distinction between statements and sentences. The latter are either true or false, whereas the former can be true, false or nonsense. If a statement makes sense, it is necessarily true or false. These underlying ideas led Bochvar to the construction of two three-valued logical systems, defined by the matrices

(a) $\mathfrak{M}B_3^{I_1} = \langle\{0,1,2\},\{2\},\sim,\wedge,\vee,\Rightarrow,\Leftrightarrow\rangle$,

(b) $\mathfrak{M}B_3^{I_2} = \langle\{0,1,2\},\{1,2\},\sim,\wedge,\vee,\Rightarrow,\Leftrightarrow\rangle$,

$(0,1,2$ denoting respectively falsity, nonsense, truth), and a third system

(c) $\mathfrak{M}B_3^{E} = \langle\{0,1,2\},\{2\},\sim,\wedge,\vee,\Rightarrow,\Leftrightarrow\rangle$.

The connectives in(a) and (b) are characterized by the tables

x	$\sim x$
0	2
1	1
2	0

$\wedge$	0	1	2
0	0	1	0
1	1	1	1
2	0	1	2

$\vee$	0	1	2
0	0	1	2
1	1	1	1
2	2	1	2

$\Rightarrow$	0	1	2
0	2	1	2
1	1	1	1
2	0	1	2

$\Leftrightarrow$	0	1	2
0	2	1	0
1	1	1	1
2	0	1	2

The connectives of (c) are given by

x	$\sim x$
0	2
1	2
2	0

$\wedge$	0	1	2
0	0	0	0
1	0	0	0
2	0	0	2

$\vee$	0	1	2
0	0	0	2
1	0	0	2
2	2	2	2

$\Rightarrow$	0	1	2
0	2	2	2
1	2	2	2
2	0	0	2

$\Leftrightarrow$	0	1	2
0	2	2	0
1	2	2	0
2	0	0	2

The set of tautologies of the calculus defined by $\mathfrak{M}B_3^{I_1}$ is empty. The calculi defined by $\mathfrak{M}B_3^{I_2}$ and $\mathfrak{M}B_3^{E}$ have their sets of tautologies identical with the classical two-valued system C_2, defined by the matrix

$$\mathfrak{M}C_2 = \langle\{0,1\},\{1\},\sim,\wedge,\vee,\Rightarrow,\Leftrightarrow\rangle,$$

with connectives

x	$\sim x$
0	1
1	0

$\wedge$	0	1
0	0	0
1	0	1

$\vee$	0	1
0	0	1
1	1	1

$\Rightarrow$	0	1
0	1	1
1	0	1

$\Leftrightarrow$	0	1
0	1	0
1	0	1

Matrices $\mathfrak{M}B_3^{I_2}$ and $\mathfrak{M}B_3^{E}$ are homomorphic with $\mathfrak{M}C_2$; hence the equality between their sets of tautologies. Moreover, the axiom systems of these two calculi in fact coincide with that of the classical two-valued propositional calculus.

3.3 The Three-Valued Calculus of Finn

Bochvar's work has been carried on by W. K. Finn [Finn 1972]. His ideas are akin to those of Bochvar; differences occur in the interpretation of variables in the matrix. Finn's primitive functors are: negation $\sim$, conjunction $\wedge$ and implication $\Rightarrow$, subject to tables identical to Bochvar's; and thus,

$$\mathfrak{M}F_3 = \langle\{0,1,2\},\{2\},\sim,\wedge,\Rightarrow\rangle.$$

Finn also considers two types of variables: propositional and sentential, the distinction being that propositional variables can take all the three logical values, while sentential variables can only have value 0 or 2 (false or true). Other functors in Finn's system are defined by

(a) $\vdash \alpha$ stands for $\alpha \Rightarrow \alpha$,

(b) $-\alpha$ stands for $\vdash\sim \alpha$,

(c) $\alpha \vee \beta$ stands for $\sim(\sim\alpha \wedge \sim\beta)$,

(d) $\alpha \supset \beta$ stands for $\sim(\alpha \wedge \sim\beta)$,

(e) $\#\alpha$ stands for $\sim (\vdash \alpha \vee -\alpha)$.

(f) $\alpha \Leftrightarrow \beta$ stands for $(\alpha \Rightarrow \beta) \wedge (\beta \Rightarrow \alpha)$.

This system has been axiomatized by the following set of axioms (imposed on the propositional variables):

(a_1) $p \Rightarrow (q \Rightarrow p)$,

(a_2) $(p \Rightarrow (q \Rightarrow r)) \Rightarrow ((p \Rightarrow a) \Rightarrow (p \Rightarrow r))$,

(a_3) $((p \Rightarrow q) \Rightarrow p) \Rightarrow p$,

(a_4) $p \wedge q \Rightarrow p$,

(a_5) $p \wedge q \Rightarrow q \wedge p$,

(a_6) $(p \Rightarrow q) \Rightarrow ((p \Rightarrow r) \Rightarrow (p \Rightarrow q \wedge r))$,

(a_7) $(p \Rightarrow r) \Rightarrow ((q \Rightarrow r) \Rightarrow (p \vee q \Rightarrow r))$,

(a_8) $p \vee q \Rightarrow q \vee p$,

(a_9) $(p \vee q) \vee r \Rightarrow p \vee (q \vee r)$,

(a_{10}) $p \wedge q \Rightarrow p \vee q$,

(a_{11}) $(p \vee q) \wedge r \Leftrightarrow (p \wedge r) \vee (q \wedge r)$,

(a_{12}) $r \wedge (p \vee q) \Leftrightarrow (r \wedge p) \vee (r \wedge q)$,

(a_{13}) $p \Leftrightarrow \sim\sim p$,

(a_{14}) $\sim p \Rightarrow (p \Rightarrow q)$,

(a_{15}) $p \Rightarrow (\sim q \Rightarrow \sim (p \Rightarrow q))$,

(a_{16}) $\sim p \Rightarrow (p \vee q \Leftrightarrow q)$,

(a_{17}) $\sim p \wedge \sim q \Leftrightarrow \sim (p \vee q)$,

(a_{18}) $\sim p \vee \sim q \Leftrightarrow \sim (p \wedge q)$,

(a_{19}) $p \Rightarrow x \vee p$,

(a_{20}) $(\sim x \Rightarrow \sim p) \Rightarrow (p \Rightarrow x)$,

(a_{21}) $\sim \# x$,

(a_{22}) $\# p \Rightarrow \# (p \vee q)$,

(a_{23}) $\vdash p \wedge \# q \Rightarrow \sim (p \Rightarrow q)$.

Primitive rules are:

(rd) $$\frac{\alpha, \alpha \Rightarrow \beta}{\beta}$$

(the rule of detachment) and two rules of substitution, for propositional and sentential variables.

3.4 The Three-Valued Calculus of Hallden

In Hallden's three-valued calculus [Hallden 1949b], the third logical value (that comes along with truth and falsity) is understood as something like meaninglessness. Regarding the philosophical premises of propositional calculi, it is a controversial question whether the lack of meaning (or no-sense-making) can be justifiably given the status of a logical value, yet more, of a designated value. Without going into that subtlety, we describe Hallden's calculus formally. It is defined by the matrix

$$\mathfrak{M}H = \langle \{0,1,2\}, \{1,2\}, \vdash, \sim, \wedge \rangle,$$

$0, 2$ and 1 assigned respectively to false, true and meaningless statements. The functors $\vdash, \sim, \wedge$ are defined by the tables

x	$\vdash x$
0	2
1	0
2	2

x	$\sim x$
0	2
1	1
2	0

$\wedge$	0	1	2
0	0	1	0
1	1	1	1
2	0	1	2

The expression $\vdash \alpha$ is to be understood as the statement that α is meaningful. Functors $\sim$ and $\wedge$ are counterparts of negation and conjunction; it is assumed that the conjunction of a meaningless statement with any statement is meaningless. Furthermore,

(a) $\alpha \vee \beta$ stands for $\sim (\sim \alpha \wedge \sim \beta)$,

(b) $\alpha \Rightarrow \beta$ stands for $\sim \alpha \vee \beta$,

(c) $\alpha \Leftrightarrow \beta$ stands for $(\alpha \Rightarrow \beta) \wedge (\beta \Rightarrow \alpha)$,

(d) $-\alpha$ stands for $\sim \vdash \alpha$.

Thus $-\alpha$ says that α is meaningless.

Hallden assumes the following system of axioms (imposed on propositional variables):

(a_1) $(\sim p \Rightarrow p) \Rightarrow p$,

(a_2) $p \Rightarrow (\sim p \Rightarrow q)$,

(a_3) $(p \Rightarrow q) \Rightarrow ((q \Rightarrow r) \Rightarrow (p \Rightarrow r))$,

(a_4) $\vdash p \Leftrightarrow \vdash \sim p$,

(a_5) $\vdash (p \wedge q) \Leftrightarrow (\vdash p \wedge \vdash q)$,

(a_6) $p \Rightarrow \vdash p$.

The rules of detachment and of substitution are the primitives; both are analogous to classical logic. Theorems on consistency and completeness hold for this axiomatic theory.

3.5 The Three-Valued Calculus of Åqvist

Taking Hallden's calculus as a starting point, L. Åqvist [Åqvist 1962] has created a three-valued calculus that differs from Hallden's in essential features. Along with truth and falsity, Åqvist admits statements which he calls normative, and to which neither truth nor falsity status is ascribed. He considers false statements, meaningful statements and meaningless ones. In his calculus, 2 is the only designated value. The calculus is defined by the matrix

$$\mathfrak{M}A = \langle \{0,1,2\}, \{2\}, \#, \sim, \vee \rangle$$

with connectives $\#, \sim, \vee$ characterized as follows:

x	$\#x$
0	0
1	0
2	2

x	$\sim x$
0	2
1	1
2	0

$\vee$	0	1	2
0	0	1	2
1	1	1	2
2	2	2	2

Further functors are defined by

(a) $F\alpha$ stands for $\# \sim \alpha$,

(b) $L\alpha$ stands for $\# \alpha \vee F\alpha$,

(c) $M\alpha$ stands for $\sim L\alpha$,

(d) $\alpha \Rightarrow \beta$ stands for $\sim \alpha \vee \beta$,

(e) $\alpha \Leftrightarrow \beta$ stands for $(\alpha \Rightarrow \beta) \wedge (\beta \Rightarrow \alpha)$,

(f) $\alpha \wedge \beta$ stands for $\sim (\sim \alpha \vee \sim \beta)$.

Loosely speaking, $F\alpha$ says "α is false;" $L\alpha$ says "α is meaningful;" $M\alpha$ says "α is meaningless." Åqvist has given a system of six axioms for this calculus and proved consistency and completeness theorems. For further information we refer to the source work, [Piróg-Rzepecka 1977].

3.6 The Three-Valued Calculi of Segerberg

K. Segerberg [Segerberg 1965], like Åqvist, has taken up Hallden's idea and created a set of three-valued propositional calculi. Primitive functors are: $\sim, \#, \vdash, \wedge$. Segerberg has defined his systems in terms of matrices and has axiomatized them, with proofs of consistency and completeness.

Here we restrict ourselves to matrix characterization of Segerberg's calculi S_1, S_2 and S_3. First,

$$\mathfrak{M}S_1 = \langle \{0,1,2\}, \{1,2\}, \sim, \#, \wedge \rangle$$

with connectives $\sim, \#, \wedge$ subject to

x	$\sim x$
0	2
1	1
2	0

x	$\# x$
0	0
1	0
2	2

$\wedge$	0	1	2
0	0	1	0
1	1	1	1
2	0	1	2

Other functors are defined by

(a) $\alpha \vee \beta$ stands for $\sim (\sim \alpha \wedge \sim \beta)$,

(b) $\alpha \Rightarrow \beta$ stands for $-\alpha \vee \beta$,

(c) $\alpha \Leftrightarrow \beta$ stands for $(\alpha \Rightarrow \beta) \wedge (\beta \Rightarrow \alpha)$,

(d) $\| \alpha$ stands for $\# \sim \alpha$,

(e) $= \alpha$ stands for $\sim (\sim \# \alpha \wedge \sim \# \sim \alpha)$,

(f) $-\alpha$ stands for $\sim \# \alpha \wedge \sim \# \sim \alpha$.

The symbols used in the last three axioms are to be read:

$\| \alpha :$ α is false,

$= \alpha :$ α makes sense,

$-\alpha :$ α makes no sense.

Axiomatization for this calculus is rather complicated, but the consistency and completeness theorems hold for it.

The two other calculi of Segerberg, also functionally complete, are given by the following matrices $\mathfrak{M}S_2$ and $\mathfrak{M}S_3$, with the corresponding tables imposed on the connectives:

$$\mathfrak{M}S_2 = \langle\{0,1,2\},\{1,2\},\vdash,\wedge\rangle,$$

x	$\vdash$
0	2
1	0
2	1

$\wedge$	0	1	2
0	0	1	0
1	1	1	1
2	0	1	2,

and

$$\mathfrak{M}S_3 = \langle\{0,1,2\},\{2\},\sim,\wedge\rangle,$$

x	$\sim x$
0	2
1	0
2	1

$\wedge$	0	1	2
0	0	0	0
1	0	1	1
2	0	1	2

Note that $\mathfrak{M}S_3$ coincides with the three-valued Post calculus, with one value designated.

3.7 The Three-Valued Calculus of Piróg-Rzepecka

This calculus, defined by K. Piróg-Rzepecka [Piróg-Rzepecka 1977], matrix-wise and axiomatically, is judged by its author to be another member of the class of so-called nonsense-logics.

The defining matrix is

$$\mathfrak{M}PR = \langle\{0,1,2\},\{2\},\sim,\wedge,\Rightarrow\rangle.$$

To these primitive functors $(\sim,\wedge,\Rightarrow)$, three further functors $(-,\vee,\Leftrightarrow)$ are adjoined. They are jointly characterized by the system of tables

x	$\sim x$
0	2
1	1
2	0

x	$-x$
0	2
1	2
2	0

$\wedge$	0	1	2
0	0	1	0
1	1	1	1
2	0	1	2

$\vee$	0	1	2
0	0	1	2
1	1	1	1
2	2	1	2

$\Rightarrow$	0	1	2
0	2	2	2
1	2	2	2
2	0	0	2

$\Leftrightarrow$	0	1	2
0	2	2	0
1	2	2	0
2	0	0	2

Plainly, the non-primitive functors are defined in terms of the primitives as follows:

(a) $\alpha \vee \beta$ stands for $\sim(\sim \alpha \wedge \sim \beta)$,

(b) $\alpha \Leftrightarrow \beta$ stands for $(\alpha \Rightarrow \beta) \wedge (\beta \Rightarrow \alpha)$,

(c) $-\alpha$ stands for $(\alpha \Rightarrow \sim \alpha)$.

Further,

(d) $\alpha \# \beta$ stands for $-\alpha \Rightarrow \beta$,

(e) $\alpha \circ \beta$ stands for $-(-\alpha \# -\beta)$,

and we have

(f) $\alpha \Leftrightarrow \beta$ if and only if $(\alpha \Rightarrow \beta) \circ (\beta \Rightarrow \alpha)$.

The calculus is axiomatized through the following system of axioms (imposed on propositional variables):

(a_1) $(p \Rightarrow q) \Rightarrow ((q \Rightarrow r) \Rightarrow (p \Rightarrow r))$,

(a_2) $p \Rightarrow (q \Rightarrow p)$,

(a_3) $((p \Rightarrow q) \Rightarrow p) \Rightarrow p$,

(a_4) $p \wedge q \Rightarrow p$,

(a_5) $p \wedge q \Rightarrow q$,

(a_6) $p \Rightarrow (q \Rightarrow p \wedge q)$,

(a_7) $p \Rightarrow (\sim p \Rightarrow q)$,

(a_8) $\sim (\sim p) \Rightarrow p$,

(a_9) $p \Rightarrow \sim (\sim p)$,

(a_{10}) $p \Rightarrow ((q \Rightarrow \sim q) \Rightarrow \sim (p \Rightarrow q))$,

(a_{11}) $\sim (p \wedge q) \Rightarrow \sim (q \wedge p)$,

(a_{12}) $\sim (p \wedge q) \Rightarrow ((p \Rightarrow \sim p) \Rightarrow \sim p)$,

(a_{13}) $p \wedge \sim q \Rightarrow \sim (p \wedge q)$,

(a_{14}) $\sim p \wedge \sim q \Rightarrow \sim (p \wedge q)$.

Primitive rules of inference are the rule of detachment and the rule of substitution (for propositional variables). The consistency theorem and the completeness theorem hold too, with respect to matrix $\mathfrak{M}PR$.

3.8 The Three-Valued Calculus of Heyting

In the classical two-valued logic the statement

$$(\alpha \Rightarrow \sim\sim \alpha) \wedge (\sim\sim \alpha \Rightarrow \alpha)$$

is tautologous. In other words, it is accepted that α twice negated is exactly the same as α.

The idea of rejecting the second member of this conjunction, while retaining the first, goes back to L. E. J. Brouwer. It was A. Heyting who, inspired by this idea, constructed the calculus with matrix

$$\mathfrak{M}He = \langle \{0, 1, 2\}, \{2\}, \sim, \Rightarrow, \wedge, \vee \rangle$$

plus characterizing tables

x	$\sim x$
0	2
1	0
2	0

$\Rightarrow$	0	1	2
0	2	2	2
1	0	2	2
2	0	1	2

$\wedge$	0	1	2
0	0	0	0
1	0	1	1
2	0	1	2

$\vee$	0	1	2
0	0	1	2
1	1	1	2
2	2	2	2

It is readily verified that $\sim\sim \alpha \Rightarrow \alpha$ is not a tautology, and neither is the law of excluded middle $\alpha \vee \sim \alpha$; whereas the expressions $\alpha \Rightarrow\sim\sim \alpha$ and $\sim\sim \alpha \vee \sim \alpha$ are tautologous. Attempts to devise an axiom system that would have $\mathfrak{M}He$ for an adequate matrix have led to the intuitionistic propositional calculus. Its axioms are satisfied in the matrix $\mathfrak{M}He$; however, there exist expressions which are satisfied in $\mathfrak{M}He$ and yet are not derivable from the axioms. Thus $\mathfrak{M}He$ has turned out to be inadequate for the intuitionistic calculus.

3.9 The Three-Valued Calculus of Kleene

S. C. Kleene [Kleene 1952] created his three-valued calculus in order to cope with problems involving partial recursive functions and arising when the concept of indefiniteness comes in. Beside truth and falsity, Kleene admits a third logical value for indefinite statements. Truth is the designated value. One of the guidelines assumed by Kleene is that, for instance, expression $a(x) \vee b(x)$ could make sense (take one of the two definite values) in certain situations where $a(x)$ or $b(x)$ is indefinite.

In fact, Kleene has introduced two three-valued calculi, which we denote by Kl_1 and Kl_2. They are defined by the following matrices and function-tables characterising their connectives:

$$\mathfrak{M}Kl_1 = \langle \{0,1,2\}, \{2\}, \sim, \vee, \wedge, \Rightarrow, \Leftrightarrow \rangle$$

with

x	$\sim x$
0	2
1	1
2	0

$\vee$	0	1	2
0	0	1	2
1	1	1	2
2	2	2	2

$\wedge$	0	1	2
0	0	0	0
1	0	1	1
2	0	1	2

$\Rightarrow$	0	1	2
0	2	2	2
1	1	1	2
2	0	1	2

$\Leftrightarrow$	0	1	2
0	2	1	0
1	1	1	1
2	0	1	2

and

$$\mathfrak{M}Kl_2 = \langle \{0,1,2\}, \{2\}, \sim, \wedge, \vee, \Rightarrow, \Leftrightarrow, \equiv \rangle$$

with $\sim, \wedge, \vee$ as above and

$\Rightarrow$	0	1	2
0	2	2	2
1	1	2	2
2	0	1	2

$\Leftrightarrow$	0	1	2
0	2	1	0
1	1	2	1
2	0	1	2

$\equiv$	0	1	2
0	2	0	0
1	0	2	0
2	0	0	2

The two calculi, Kl_1 and Kl_2, are respectively referred to as strong and weak.

The calculus defined by matrix $\mathfrak{M}Kl_1$ has a void set of tautologies. However, if one takes 2 and 1 for designated values, then matrix $\mathfrak{M}Kl_1$ becomes homomorphic to the classical two-valued matrix, the set of tautologies coinciding with that of classical logic.

The calculus defined by $\mathfrak{M}Kl_2$ is a slight extension of the three-valued Łukasiewicz calculus.

3.10 The Three-Valued Calculus of Reichenbach

H. Reichenbach [Reichenbach 1946] created his calculus in 1946, in an attempt to overcome certain philosophical and logical difficulties that arise in quantum mechanics. Along with conjunction and disjunction, he introduced three negation functors, three implications and two equivalences. The matrix is

$$\mathfrak{M}R = \langle \{0,1,2\}, \{2\}, \sim, -, \neg, \vee, \wedge, \supset, \Rightarrow, \rightarrow, \equiv, \Leftrightarrow \rangle,$$

with connectives characterized by the tables

x	$\sim x$
0	1
1	2
2	0

x	$-x$
0	2
1	1
2	0

x	$\neg x$
0	2
1	2
2	1

$\vee$	0	1	2
0	0	1	2
1	1	1	1
2	2	2	2

$\wedge$	0	1	2
0	0	0	0
1	0	1	1
2	0	1	2

$\supset$	0	1	2
0	2	2	2
1	1	2	2
2	0	1	2

$\Rightarrow$	0	1	2
0	2	2	2
1	2	2	2
2	0	0	2

$\rightarrow$	0	1	2
0	1	1	1
1	1	1	1
2	0	1	2

$\equiv$	0	1	2
0	2	1	0
1	1	2	1
2	0	1	2

$\Leftrightarrow$	0	1	2
0	2	0	0
1	0	2	0
2	0	0	2

Reichenbach calls his three negations $\sim, -, \neg$ respectively: cyclic, diametrical and full negation.

The calculus is readily seen to be functionally complete.

3.11 The Three-Valued Calculus of Słupecki

In his paper [Słupecki 1946], J. Słupecki formulated yet another variant of a functionally complete three-valued propositional calculus. It is defined by the matrix

$$\mathfrak{M}Sl = \langle \{0,1,2\}, \{2\}, \sim, \neg, \Rightarrow \rangle$$

with connectives

x	$\sim x$
0	2
1	2
2	0

x	$\neg x$
0	2
1	2
2	1

$\Rightarrow$	0	1	2
0	2	2	2
1	2	2	2
2	0	1	2

The motive was purely logical: to construct a three-valued calculus which was functionally complete and fulfilled certain additional requirements.

Słupecki has given an adequate axiom system for his calculus. His axioms are:

(a_1) $(p \Rightarrow q) \Rightarrow ((q \Rightarrow r) \Rightarrow (p \Rightarrow r))$,

(a_2) $(\neg p \Rightarrow p) \Rightarrow p$,

(a_3) $p \Rightarrow (\neg p \Rightarrow q)$,

(a_4) $\sim p \Rightarrow \neg p$,

(a_5) $\sim (p \Rightarrow q) \Rightarrow \sim q$,

(a_6) $p \Rightarrow (\sim q \Rightarrow \sim (p \Rightarrow q))$,

(a_7) $\sim\sim p \Rightarrow p$,

(a_8) $p \Rightarrow \sim\sim p$,

(a_9) $\neg \sim \neg p \Rightarrow \sim p$.

Primitive laws are the rule of detachment (modus ponens) and the rule of substitution for propositional variables.

3.12 The Three-Valued Calculus of Sobociński

B. Sobociński [Sobociński 1936] defines his three-valued calculus with two designated values by the matrix

$$\mathfrak{M}Sb = \langle \{0,1,2\}, \{1,2\}, \sim, \wedge, \vee, \Rightarrow, \Leftrightarrow \rangle,$$

characterized by the tables of connectives:

x	$\sim x$
0	1
1	2
2	0

$\wedge$	0	1	2
0	0	0	0
1	0	1	2
2	0	2	2

$\vee$	0	1	2
0	0	0	2
1	0	1	2
2	2	2	2

$\Rightarrow$	0	1	2
0	2	1	2
1	0	2	2
2	0	1	2

$\Leftrightarrow$	0	1	2
0	2	0	0
1	0	2	1
2	0	1	2

The calculus is axiomatized by the following system of axioms:

(a_1) $((p \Rightarrow q) \Rightarrow (r \Rightarrow s)) \Rightarrow (t \Rightarrow ((s \Rightarrow p) \Rightarrow (r \Rightarrow p)))$,

(a_2) $\sim\sim\sim p \Rightarrow p$,

(a_3) $p \Rightarrow \sim\sim\sim p$,

(a_4) $\sim\sim (p \Rightarrow q) \Rightarrow (\sim\sim p \Rightarrow \sim\sim q)$,

(a_5) $(\sim\sim p \Rightarrow \sim\sim q) \Rightarrow \sim\sim (p \Rightarrow q)$,

(a_6) $\sim (p \Rightarrow q) \Rightarrow \sim q$,

(a_7) $\sim (p \Rightarrow q) \Rightarrow ((p \Rightarrow q) \Rightarrow q)$,

(a_8) $\sim (p \Rightarrow q) \Rightarrow (p \Rightarrow (\sim p \Rightarrow \sim\sim q))$,

(a_9) $(p \Rightarrow (\sim p \Rightarrow \sim\sim q)) \Rightarrow (\sim q \Rightarrow ((p \Rightarrow q) \Rightarrow q) \Rightarrow \sim (p \Rightarrow q))$,

(a_{10}) $((p \Rightarrow q) \Rightarrow r) \Rightarrow (\sim p \Rightarrow (\sim\sim p \Rightarrow r))$,

(a_{11}) $(q \Rightarrow r) \Rightarrow ((\sim p \Rightarrow (\sim\sim p \Rightarrow r)) \Rightarrow ((p \Rightarrow q) \Rightarrow r))$,

(a_{12}) $\sim\sim (p \Rightarrow p)$,

(a_{13}) $p \Rightarrow (\sim p \Rightarrow (\sim\sim p \Rightarrow q))$.

with modus ponens and the law of substitution (for propositional variables) as primitive rules.

The above axiom system involves $\sim$ and $\Rightarrow$ alone. The remaining connectives are defined in terms of these two.

4 Some n-valued Propositional Calculi: A Selection

So far we have presented, in more or less detail, the n-valued calculi of Łukasiewicz and Post. The Łukasiewicz calculus differs from the Post calculus in that it is not functionally complete (at least as long as we are considering the original version).

In the rest of this book we ignore calculi defined by matrices whose contents coincide with those of the classical two-valued logic. We also do not present calculi given by matrices of cardinality $\aleph_0$; these will be the subject of a separate treatise. We just make a survey of n-valued calculi other than those of Łukasiewicz and Post.

4.1 The Many-Valued Calculus of Słupecki

In 1938, J. Słupecki [Słupecki 1938] defined a very extensive class of n-valued logics. Those calculi are complete and functionally complete, and are given by the matrices

$$\mathfrak{M}Slnr = \langle\{0,1,\ldots,n-1\},\{r,r+1,\ldots,n-1\},\Rightarrow,\sim,\neg\rangle$$

with functors characterized by the tables:

$\Rightarrow$	0	1	2	...	$r-1$	r	$r+1$	...	$n-1$
0	$n-1$	$n-1$	$n-1$	...	$n-1$	$n-1$	$n-1$	...	$n-1$
1	$n-1$	$n-1$	$n-1$	...	$n-1$	$n-1$	$n-1$	...	$n-1$
2	$n-1$	$n-1$	$n-1$	...	$n-1$	$n-1$	$n-1$	...	$n-1$
$\vdots$	$\vdots$	$\vdots$	$\vdots$		$\vdots$	$\vdots$	$\vdots$		$\vdots$
$r-1$	$n-1$	$n-1$	$n-1$	...	$n-1$	$n-1$	$n-1$	...	$n-1$
r	0	1	2	...	$r-1$	r	$r+1$	...	$n-1$
$r+1$	0	1	2	...	$r-1$	r	$r+1$	...	$n-1$
$\vdots$	$\vdots$	$\vdots$	$\vdots$		$\vdots$	$\vdots$	$\vdots$		$\vdots$
$n-1$	0	1	2	...	$r-1$	r	$r+1$	...	$n-1$

x	$\sim x$
0	$n-1$
1	0
2	1
$\vdots$	$\vdots$
$r-1$	$r-2$
r	$r-1$
$r+1$	r
$\vdots$	$\vdots$
$n-3$	$n-4$
$n-2$	$n-3$
$n-1$	$n-2$

x	$\neg x$
0	$n-1$
1	$n-1$
2	$n-1$
$\vdots$	$\vdots$
$r-1$	$n-1$
r	$n-1$
$r+1$	$n-1$
$\vdots$	$\vdots$
$n-3$	$n-1$
$n-2$	$n-1$
$n-1$	$n-2$

He has axiomatized his calculi and has proved their completeness and decidability.

The axiom system for the Słupecki calculus is not easy to formulate. We have to prepare several auxiliary definitions.

Let α be a formula. Deleting from α all one-argument connectives we obtain a formula μ, which we call the core of α. Thus the core of α is either a single variable or a formula involving no connectives other than implication.

A formula whose core is a single variable has order 1. A formula is said to be of order m, $m > 1$, if its core has the form

$$\mu_1 \Rightarrow (\mu_2 \Rightarrow (\mu_3 \Rightarrow (\ldots \Rightarrow (\mu_l \Rightarrow \nu_l)\ldots))),$$

where all μ_i are of order less than or equal to $m-1$, with equality for at least one i.

Let $A_1 = (0, 1, 2, \ldots, n-1)$ and let A_i, $i = 1, 2, \ldots, n!$, be all permutations of the sequence A_1.

Suppose a formula α contains propositional variables $p_1, p_2, \ldots, p_s$. The symbol $w(\alpha, c)$ will denote the value of α on the sequence $c \in \bar{n}^{\bar{s}}$ where $\bar{n} = \{0, 1, 2, \ldots, n-1\}$, $\bar{s} = \{1, 2, 3, \ldots, s\}$. Let $i^* = \{1, 2, 3, \ldots, n!\}$. For each $i \in i^*$ there exist formulas α of order 1 such that, for any value $w(p, c)$, $w(\alpha, w(p, c)) = \alpha_i \in A_i$; this follows from a theorem of Picard [Picard 1935].

Choose any one of those formulas, having no more variables than the remaining ones, and denote it by $\alpha_i p$. The number of its variables will be denoted by j_i. Define $j = \max\{j_i : i \in i^*\}$.

For $m = 1, 2, 3, \ldots, \sum_{l=0}^{j} 2^l$, we will denote by $\beta_m p$ an arbitrary formula of order 1 that contains variable p preceded by a combination of functors $\sim$ and $\neg$ (in any order), their joint number not exceeding j.

Evidently, for every formula $\beta_m p$ there exists i such that

$$w(\beta_m p, c) = w(\alpha_i p, c).$$

Any formula with this property will be denoted by the symbol $\beta_l^i p$, with subscript l ranging over a subsequence of $1, 2, 3, \ldots, \sum_{s=0}^{j} 2^s$.

Further, define

$$\begin{aligned} \Gamma &= \{\alpha_i p : w(\alpha_i p, n-1) < r,\ i = 1,2,3,\ldots,(n-1)!(n-k)\} \\ \Delta &= \{\alpha_i p : w(\alpha_i p, n-1) \geq r,\ i = 1,2,3,\ldots,(n-1)!r\} \\ \Phi &= \{\alpha_i p : w(p,t) < r, t \geq n-r,\ w(\alpha_i p, r+1) = n-1, \\ &\qquad i = 0,1,2,\ldots,n-r-1\} \end{aligned}$$

In what follows we will write $\gamma_i p$, $\delta_i p$, $\varphi_i p$ and $\psi_i p$ for formulas belonging in Γ, Δ, Φ and Ψ, respectively.

With this notation, we are able to formulate the axioms of the n-valued Słupecki calculus $Slnr$ as follows:

$$\Psi = \{\alpha_i p : w(p,t) \geq r,\ t < n-k,\ w(\alpha_i p, i) = 0,\ i = 0,1,2,\ldots,r\}$$

(a_1) $((p \Rightarrow q) \Rightarrow z) \Rightarrow ((z \Rightarrow p) \Rightarrow (s \Rightarrow p))$,

(a_2il) $\alpha_i p \Rightarrow \beta_l^i p;\ i,l = 1,2,3,\ldots,n$,

(a_3il) $\beta_l^i p \Rightarrow \alpha_i p;\ i,l = 1,2,3,\ldots,n$,

(a_4i) $\gamma_i(p \Rightarrow q) \Rightarrow p;\ i = 1,2,3,\ldots,(n-1)!(n-r)$,

(a_5i) $\gamma_i(p \Rightarrow q) \Rightarrow \gamma_i q;\ i = 1,2,3,\ldots,(n-1)!(n-r)$,

(a_6i) $p \Rightarrow (\gamma_i q \Rightarrow \gamma_i(p \Rightarrow q));\ i = 1,2,3,\ldots,(n-1)!(n-r)$,

(a_7i) $\delta_i(p \Rightarrow q) \Rightarrow (p \Rightarrow \delta_i q);\ i = 1,2,3,\ldots,k(n-r)!$,

(a_8i) $(p \Rightarrow \delta_i q) \Rightarrow \delta_i(p \Rightarrow q);\ i = 1,2,3,\ldots,k(n-r)!$,

(a_9i) $((p \Rightarrow q) \Rightarrow z) \Rightarrow \phi_i(p \Rightarrow z);\ i = 1,2,3,\ldots,n-t, t < n-r$,

(a_{10}) $(q \Rightarrow z) \Rightarrow ((\phi_1 p \Rightarrow z) \Rightarrow ((\phi_2 p \Rightarrow z) \Rightarrow \ldots \Rightarrow ((\phi_{n-t} p \Rightarrow z) \Rightarrow ((p \Rightarrow q) \Rightarrow z))\ldots));\ t < n-r$,

(a_{11}) $((p \Rightarrow q) \Rightarrow z) \Rightarrow (\psi_1 p \Rightarrow (\psi_2 p \Rightarrow (\ldots \Rightarrow (\psi_t p \Rightarrow z)\ldots)));\ t > n-r$,

(a_{12}) $(q \Rightarrow z) \Rightarrow (\psi_1 p \Rightarrow (\psi_2 p \Rightarrow (\psi_3 p \Rightarrow \ldots \Rightarrow (\psi_{t-1} p \Rightarrow (\psi_t(p \Rightarrow z) \Rightarrow ((p \Rightarrow q) \Rightarrow z)))\ldots)));\ t > n-r$, (a13i) $\alpha_{i_1} \Rightarrow (\alpha_{i_2}p \Rightarrow (\alpha_{i_3}p \Rightarrow (\ldots \Rightarrow (\alpha_{i_m}p \Rightarrow z)\ldots)))$;

where z either is of the form $\alpha_i p$ or is a single propositional variable,, the sequence $i_1, i_2, i_3, \ldots, i_m$ is any increasing subsequence of $1, 2, 3, \ldots, n!$ and $i = 1, 2, 3, \ldots, (n!+1)\sum_{j=1}^{n!}\binom{n!}{j}$.

The primitive rules are the rule of substitution and the rule of detachment.

One could say that, in a sense, this device puts an end to the problem of construction of finitely-valued logics. However, this does not mean that there are no further possibilities of a different characterization of propositional connectives and their axiomatization, especially when the systems that arise fail to be functionally complete.

4.2 The Many-Valued Calculus of Sobociński

An interesting class of n-valued calculi was invented by B. Sobociński in his paper [Sobociński 1936]. His n-valued calculus is given by the matrix

$$\mathfrak{M}Sbn = \langle\{0, 1, \ldots, n-1\}, \{1, 2, \ldots, n-1\}, \Rightarrow, \sim\rangle$$

with functors defined by

$\Rightarrow$	0	1	2	3	...	$n-2$	$n-1$
0	$n-1$	1	2	3	...	$n-2$	$n-1$
1	0	$n-1$	2	3	...	$n-2$	$n-1$
2	0	1	$n-1$	3	...	$n-2$	$n-1$
3	0	1	2	$n-1$	...	$n-2$	$n-1$
$\vdots$	$\vdots$	$\vdots$	$\vdots$	$\vdots$		$\vdots$	$\vdots$
$n-2$	0	0	2	3	...	$n-1$	$n-1$
$n-1$	0	1	2	3	...	$n-2$	$n-1$

x	$\sim x$
0	1
1	2
2	3
3	4
$\vdots$	$\vdots$
$n-2$	$n-1$
$n-1$	0

Obviously, the calculi thus defined (for a finite n) are functionally complete and decidable.

Sobociński imposes on his n-valued calculus the following axiom system.

Writing $\overset{1}{\sim} p = \sim p$, $\overset{m+1}{\sim} p = \sim (\overset{m}{\sim} p)$,

(a_1) $(p \Rightarrow q) \Rightarrow ((r \Rightarrow s) \Rightarrow (t \Rightarrow ((s \Rightarrow p) \Rightarrow (r \Rightarrow p))))$,

(a_2) $\overset{n}{\sim} p \Rightarrow p;\ n \geq 2$,

(a_3) $p \Rightarrow \overset{n}{\sim} p;\ n \geq 2$,

(a_4) $\overset{k}{\sim} (p \Rightarrow q) \Rightarrow (\overset{k}{\sim} p \Rightarrow \overset{k}{\sim} q)$,

(a_5) $\overset{k}{\sim} p \Rightarrow \overset{k}{\sim} q \Rightarrow \overset{k}{\sim} (p \Rightarrow q)$,

(a_6) $\sim (p \Rightarrow q) \Rightarrow \sim q$,

(a_7) $\sim (p \Rightarrow q) \Rightarrow ((p \Rightarrow q) \Rightarrow q)$,

(a_8) $((p \Rightarrow q) \Rightarrow r) \Rightarrow (\sim p \Rightarrow (\overset{2}{\sim} p \Rightarrow (\overset{3}{\sim} p \Rightarrow (\ldots$
$\Rightarrow \overset{n-3}{\sim} p \Rightarrow (\overset{n-2}{\sim} p \Rightarrow r)\ldots))))$,

(a_9) $(q \Rightarrow r) \Rightarrow (\sim p \Rightarrow (\overset{2}{\sim} p \Rightarrow (\overset{3}{\sim} p \Rightarrow (\ldots$
$\Rightarrow (\overset{n-3}{\sim} p \Rightarrow (\overset{n-2}{\sim} p \Rightarrow r))\ldots)))) \Rightarrow ((p \Rightarrow q) \Rightarrow r))$,

(a_{10^1}) $\overset{2}{\sim} (p \Rightarrow p)$,

(a_{10^2}) $\overset{3}{\sim} (p \Rightarrow p)$,

. . .

($a_{10^{n-3}}$) $\overset{n-2}{\sim} (p \Rightarrow p)$,

(a_{11}) $p \Rightarrow (\sim p \Rightarrow (\overset{2}{\sim} p \Rightarrow (\overset{3}{\sim} p \Rightarrow (\ldots \Rightarrow (\overset{n-2}{\sim} p \Rightarrow q)\ldots))))$,

(a_{12}j) $\sim (p \Rightarrow q) \Rightarrow (p \Rightarrow (\sim p \Rightarrow (\overset{2}{\sim} p \Rightarrow (\overset{3}{\sim} p \Rightarrow \ldots \Rightarrow (\overset{j-1}{\sim} p \Rightarrow (\overset{j+1}{\sim} p \Rightarrow$
$(\overset{j+2}{\sim} p \Rightarrow \ldots \Rightarrow (\overset{n-4}{\sim} p \Rightarrow (\overset{n-3}{\sim} p \Rightarrow (\overset{n-2}{\sim} p \Rightarrow \overset{j}{\sim} q)))\ldots)))\ldots))))$
for $1 \leq j \leq n-3$,

(a_{13}) $(p \Rightarrow (\sim p \Rightarrow (\overset{3}{\sim} p \Rightarrow (\overset{4}{\sim} p \Rightarrow \ldots (\overset{n-3}{\sim} p \Rightarrow (\overset{n-2}{\sim} p \Rightarrow \overset{2}{\sim} q))\ldots))) \Rightarrow (p \Rightarrow$
$(\sim p \Rightarrow (\overset{2}{\sim} p \Rightarrow (\overset{4}{\sim} p \Rightarrow \ldots \Rightarrow (\overset{n-3}{\sim} p \Rightarrow (\overset{n-2}{\sim} p \Rightarrow \overset{3}{\sim} q))\ldots)))) \Rightarrow (p \Rightarrow$
$(\sim p \Rightarrow (\overset{2}{\sim} p \Rightarrow (\overset{3}{\sim} p \Rightarrow (\overset{5}{\sim} p \Rightarrow \ldots \Rightarrow (\overset{n-3}{\sim} p \Rightarrow (\overset{n-2}{\sim} p \Rightarrow \overset{4}{\sim} q))\ldots))))) \Rightarrow$
$\ldots \Rightarrow (p \Rightarrow (\sim p \Rightarrow (\overset{2}{\sim} p \Rightarrow (\overset{3}{\sim} p \Rightarrow \ldots \Rightarrow (\overset{n-4}{\sim} p \Rightarrow (\overset{n-2}{\sim} p \Rightarrow$
$\overset{n-3}{\sim} q))\ldots)))) \Rightarrow ((p \Rightarrow (\sim p \Rightarrow (\overset{2}{\sim} p \Rightarrow (\overset{3}{\sim} p \Rightarrow \ldots \Rightarrow (\overset{n-4}{\sim} p \Rightarrow (\overset{n-3}{\sim} p \Rightarrow$
$\overset{n-2}{\sim} q))\ldots)))) \Rightarrow ((\sim q \Rightarrow ((((p \Rightarrow q) \Rightarrow q) \Rightarrow \sim (p \Rightarrow q))))))$.

As usual, the rules of substitution and detachment are the primitives.

The proofs of completeness and consistency can be found in Sobociński's paper mentioned above.

4.3 The Many-Valued Calculi of Gödel

Another class of n-valued logics for a finite n has been defined by Kurt Gödel [Gödel 1930] by the matrix

$$\mathfrak{M}Gn = \langle\{0, 1, \ldots, n-1\}, \{n-1\}, \sim, \wedge, \vee, \Rightarrow, \Leftrightarrow\rangle.$$

Its propositional functors are characterized by the following equalities:

$$\sim x = \begin{cases} n-1 & \text{if } x = 0, \\ 0 & \text{if } x \neq 0, \end{cases}$$

$$x \wedge y = \min(x, y),$$

$$x \vee y = \max(x, y),$$

$$x \Rightarrow y = \begin{cases} n-1 & \text{if } x \leq y, \\ y & \text{if } x > y, \end{cases}$$

$$x \Leftrightarrow y = \begin{cases} n-1 & \text{if } x = y, \\ \min(x, y) & \text{if } x \neq y. \end{cases}$$

The case of $n = 3$ deserves special attention. The matrix then involves all assertions of the intuitionistic propositional calculus. Gödel's axiomatization coincides with the usual axiom system of the intuitionistic calculus, with

$$(\sim \alpha \Rightarrow \beta) \Rightarrow (((\beta \Rightarrow \alpha) \Rightarrow \beta) \Rightarrow \beta)$$

added. Thus, the axiom system for the three-valued calculus of Gödel is the following:

(a_1) $(\alpha \wedge \beta) \Rightarrow \alpha$,

(a_2) $(\alpha \wedge \beta) \Rightarrow \beta$,

(a_3) $\alpha \Rightarrow (\alpha \vee \beta)$,

(a_4) $\beta \Rightarrow (\alpha \vee \beta)$,

(a_5) $\alpha \Rightarrow (\beta \Rightarrow \alpha)$,

(a_6) $\alpha \Rightarrow (\sim \alpha \Rightarrow \beta)$,

(a_7) $\alpha \Rightarrow (\beta \Rightarrow (\alpha \wedge \beta))$,

(a_8) $(\alpha \Rightarrow \sim \beta) \Rightarrow (\beta \Rightarrow \sim \alpha)$,

(a_9) $(\alpha \Rightarrow (\beta \Rightarrow \gamma)) \Rightarrow ((\alpha \Rightarrow \beta) \Rightarrow (\alpha \Rightarrow \gamma))$,

(a_{10}) $(\alpha \Rightarrow \gamma) \Rightarrow ((\beta \Rightarrow \gamma) \Rightarrow ((\alpha \vee \beta) \Rightarrow \gamma))$,

(a_{11}) $(\sim \alpha \Rightarrow \beta) \Rightarrow (((\beta \Rightarrow \alpha) \Rightarrow \beta) \Rightarrow \beta)$.

Its primitive rule is modus ponens (rule of detachment).

4.4 The Many-Valued Calculus *Cnr*

We close our survey of n-valued logics with a detailed discussion of the *Cnr* calculus, a version of the n-valued Post calculus. We display the matrix and the axiom system, and then we present the proof of completeness. The method is due to Kalmar; the proof is given in the form elaborated by J. Surma.

The matrix is

$$\mathfrak{M}Cnr = \langle \{0, 1, 2, \ldots, n-1\}, \{r, r+1, \ldots, n-1\}, \sim, \Rightarrow \rangle,$$

where $0 < r \leq n-1$; the functors are characterized by the equalities

$$x \Rightarrow y = \begin{cases} n-1 & \text{if } x \leq y, \\ y & \text{if } x > y \text{ and } r \leq x, \\ n-1-x+y & \text{if } x > y \text{ and } r > x, \end{cases}$$

$$\sim x = \begin{cases} x-1 & \text{if } x \neq 0, \\ n-1 & \text{if } x = 0. \end{cases}$$

Thus in fact, $\Rightarrow$ is nothing else than the Post implication. If $r = n-1$ (that is, when $n-1$ alone is a designated value), we obtain the Lukasiewicz implication. The negation $\sim$ is just Post's cyclic negation.

It will be convenient to introduce some notation:

(a) $\overset{0}{\sim} \alpha = \alpha$;

(b) $\overset{i+1}{\sim} \alpha = \overset{i}{\sim} (\alpha)$;

(c) $\alpha \overset{n-1}{\Rightarrow} \beta = \alpha \Rightarrow (\alpha \Rightarrow (\ldots(\alpha \Rightarrow \beta)\ldots))$;

with $n-1$ implication signs on the right-hand side;

(d) $B^i_\alpha = \left\{ \overset{i}{\sim} \alpha, \overset{i+1}{\sim} \alpha, \ldots, \overset{i+s-1}{\sim} \alpha \right\}$,

where s denotes the number of designated values;

(e) $A^i_\alpha(\beta) = \overset{i+s-1}{\sim} \alpha \overset{n-1}{\Rightarrow} \left(\overset{i+s-2}{\sim} \alpha \overset{n-1}{\Rightarrow} (\ldots (\overset{i}{\sim} \alpha \Rightarrow \beta)\ldots) \right)$,

with s as above.

Examples

Let $n = 3$, $s = 1$. Then

$$B^i_\alpha = \{\overset{i}{\sim} \alpha\},$$

$$A^i_\alpha = \overset{i}{\sim} \alpha \Rightarrow (\overset{i}{\sim} \alpha \Rightarrow \beta) = \overset{i}{\sim} \alpha \overset{2}{\Rightarrow} \beta.$$

Now let $n = 3$, $s = 2$. Then

$$B^i_\alpha = \{\overset{i}{\sim} \alpha, \overset{i+1}{\sim} \alpha\},$$

$$A^i_\alpha(\beta) = \overset{i+1}{\sim} \alpha \Rightarrow (\overset{i+1}{\sim} \alpha \Rightarrow (\overset{i}{\sim} \alpha \Rightarrow (\overset{i}{\sim} \alpha \Rightarrow \beta))) = \overset{i+1}{\sim} \alpha \overset{2}{\Rightarrow} (\overset{i}{\sim} \alpha \overset{2}{\Rightarrow} \beta).$$

Let $LCnr$ denote the axiomatic setting for this n-valued calculus. The following formulas are its axioms:

(a$_1$) $\alpha \Rightarrow (\beta \Rightarrow \alpha)$,

(a$_2$) $(\alpha \overset{n-1}{\Rightarrow} (\beta \Rightarrow \gamma)) \Rightarrow ((\alpha \overset{n-1}{\Rightarrow} \beta) \Rightarrow (\alpha \overset{n-1}{\Rightarrow} \gamma))$,

(a$_3$) $A^1_\alpha(\beta) \Rightarrow (A^2_\alpha(\beta) \Rightarrow (\ldots \Rightarrow (A^{n-1}_\alpha(\beta) \Rightarrow \beta)\ldots))$,

(a$_4$) $\overset{n}{\sim} \alpha \Rightarrow \alpha$,

(a$_5$) $\alpha \Rightarrow \overset{n}{\sim} \alpha$,

(a$_6$) For every i, j, k such that $i \Rightarrow j = k$, the formulas

$A^i_\alpha(A^j_\beta(\overset{k}{\sim} (\alpha \Rightarrow \beta)))$,

$A^i_\alpha(A^j_\beta(\overset{k+1}{\sim} (\alpha \Rightarrow \beta)))$,

...

$A^i_\alpha(A^j_\beta(\overset{k+s-1}{\sim} (\alpha \Rightarrow \beta)))$

are axioms; as before, s is the number of designated values.

The only primitive rule is the rule of detachment for $\Rightarrow$, with the usual pattern

(rd) $$\frac{\alpha, \alpha \Rightarrow \beta}{\beta}.$$

This means that if $\alpha \in LCnr$ and $\alpha \Rightarrow \beta \in LCnr$ then also $\beta \in LCnr$.

Explicit axiomatization of LC32 (hence, $s = 1$).

(a_1) $\alpha \Rightarrow (\beta \Rightarrow \alpha)$,

(a_2) $(\alpha \overset{2}{\Rightarrow} (\beta \Rightarrow \gamma)) \Rightarrow ((\alpha \overset{2}{\Rightarrow} \beta) \Rightarrow (\alpha \overset{2}{\Rightarrow} \gamma))$,

(a_3) $(\sim \alpha \Rightarrow (\sim \alpha \Rightarrow \beta)) \Rightarrow ((\overset{2}{\sim} \alpha \Rightarrow (\overset{2}{\sim} \alpha \Rightarrow \beta))$
$\Rightarrow (\overset{3}{\sim} \alpha \Rightarrow (\overset{3}{\sim} \alpha \Rightarrow \beta) \Rightarrow \beta))$,

(a_4) $\overset{3}{\sim} \alpha \Rightarrow \alpha$,

(a_5) $\alpha \Rightarrow \overset{3}{\sim} \alpha$,

(a_6) $\sim \alpha(\sim \alpha \Rightarrow (\sim \beta \Rightarrow \overset{3}{\sim} (\alpha \Rightarrow \beta)))$,

(a_7) $\sim \alpha \Rightarrow (\sim \alpha \Rightarrow (\overset{2}{\sim} \beta \Rightarrow \overset{3}{\sim} (\alpha \Rightarrow \beta)))$,

(a_8) $\sim \alpha \Rightarrow (\sim \alpha \Rightarrow (\overset{3}{\sim} \beta \Rightarrow \overset{3}{\sim} (\alpha \Rightarrow \beta)))$,

(a_9) $\overset{2}{\sim} \alpha \Rightarrow (\overset{2}{\sim} \alpha \Rightarrow (\overset{1}{\sim} \beta \Rightarrow \overset{2}{\sim} (\alpha \Rightarrow \beta)))$,

(a_{10}) $\overset{2}{\sim} \alpha \Rightarrow (\overset{2}{\sim} \alpha \Rightarrow (\overset{2}{\sim} \beta \Rightarrow \overset{3}{\sim} (\alpha \Rightarrow \beta)))$,

(a_{11}) $\overset{2}{\sim} \alpha \Rightarrow (\overset{2}{\sim} \Rightarrow (\overset{3}{\sim} \beta \Rightarrow \overset{3}{\sim} (\alpha \Rightarrow \beta)))$,

(a_{12}) $\overset{3}{\sim} \alpha \Rightarrow (\overset{3}{\sim} \alpha \Rightarrow (\overset{1}{\sim} \beta \Rightarrow \sim (\alpha \Rightarrow \beta)))$,

(a_{13}) $\overset{3}{\sim} \alpha \Rightarrow (\overset{3}{\sim} \alpha \Rightarrow (\overset{2}{\sim} \beta \Rightarrow \overset{2}{\sim} (\alpha \Rightarrow \beta)))$,

(a_{14}) $\overset{3}{\sim} \alpha \Rightarrow (\overset{3}{\sim} \alpha \Rightarrow (\overset{3}{\sim} \beta \Rightarrow \overset{3}{\sim} (\alpha \Rightarrow \beta)))$.

Let Cn denote the operation of consequence in the calculus $LCnr$. That is to say, if X is any set of formulas, then

$$CnX = \{\alpha : \exists \alpha_1, \alpha_2, \ldots, \alpha_n(\alpha_n = \alpha \wedge \forall k \leq n(\alpha_k \in LCnr \cup X \vee \exists i, j(i < k \wedge j < k \wedge \alpha_i = \alpha_j \Rightarrow \alpha_k)))\}$$

The operation thus defined has all properties needed.

Theorem 4.1 (Deduction Theorem)
If $\alpha \in Cn(X \cup \{\beta\})$ then $\beta \overset{n-1}{\Rightarrow} \alpha \in CnX$.

Proof

Assume $\alpha \in Cn(X \cup \{\beta\})$. Then there exists a finite sequence $\alpha_1, \alpha_2, \ldots, \alpha_n$ fulfilling the conditions stated in the definition of $Cn(\cdot)$, with $\alpha = \alpha_n$. We now show by induction that $\beta \overset{n-1}{\Rightarrow} \alpha_i \in CnX$ for each i, $1 \le i \le n$.

If $\alpha_1 = \beta$ then by axioms (a_1) and (a_2), $\beta \overset{n-1}{\Rightarrow} \alpha_1 \in CnX$. If $\alpha_1 \in LCnr$ then, applying (a_1) $(\alpha_1 \Rightarrow (\beta \Rightarrow \alpha_1))$ with modus ponens we get $\beta \Rightarrow \alpha_1 \in CnX$. Repeating the argument $n-1$ times we obtain

$$\beta \overset{n-1}{\Rightarrow} \alpha_1 \in CnX.$$

Now, if $\alpha_1 \in X$ then of course $\alpha_1 \in CnX$ and hence, just as above,

$$\beta \overset{n-1}{\Rightarrow} \alpha_1 \in CnX.$$

Assume $\beta \overset{n-1}{\Rightarrow} \alpha_1 \in CnX$ for $i < k$. We have to show that $\beta \overset{n-1}{\Rightarrow} \alpha_k \in CnX$.

Formula α_k either belongs to $LCnr \cup X \cup \{\beta\}$ or not. In the first case the proof is obtained in exactly the same way as for α_1. In the second case there are i, j with $i < k$, $j < k$, such that $\alpha_i = (\alpha_j \Rightarrow \alpha_k)$. Since $\beta \overset{n-1}{\Rightarrow} \alpha_i \in CnX$ and $\beta \overset{n-1}{\Rightarrow} \alpha_j \in CnX$, we infer

$$\beta \overset{n-1}{\Rightarrow} (\alpha_j \Rightarrow \alpha_k) \in CnX.$$

According to axiom (a_2),

$$(\beta \overset{n-1}{\Rightarrow} (\alpha_j \Rightarrow \alpha_k)) \Rightarrow ((\beta \overset{n-1}{\Rightarrow} \alpha_j) \Rightarrow (\beta \overset{n-1}{\Rightarrow} \alpha_k)).$$

Consequently

$$\beta \overset{n-1}{\Rightarrow} \alpha_k \in CnX,$$

completing the induction.

We have thus shown that $\beta \overset{n-1}{\Rightarrow} \alpha_i \in CnX$ for all i, $1 \le i \le n$. Taking in particular $i = n$ we get $\beta \overset{n-1}{\Rightarrow} \alpha \in CnX$, as claimed. ∎

Lemma 4.1

Let X be a set of formulas, closed with respect to the rule of detachment (modus ponens). If $\alpha \in X$ and $\alpha \overset{n-1}{\Rightarrow} \beta \in X$ then $\beta \in X$.

Proof

Apply modus ponens $n-1$ times. ∎

Lemma 4.2

If a set of formulas Z is closed with respect to modus ponens and if $A^i_\alpha(\beta) \in Z$ and $B^i_\alpha \subseteq Z$, then $\beta \in Z$.

Proof

All the premises of $A^i_\alpha(\beta)$ are in Z; apply Lemma 4.1 r times. ∎

Now let V denote the set of propositional variables and let v be any valuation of V in $\{0, 1, \ldots, n-1\}$; i.e.,

$$v : V \to \{0, 1, \ldots, n-1\}.$$

Recall that a formula $\alpha \in S$ is a tautology over the matrix $\mathfrak{M}Cnr$ if and only if

$$\mid \alpha \mid (v) \in \{r, r+1, \ldots, n-1\}$$

holds for every

$$v : V \to \{0, 1, \ldots, r, r+1, \ldots, n-1\};$$

$\mid \cdot \mid (v)$ denotes the extension of v to the whole of S. Let

$$E(\mathfrak{M}Cnr) = \{\alpha \in S : \alpha \text{ is a tautology over } \mathfrak{M}Cnr\}.$$

Lemma 4.3

Suppose Z is a set of formulas, closed with respect to modus ponens; i.e., suppose $B_\alpha^{|\alpha|(v)} \subseteq Z$. Then $B_{\sim\alpha}^{|\sim\alpha|(v)} \subseteq Z$.

Proof

If $i > 0$ and $\mid \alpha \mid (v) = i$, then $\mid \sim \alpha \mid (v) = i - 1$. If $i = 0$ then $\mid \sim \alpha \mid (v) = n - 1$. Therefore

$$B_{\sim\alpha}^{i-1} = B_\alpha^i$$

for $i > 0$. When $i = 0$, it suffices to apply axiom (a_5) with modus ponens to formulas of the set B_α^0, thus obtaining B_α^{n-1}. This ends the proof. ■

Theorem 4.2 (The generalized Kalmar's Lemma)

Let $\alpha \in S$ be a formula involving propositional variables $p_1, p_2, \ldots, p_m$ and let v be any valuation of V in $\{0, 1, \ldots, n-1\}$. Then

$$B_\alpha^{|\alpha|(v)} \subseteq Cn\left(B_{p_1}^{v(p_1)} \cup B_{p_2}^{v(p_2)} \cup \ldots \cup B_{p_m}^{v(p_m)}\right).$$

Proof

We use induction on the "architecture" of α. When α is just a single variable, $\alpha = p_i$, we note that

$$B_{p_i}^{|p_i|(v)} = B_{p_i}^{v(p_i)} \subseteq Cn\left(B_{p_i}^{v(p_i)}\right),$$

proving the claim.

Assume the claim for a certain formula β and consider $\alpha = \sim \beta$. Since

$$B_\beta^{|\beta|(v)} \subseteq Cn\left(\bigcup_{i=1}^{m} B_{p_i}^{v(p_i)}\right),$$

we get, by Lemma 4.3,

$$B_{\alpha}^{|\alpha|(v)} = B_{\sim\beta}^{|\sim\beta|(v)} \subseteq Cn\left(\bigcup_{i=1}^{m} B_{p_i}^{v(p_i)}\right),$$

as needed.

Now consider $\alpha = (\beta \Rightarrow \gamma)$. Suppose β is formed from variables $p_1, p_2, \ldots, p_t \in \{p_1, p_2, \ldots, p_m\}$ and γ is formed from $p_{t+1}, p_{t+2}, \ldots, p_m$. Suppose

$$B_{\beta}^{|\beta|(v)} \subseteq Cn\left(\bigcup_{s=1}^{m} B_{p_s}^{v(p_s)}\right),$$

$$B_{\gamma}^{|\gamma|(v)} \subseteq Cn\left(\bigcup_{s=1}^{m} B_{p_s}^{v(p_s)}\right).$$

Further assume

$$|\,\beta\,|\,(v) = i, \quad |\,\gamma\,|\,(v) = j, \quad |\,\beta \Rightarrow \gamma\,|\,(v) = k.$$

The values i, j, k satisfy $i \Rightarrow j = k$ and

$$B_{\beta}^{i} \subseteq Cn\left(\bigcup_{s=1}^{m} B_{p_s}^{v(p_s)}\right),$$

$$B_{\gamma}^{j} \subseteq Cn\left(\bigcup_{s=1}^{m} B_{p_s}^{v(p_s)}\right).$$

The set $Cn\left(\bigcup_{s=1}^{m} B_{p_s}^{v(p_s)}\right)$ includes, in particular, the following formulas (which are specific cases of axiom (a_6)):

$$A_{\beta}^{i}(A_{\gamma}^{j}(\overset{k}{\sim}(\beta \Rightarrow \gamma)))$$
$$A_{\beta}^{i}(A_{\gamma}^{j}(\overset{k+1}{\sim}(\beta \Rightarrow \gamma)))$$
$$\ldots$$
$$A_{\beta}^{i}(A_{\gamma}^{j}(\overset{k+s-1}{\sim}(\beta \Rightarrow \gamma)))$$

Since B_{β}^{i} and B_{γ}^{j} are subsets of $Cn\left(\bigcup_{s=1}^{m} B_{p_s}^{v(p_s)}\right)$, we conclude by Lemma 4.2 that the following formulas belong to this set:

$$\overset{k}{\sim}(\beta \Rightarrow \gamma)$$
$$\overset{k+1}{\sim}(\beta \Rightarrow \gamma)$$
$$\ldots$$
$$\overset{k+s-1}{\sim}(\beta \Rightarrow \gamma).$$

These formulas are elements of the set $B^k_{\beta\Rightarrow\gamma}$ as well. Therefore

$$B^k_\alpha = B^k_{\beta\Rightarrow\gamma} \subseteq Cn\left(\bigcup_{s=1}^{m} B^{v(p_s)}_{p_s}\right).$$

This completes the proof of the theorem. ∎

Lemma 4.4

Suppose α is a tautology over $\mathfrak{M}Cnr$, formed of variables $p_1, p_2, \ldots, p_m$. Then

$$\alpha \in Cn\left(\bigcup_{i=1}^{m} B^{v(p_i)}_{p_i}\right)$$

for every valuation $v : V \to \{0, 1, \ldots, n-1\}$.

Proof

Let v be any valuation and let $\mid \alpha \mid (v) = k$. Since α is a tautology, $k \geq r$. By Theorem 4.2,

$$B^k_\alpha \subseteq \left(\bigcup_{i=1}^{m} B^{v(p_i)}_{p_i}\right).$$

Writing, as before, $s = n - r$ we have $k + s - 1 \geq n - 1$, and so $\overset{m}{\sim} \alpha$ is an element of B^k_α. Hence, by axiom (a_4) and the rule of detachment,

$$\alpha \in Cn\left(\bigcup_{i=1}^{m} B^{v(p_i)}_{p_i}\right). \quad ∎$$

Lemma 4.5

Let Z be a set of formulas, closed with respect to detachment. Suppose $A^i_\alpha(\beta) \in Z$ for all i, $0 \leq i \leq n-1$; then $\beta \in Z$.

Proof

Obvious by detachment of the successive premises, in accordance with axiom (a_3). ∎

Lemma 4.6

If $\beta \in Cn(B^i_\alpha \cup B)$ then $A^i_\alpha(\beta) \in Cn(B)$.

Proof

Results by s-fold application of the Deduction Theorem, where s is the number of designated variables. ∎

Theorem 4.3 (Completess Theorem)

If a formula $\alpha \in S$ is a tautology over $\mathfrak{M}Cnr$, then α is provable in $LCnr$; i.e.,

$$E(\mathfrak{M}Cnr) \subseteq LCnr.$$

Proof

Suppose α is tautologous, $\alpha \in E(\mathfrak{M}Cnr)$, and let $v : V \to \{0, 1, \ldots, n-1\}$ be any valuation. Denote by $\overset{i}{\sim} v^p$ the following valuation:

$$\overset{i}{\sim} v^p = \begin{cases} \overset{i}{\sim} v(p) & \text{if} \quad q = p, \\ v(p) & \text{if} \quad q \neq p. \end{cases}$$

Applying Lemma 4.4 to each one of the chain of valuations $v, \overset{1}{\sim} v^{p_1}, \overset{2}{\sim} v^{p_1}, \ldots, \overset{n-2}{\sim} v^{p_1}$ we infer

$$\alpha \in Cn\left(\bigcup_{i=1}^{m} B_{p_i}^{v(p_i)}\right),$$

$$\alpha \in Cn\left(\bigcup_{i=1}^{m} B_{p_i}^{\overset{1}{\sim} v^{p_1}(p_i)}\right),$$

$$\ldots$$

$$\alpha \in Cn\left(\bigcup_{i=1}^{m} B_{p_i}^{\overset{n-2}{\sim} v^{p_1}(p_i)}\right).$$

To each of the expressions above we now apply Lemma 4.6, thus obtaining

$$A_{p_1}^{v(p_1)}(\alpha) \in Cn\left(\bigcup_{i=2}^{m} B_{p_1}^{v(p_1)}\right),$$

$$A_{p_1}^{\overset{1}{\sim} v(p_1)}(\alpha) \in Cn\left(\bigcup_{i=2}^{m} B_{p_i}^{\overset{1}{\sim} v^{p_1}(p_i)}\right),$$

$$A_{p_1}^{\overset{2}{\sim} v(p_1)}(\alpha) \in Cn\left(\bigcup_{i=2}^{m} B_{p_i}^{\sim v^{p_1}(p_1)}\right),$$

$$\ldots$$

$$A_{p_1}^{\overset{n-2}{\sim} v(p_1)}(\alpha) \in Cn\left(\bigcup_{i=2}^{m} B_{p_i}^{\overset{n-2}{\sim} v^{p_1}(p_i)}\right).$$

It is easy to see that the values $v(p_1)$, $\sim v(p_1)$, $\overset{2}{\sim} v(p_1), \ldots, \overset{n-2}{\sim} v(p_1)$ exhaust the whole set $\{0, 1, \ldots, n-1\}$. So

$$A_{p_1}^{i}(\alpha) \in Cn\left(\bigcup_{i=2}^{m} B_{p_i}^{v(p_i)}\right)$$

for all i, $0 \leq i \leq n-1$. Hence by Lemma 4.5,

$$\alpha \in Cn\left(\bigcup_{i=2}^{m} B_{p_i}^{v(p_i)}\right).$$

The argument can be repeated with regard to the successive variables $p_2, p_3, \ldots, p_m$, occurring in α. Thus, finally,

$$\alpha \in Cn(\emptyset) = LCnr,$$

proving the claim. ∎

Lemma 4.7

For $i, j \in \{0, 1, \ldots, n-1\}$,

$$i \overset{n-1}{\Rightarrow} j = \begin{cases} m & \text{if } \ i \leq j \ \text{ or } \ j < i < r, \\ j & \text{if } \ j < i \ \text{ and } \ i \geq r. \end{cases}$$

Proof

We have to examine the value of

$$i \Rightarrow (i \Rightarrow (\ldots \Rightarrow (i \Rightarrow (i \Rightarrow j))\ldots)).$$

Denote this value by w. If $i \leq j$, then $w = n-1$. If $j < i$ and $r \leq i$, then $w = j$, in view of the definition of the value of $i \Rightarrow j$. If $j < i < r$, then the value of $i \Rightarrow j$ equals $n - 1 - i + j$. Now, if $n - 1 - i + j < i$, then the value of $i \Rightarrow (i \Rightarrow j)$ is $2n - 2 - 2i + j$. If this number is less than i, we consider $i \Rightarrow (i \Rightarrow (i \Rightarrow j))$, which evaluates to $3n - 3 - 3i + j$, and so on, until we reach $(n-3)(n-1) - (n-3)i + j$, still less than i. The value of w will then be $(n-2)(n-1) - (n-2)i + j$. Since the sequence increases with each step of the procedure, either we exceed i at some instant, and then clearly w equals $n-1$, or we stop at $w = (n-2)(n-1) - (n-2)i + j$. This, after slightly recasting and taking into account that $j < i < r$, is also seen to be equal to $n-1$. This ends the proof. ∎

Lemma 4.8

Let $\mid \alpha \mid (v) = i$ and $\mid \beta \mid (v) = j$. Then

$$\mid A^k_\alpha(\beta) \mid (v) = \begin{cases} n-1 & \text{if } \ k \neq i \ \text{ or } \ r \leq j, \\ j & \text{if } \ k = i \ \text{ and } \ r > j. \end{cases}$$

Proof

The proof consists in the evaluation of

$$\overset{k}{\sim} i \overset{n-2}{\Rightarrow} \left(\overset{k+1}{\sim} i \overset{n-2}{\Rightarrow} \left(\ldots \overset{n-2}{\Rightarrow} \left(\overset{k+s-1}{\sim} i \overset{n-2}{\Rightarrow} j\right)\ldots\right)\right).$$

This expression has value $n-1$ when at least one of the premises equals 0. The cases $i - k - s + 1 > 0$, $i - k \leq 0$ and $i - k - s + 1 > -n + 1$ are obvious, whereas the case $i - j \leq -n + 1$ is contradictory. ∎

Theorem 4.4 (Consistency Theorem)

If a formula has a proof in $LCnr$, then it is a tautology over $\mathfrak{M}Cnr$. Thus

$$LCnr \subseteq E(\mathfrak{M}Cnr).$$

Proof

First, one shows that the axioms belong to $E(\mathfrak{M}Cnr)$. This is obvious for (a_1), (a_2), (a_4), (a_5). Axiom (a_3) assumes value $n-1$, on account of Lemma 4.8. Simple manipulation shows that also (a_6) is in $E(\mathfrak{M}Cnr)$.

It remains to notice that $E(\mathfrak{M}Cnr)$ is closed under modus ponens. ∎

Corollary 4.1

The sets $LCnr$ and $E(\mathfrak{M}Cnr)$ are identical.

5 Intuitionistic Propositional Calculus

5.1 The Intuitionistic Propositional Logic in an Axiomatic Setting

Intuitionism is deemed to be the most important non-classical logic calculus. There is not much exaggeration in saying that it has arisen accidentally, as an attempt to axiomatize a certain three-valued logic. In fact, it should be rather regarded as the result of a programme of putting constraints on the laws and rules of classical logic.

Various motivations can be traced in the historical background of intuitionistic calculi. Intuitionists do not accept the indirect proofs (reductio ad absurdum) so frequently used in classical logic, where it is assumed that a contradiction resulting from the negation of a statement proves the statement. From the intuitionistic standpoint, such a contradiction shows that the negation of the statement is false, and nothing more. The falsity of the negation yields the lack of its truth but fails to imply the truth of the statement. Thus the laws of double negation and of excluded middle are rejected; this forces further rejection of many assertions and rules accepted in the classical two-valued logic, and in particular the reductio ad absurdum principle.

As examples we list here some theses of classical logic that are unprovable in the intuitionistic propositional calculus:

(n_1) $(\sim \alpha \Rightarrow \sim \beta) \Rightarrow (\beta \Rightarrow \alpha)$,

(n_2) $(\sim \alpha \Rightarrow \beta) \Rightarrow (\sim \beta \Rightarrow \alpha)$,

(n_3) $\sim\sim \alpha \Rightarrow \alpha$,

(n_4) $(\sim \alpha \Rightarrow \alpha) \Rightarrow \alpha$,

(n_5) $(\sim \alpha \Rightarrow \beta) \Rightarrow ((\alpha \Rightarrow \beta) \Rightarrow \alpha)$,

(n_6) $((\alpha \Rightarrow \beta) \Rightarrow \alpha) \Rightarrow \alpha$,

(n_7) $\alpha \vee \sim \alpha$,

(n_8) $\sim (\sim \alpha \vee \sim \beta) \Rightarrow (\alpha \wedge \beta)$,

(n_9) $\sim (\sim \alpha \wedge \sim \beta) \Rightarrow (\alpha \vee \beta)$,

(n_{10}) $(\alpha \Rightarrow \beta) \Rightarrow (\sim \alpha \vee \beta)$,

(n_{11}) $\sim (\alpha \Rightarrow \sim \beta) \Rightarrow (\alpha \wedge \beta)$,

(n_{12}) $\sim (\alpha \wedge \sim \beta) \Rightarrow (\alpha \Rightarrow \beta)$.

The language of intuitionistic calculus is defined similarly to classical logic; that is,

$$L = (S, \sim, \Rightarrow, \vee, \wedge, \Leftrightarrow)$$

with S defined in the standard way. The operation of consequence Cn_I is the same as that in the classical logic (denoted Cn), just restricted to the axiom system given below.

The following set of formulas constitutes an axiomatization of the intuitionistic calculus:

(i_1) $\alpha \Rightarrow (\beta \Rightarrow \alpha)$,

(i_2) $(\alpha \Rightarrow (\beta \Rightarrow \gamma)) \Rightarrow ((\alpha \Rightarrow \beta) \Rightarrow (\alpha \Rightarrow \gamma))$,

(i_3) $(\alpha \vee \beta) \Rightarrow (\beta \vee \alpha)$,

(i_4) $(\alpha \wedge \beta) \Rightarrow (\beta \wedge \alpha)$,

(i_5) $\alpha \Rightarrow \alpha \vee \beta$,

(i_6) $\alpha \wedge \beta \Rightarrow \alpha$,

(i_7) $\alpha \Rightarrow (\beta \Rightarrow (\alpha \wedge \beta))$,

(i_8) $(\alpha \Rightarrow \gamma) \Rightarrow ((\beta \Rightarrow \gamma) \Rightarrow ((\alpha \Rightarrow \beta) \Rightarrow \gamma))$,

(i_9) $(\alpha \Rightarrow (\beta \wedge \sim \beta)) \Rightarrow \sim \alpha$,

(i_{10}) $(\alpha \wedge \sim \alpha) \Rightarrow \beta$,

(i_{11}) $(\alpha \Leftrightarrow \beta) \Rightarrow (\alpha \Rightarrow \beta)$,

(i_{12}) $(\alpha \Leftrightarrow \beta) \Rightarrow (\beta \Rightarrow \alpha)$,

(i_{13}) $(\alpha \Rightarrow \beta) \Rightarrow ((\beta \Rightarrow \alpha) \Rightarrow (\alpha \Leftrightarrow \beta))$,

with modus ponens as the only primitive rule:

$$\frac{\alpha, \alpha \Rightarrow \beta}{\beta}.$$

It is easily observed that, by the non-provability of formulas (n_1)–(n_{12}), no one of the four connectives $\Rightarrow, \vee, \wedge, \sim$ can be defined in terms of the other three (as is the case in the two-valued logic).

The interrelation between the classical and intuitionistic logics is expressed by the following two theorems.

Theorem 5.1 (Glivenko)

Let $Cn(\emptyset)$ and $Cn_I(\emptyset)$ denote respectively the set of assertions of classical logic and of intuitionistic logic.

If $\alpha \in Cn(\emptyset)$ then $\sim\sim \alpha \in Cn_I(\emptyset)$.

Proof

Suppose we attach the law of excluded middle $(\alpha \vee \sim \alpha)$ to the axioms (i_1)–(i_{13}). What results is the usual axiom system of classical logic. The formula $\alpha \Rightarrow \sim\sim \alpha$ is an assertion of intuitionism, on account of axiom (i_9). Consequently, the two-fold negation of each one of the axioms (i_1)–(i_8) is an assertion of intuitionistic logic.

In view of the deduction theorem and axiom (i_9), the following law is valid:

$$(\alpha \Rightarrow \beta) \Rightarrow (\sim \beta \Rightarrow \sim \alpha).$$

Hence,

$$\sim (\alpha \vee \beta) \Rightarrow (\sim \alpha \wedge \sim \beta)$$

and

$$\sim (\sim \alpha \wedge \sim\sim \alpha) \Rightarrow \sim\sim (\alpha \vee \sim \alpha)$$

are assertions of intuitionism. So is the formula $\sim\sim (\alpha \vee \sim \alpha)$, by the rule of detachment. It follows that all axioms of classical two-valued logic, doubly negated, become assertions of intuitionistic logic.

Now suppose a formula β is an assertion of classical logic. It is provable by detachment. We proceed by induction. Suppose that the detachment argument

$$\frac{\alpha, \alpha \Rightarrow \beta}{\beta}$$

is the final link in the proof of β and that the formulas $\alpha, \alpha \Rightarrow \beta$ satisfy the claim. Thus $\sim\sim \alpha, \sim\sim (\alpha \Rightarrow \beta)$ are assertions of intuitionism. So are the formulas $\sim\sim (\alpha \Rightarrow \beta) \Rightarrow (\sim\sim \alpha \Rightarrow \sim\sim \beta)$ and $\alpha \Rightarrow \sim\sim \alpha$, hence also $\sim\sim \beta$.

Induction now shows that double negation makes every formula which is an assertion of classical logic into an assertion of intuitionistic logic, as claimed. ∎

Theorem 5.2 (Gödel)

Suppose a formula $\alpha \in Cn(\emptyset)$ is formed of propositional variables and connectives $\wedge, \sim$ alone. Then $\alpha \in Cn_I(\emptyset)$. ∎

In the intuitionistic logic, the classical deduction theorem does hold. We formulate it as follows:

For any formulas $\alpha, \beta \in S$ and any set $X \subseteq S$, if $\beta \in Cn_I(X \cup \{\alpha\})$ then $(\alpha \Rightarrow \beta) \in Cn_I(X)$. ∎

The proof is exactly the same as in the classical logic.

Another fact worth noticing is this:

If $(\alpha \vee \beta) \in Cn_I(\emptyset)$ then $\alpha \in Cn_I(\emptyset)$ or $\beta \in Cn_I(\emptyset)$.

Let us mention here that the intuitionistic logic fails to be functionally complete and has no finite adequate matrix. Though it is non-complete in the sense of Post, it has the following useful property: if $\alpha \notin Cn(\emptyset)$ then $Cn_I(\{\alpha\}) = S$.

The intuitionistic propositional logic is decidable.

Along with the axiomatization, as given above, various other formalizations of intuitionism have been worked out. Let us mention here the natural-deduction method, the Gentzen formalism, the metasystem description of the consequence operator, Kripke's models, and the tableau method of Beth. We now briefly discuss some of these.

5.2 The Natural-Deduction Method for the Intuitionistic Propositional Logic

Let $X \subseteq S$ be a finite set and let $\alpha \in S$. The expressions $X \vdash \alpha$ and $\vdash \alpha$ will be called natural intuitionist sequents, abbreviated nis. Sequents will be denoted by small Latin characters (with or without subscripts) $s, s_1, s_2, \ldots$. The key role in the deduction process is played by rules. A rule is defined as a relation $r \subseteq P(SQ) \times SQ$, where SQ denotes the set of nis and $P(SQ)$ is its power set. A rule is written down in the form of a pattern

$$\frac{s_1, s_2, \ldots, s_n}{s},$$

sequents above the line denoting the premises and below the line – the conclusion.

The system is constituted by the axiom

(a) For every $\alpha \in S$, $\alpha \vdash \alpha$,

together with the following set of primitive sequent rules:

(r$_1$) $\dfrac{X \vdash \alpha, X \vdash \beta}{X \vdash \alpha \wedge \beta}$,

(r$_2$) $\dfrac{X \vdash \alpha \wedge \beta}{\alpha}$,

(r$_3$) $\dfrac{X \vdash \alpha \wedge \beta}{X \vdash \beta}$,

(r$_4$) $\dfrac{X \vdash \alpha}{X \vdash \alpha \vee \beta}$,

(r$_5$) $\dfrac{X \vdash \beta}{X \vdash \alpha \vee \beta}$,

(r$_6$) $\dfrac{X \cup \{\alpha\} \vdash \gamma, X \cup \{\beta\} \vdash \gamma, X \vdash \alpha \vee \beta}{X \vdash \gamma}$,

(r$_7$) $\dfrac{X \cup \{\alpha\} \vdash \beta}{X \vdash \alpha \Rightarrow \beta}$,

(r$_8$) $\dfrac{X \vdash \alpha, X \vdash \alpha \Rightarrow \beta}{X \vdash \beta}$,

(r$_9$) $\dfrac{X \cup \{\alpha\} \vdash \beta \wedge \sim \beta}{X \vdash \sim \alpha}$,

(r$_{10}$) $\dfrac{X \vdash \beta \wedge \sim \beta}{X \vdash \alpha}$,

(r$_{11}$) $\dfrac{X \vdash \alpha \Rightarrow \beta, X \vdash \beta \Rightarrow \alpha}{X \vdash \alpha \Leftrightarrow \beta}$,

(r$_{12}$) $\dfrac{X \vdash \alpha \Leftrightarrow \beta}{X \vdash \alpha \Rightarrow \beta}$,

(r$_{13}$) $\dfrac{X \vdash \alpha \Leftrightarrow \beta}{X \vdash \beta \Rightarrow \alpha}$,

(r$_{14}$) $\dfrac{X \vdash \alpha}{X \cup \{\beta\} \vdash \alpha}$,

holding for $\alpha, \beta, \gamma \in S$, $X \subseteq S$.

A sequent $X \vdash \alpha$ is provable if and only if there exists a finite chain of nis $s_1, s_2, \ldots, s_n$ such that

(d$_1$) $s_n = (X \vdash \alpha)$,

(d$_2$) for each i with $1 \leq i \leq n$, either $s_i = (\alpha \vdash \alpha)$ or $s_i = r(s_{i_1}, s_{i_2}, \ldots, s_{i_k})$ for some $i_1, i_2, \ldots, i_k < i$ and a certain $r \in \{$r$_1$, r$_2$,..., r$_{14}\}$.

Theorem 5.3

For any $X \subseteq S$ and $\alpha \in S$, the sequent $X \vdash \alpha$ is provable if and only if $\alpha \in Cn_I(X)$.

Proof (outline)

Assume at first that $\alpha \in Cn_I(X)$. This means that α has an intuitionistic proof from the set X. Induction with respect to the complexity of α shows easily that the sequent $X \vdash \alpha$ is provable.

Now assume that $X \vdash \alpha$ is provable in the natural deduction calculus of intuitionistic logic. Then there exists a finite chain of sequents $s_1, s_2, \ldots, s_n$ satisfying conditions (d_1) and (d_2). Fix i, $1 \leq i \leq n$. If s_i has the form $\beta \vdash \beta$ then of course $\beta \in Cn_I(X)$. If $s_i = r(s_{i_1}, s_{i_2}, \ldots, s_{i_k})$ for some $r = \mathrm{r}_j$, $1 \leq j \leq 14$, then we get $s_i = X \vdash \beta_i$ where $\beta_i \in Cn_I(X)$. Hence $\alpha \in Cn_I(X)$. ∎

5.3 Characterization of the Intuitionistic Propositional Logic in Terms of the Consequence Operator Cn_I

The intuitionistic logic can be defined by imposing the following conditions upon the operator Cn_I:

(c_1) $X \subseteq Cn_I(X)$,

(c_2) $X \subseteq Cn_I(Y) \Rightarrow Cn_I(X) \subseteq Cn_I(Y)$,

(c_3) $Cn_I(X) = \bigcup\{Cn_I(Y) : Y \subseteq X\}$,

(c_4) $\beta \in Cn_I(X \cup \{\alpha\}) \Leftrightarrow \alpha \Rightarrow \beta \in Cn_I(X)$,

(c_5) $\alpha \vee \beta \in Cn_I(X) \Leftrightarrow Cn_I(X \cup \{\alpha\}) \cap Cn_I(X \cup \{\beta\}) \subseteq Cn_I(X)$

(c_6) $\alpha \wedge \beta \in Cn_I(X) \Leftrightarrow Cn_I(X \cup \{\alpha\}) \cup Cn_I(X \cup \{\beta\}) \subseteq Cn_I(X)$

(c_7) $(\alpha \Leftrightarrow \beta) \in Cn_I(X) \Leftrightarrow Cn_I(X \cup \{\alpha, \beta\} \cap Cn_I(X \cup \{\sim \alpha, \sim \beta\}) \subseteq Cn_I(X)$, (c$_8$) $\sim \alpha \in Cn_I(X) \Leftrightarrow Cn_I(X \cup \{\alpha\}) = S$,

(c$_9$) Cn_I is the least consequence operation fulfilling conditions (c$_1$) – (c$_8$).

5.4 Algebraic Characterization of the Intuitionistic Propositional Logic

The algebraic semantics adequate for a treatment of intuitionistic logic is provided by the class of so-called pseudo-Boolean algebras.

A pseudo-Boolean algebra, abbreviated pba, is defined as a pair $(B, \leq)$ in which B is a nonempty set and $\leq$ is an order relation in B satisfying for every elements $a, b \in B$ the following conditions:

(p$_1$) there exists $a \cup b$, the least upper bound of a and b;

(p$_2$) there exists $a \cap b$, the greatest lower bound of a and b;

(p$_3$) there exists $a \Rightarrow b$, the pseudo-complement of a relative to b, defined as the greatest element $x \in B$ with $a \cap x \leq b$;

(p$_4$) there exists the least element in B, denoted by $\wedge$.

The symbol $-a$ stands for the element $a \Rightarrow \wedge$; so

$$-a = a \Rightarrow \wedge;$$

and we define

$$-\wedge = \vee$$

Suppose $(B, \leq)$ is a pba. As before, let S denote the set of all well formed formulas of the intuitionistic propositional calculus. By a homomorphism of S into $(B, \leq)$ we mean any function $h : S \to B$ with the following properties:

(h$_1$) $h(\alpha \wedge \beta) = h(\alpha) \cap h(\beta)$,

(h$_2$) $h(\alpha \vee \beta) = h(\alpha) \cup h(\beta)$,

(h$_3$) $h(\sim \alpha) = -h(\alpha)$,

(h$_4$) $h(\alpha \Rightarrow \beta) = h(\alpha) \Rightarrow h(\beta)$.

If $(B, \leq)$ is a pba and h is a homomorphism, we call the triple $(B, \leq, h)$ an algebraic model for the formula set S. A formula $\alpha \in S$ is algebraically true in a model $(B, \leq, h)$ when $h(\alpha) = \vee$. A formula α is a tautology if and only if it is algebraically true in every algebraic model.

Theorem 5.4

A formula $\alpha \in S$ is a tautology if and only if $\alpha \in Cn_I(\emptyset)$. ∎

A proof of this theorem can be found, e.g., in [Rasiowa and Sikorski 1970].

5.5 Kripke's Semantics for the Intuitionistic Propositional Calculus

By an intuitionist model over a calculus $L = (V, S)$ we mean a triple $(G, r, \models)$ in which G is a nonempty set, $r \subseteq G \times G$ is a relation, reflexive and transitive, and $\models \subseteq G \times S$ is a relation subject to the following requirements (as usual, α, β denote element of S):

(a_1) for every $p \in V$, if $g \models p$ and $(g, d) \in r$, then $d \models p$;

(a_2) $g \models \alpha \wedge \beta$ iff $g \models \alpha$ and $g \models \beta$;

(a_3) $g \models \alpha \vee \beta$ iff $g \models \alpha$ or $g \models \beta$;

(a_4) $g \models \sim \alpha$ iff, for every $d \in G$ with $(g, d) \in r$, $d \models \alpha$ does not hold;

(a_5) $g \models (\alpha \Rightarrow \beta)$ iff, for every $d \in G$ with $(g, d) \in r$, $d \models \alpha$ implies $d \models \beta$.

Given $g \in G$, let us agree to write g^* for any element d such that $(g, d) \in r$. Then we can restate conditions (a_4) and (a_5) in a slightly more concise form:

(a'_4) $g \models \sim \alpha$ iff $g^* \models \alpha$ fails to hold, for all g^*;

(a'_5) $g \models \alpha \Rightarrow \beta$ iff $g^* \models \alpha$ implies $g^* \models \beta$, for all g^*.

A formula $\alpha \in S$ is true in a model $(G, r, \models)$ if and only if $g \models \alpha$ for all g in G.

A formula $\alpha \in S$ is a tautology of the intuitionistic propositional calculus if and only if it is true in every model $(G, r, \models)$.

Lemma 5.1

Let $(G, r, \models)$ and $(G, r, \models')$ be two models. Suppose that, for every $g \in G$ and every propositional variable p,

$$g \models p \quad \text{iff} \quad g \models' p.$$

Then the two relations $\models$ and $\models'$ coincide.

Proof

Induction with respect to the complexity of the formula. ■

Lemma 5.2

Let G be a nonempty set, let $r \subseteq G \times G$ be a relation, which is reflexive and transitive, and let $\models \subseteq G \times V$ (V denoting the set of propositional variables). Then $\models$ can be extended to a relation $\models' \subseteq G \times S$ such that $(G, r, \models')$ is a model.

Proof

The extended relation $\models'$ is defined inductively as follows:

(b_1) if $g \models \alpha$ then, for every g^*, $g^* \models' \alpha$;

(b_2) $g \models' \alpha \wedge \beta$ iff $g \models' \alpha$ and $g \models' \beta$;

(b_3) $g \models' \alpha \vee \beta$ iff $g \models' \alpha$ or $g \models' \beta$;

(b_4) $g \models' \sim \alpha$ iff, for every g^*, $g^* \models' \alpha$ fails to hold;

(b_5) $g \models' (\alpha \Rightarrow \beta)$ iff, for every g^*, $g^* \models' \alpha$ implies $g^* \models' \beta$.

It is verified without any difficulty that the triple $(G, r, \models')$ is a model. ∎

Theorem 5.5

Let G be a nonempty set and let $r \subseteq G \times G$ be a relation, which is reflexive and transitive. Then every relation $\models \subseteq G \times V$ extends uniquely to a relation $\models' \subseteq G \times S$ such that $(G, r, \models')$ is a model.

Proof

Immediate from Lemmas 5.1 and 5.2. ∎

Theorem 5.6

Let $\alpha \in S$. Suppose $(G, r, \models)$ is a model and let $g, d \in G$. If $g \models \alpha$ and $(g, d) \in r$, then $d \models \alpha$.

Proof

Induction with respect to the complexity of α.

When α is an individual variable, the claim follows from (a_1). Assume the claim for a certain formula α; i.e., assume that the conditions $g \models \alpha$ and $(g, d) \in r$ imply $d \models \alpha$. We show that the claim holds for $\sim \alpha$.

Thus suppose $g \models \sim \alpha$ and $(g, d) \in r$. Then, for every g^*, $g^* \models \alpha$ fails to hold. But $(g, d) \in r$ and r is transitive, so each successor of d is automatically a successor of g. Therefore $d^* \models \alpha$ does not hold for any d^*, and hence $d \models \sim \alpha$.

For other connectives, the proof is obtained along the same lines. ∎

It can be proved (see [Fitting 1973]) that a formula α is a tautology over Kripke's models if and only if it is a tautology over the class of pseudo-Boolean algebras.

6 First-Order Predicate Calculus for Many-Valued Logics

The first-order predicate calculus in its modern shape entered the stage toward the end of the nineteenth century. G. Frege, C. S. Peirce and E. Schroeder are considered to be its creators.

The first formalization and axiomatization in the classical case – without a proof of completeness – was achieved by B. Russell and A. N. Whitehead in their monograph *Principia Mathematica*, Vol. I, 1910.

Unlike the classical propositional calculus, the classical predicate calculus has been given various nonequivalent axiomatizations. In general, they are lexically weaker than the well grounded axiomatizations presented in the monographs by D. Hilbert and P. Bernays, *Grundlagen der Mathematik*, Vol. I, 1934, or D. Hilbert and W. Ackermann, *Grundzüge der theoretischen Logik*, 1967, or the work of A. Mostowski, *Logika matematyczna*, 1948.

The discovery of non-classical calculi has made it clear that the predicate calculus is in fact a superstructure over a given propositional calculus and is entirely determined by the latter (within the given approach). Ever since the appearance of non-classical logics, more or less successful attempts have been made to devise a general theory of predicate calculus for those logics. There also exist neat predicate calculi for classes of many-valued logics.

The method of constructing the predicate calculus for a given many-valued logic L follows the lines of the classical case of two-valued logic.

6.1 The Language of the First-Order Predicate Calculus

Let L be a propositional calculus. Consider the ordered octuple

$$A = (V, \{F_i : i \in I_0\}, \{P_i : i \in I_1\}, L_0, L_1, L_2, Q, U)$$

where:

(i_1) I_0 is the set of nonnegative integers, I_1 is the set of positive integers;

(i_2) V is a countable set of individual variables: $V = \{x_i : i \in I_1\}$;

(i_3) each $F_i (i \in I_0)$ is a countable (possibly void) set of function symbols; each $P_i (i \in I_1)$ is a countable (possibly void) set of relation symbols (predicates);

(i_4) the set $\bigcup_{i \in I_1} P_i$ is nonempty;

(i_5) L_0 is the set of propositional constants;

(i_6) L_1 and L_2 are the sets of unary and binary connectives;

(i_7) Q is a set containing two elements, called quantifiers, the general (universal) and the particular (existential), denoted respectively by $\forall$ and $\exists$;

(i_8) U is a set of two elements, called parentheses, (and), viewed as auxiliary symbols;

(i_9) the sets $V, F_i (i \in I_0), P_i (i \in I_1), L_0, L_1, L_2, Q, U$ are all disjoint.

For each $i \in I_0$, the elements of F_i are i-argument function symbols; they will be denoted by the letters f, g, h, possibly with subscripts. Zero-argument function symbols, i.e., the elements of F_0, will be called individual variables and denoted by c_k, where k ranges over an index set K.

For each $i \in I_1$, the elements of P_i are i-argument relation symbols (predicates). To denote them, we will use the letters p, g, r, with subscripts if necessary.

The union of all sets that constitute A will be called the alphabet of the language of predicate calculus; its elements are the alphabet symbols.

We now define T, the set of terms over A, as the smallest set T' with the properties:

(t_1) $V \cup F_0 \subseteq T'$;

(t_2) if $f \in F_i$, $i > 0$, and $t_1, t_2, \dots, t_i \in T'$, then $f(t_1, t_2, \dots, t_i) \in T'$.

Terms will be denoted by t, s, u, with subscripts if the need arises.

Finally, we introduce S, the set of predicate formulas (for short, formulas) over A, defined as the smallest set S' with the properties:

(s_1) if $r \in P_i$ and $t_1, t_2, \dots, t_i \in T$, then $r(t_1, t_2, \dots, t_i) \in S'$;

(s_2) if $e \in L_0$ then $e \in S'$;

(s_3) if $o \in L_1$ and $a \in S'$ then $oa \in S'$;

(s_4) if $o \in L_2$ and $a, b \in S'$ then $aob \in S'$;

(s_5) if $\alpha(x_k) \in S'$ for some $x_k \in V$, then $\forall x_k \alpha(x_k) \in S'$ and $\exists x_k \alpha(x_k) \in S'$.

Formulas of type as in (s_1) or (s_2) are called atomic, or elementary. Thus, propositional constants are atomic formulas.

6.2 Free Variables and Bound Variables

The set of free individual variables occuring in a formula α (or in a term t), denoted by $zw(\alpha)$(or $zw(t)$, respectively), is defined as follows:

(w₁) $zw(x_i) = x_i$, $zw(c_k) = \emptyset$, for $i \in N, k \in K$;

(w₂) $zw(f(t_1, t_2, \ldots, t_i)) = zw(t_1) \cup zw(t_2) \cup \ldots \cup zw(t_i)$;

(w₃) $zw(r(t_1, t_2, \ldots, t_i)) = zw(t_1) \cup zw(t_2) \cup \ldots \cup zw(t_i)$;

(w₄) $zw(o\alpha) = zw(\alpha)$, for $o \in L_1$;

(w₅) $zw(\alpha o b) = zw(\alpha) \cup zw(\beta)$, for $o \in L_2$;

(w₆) $zw(\forall x_k \alpha) = zw(\exists x_k \alpha) = zw(\alpha) - \{x_k\}$.

Similarly, we define the set of individual constants occurring in a term t or in a formula α, in symbols: $ct(t), ct(\alpha)$:

(c₁) $ct(x_i) = \emptyset, ct(c_k) = c_k$, for $i \in N, k \in K$;

(c₂) $ct(f(t_1, t_2, \ldots, t_i)) = ct(t_1) \cup ct(t_2) \cup \ldots \cup ct(t_i)$;

(c₃) $ct(r(t_1, t_2, \ldots, t_i)) = ct(t_1) \cup ct(t_2) \cup \ldots \cup ct(t_i)$;

(c₄) $ct(o\alpha) = ct(\alpha)$, for $o \in L_1$;

(c₅) $ct(\alpha o \beta) = ct(\alpha) \cup ct(\beta)$, for $o \in L_2$;

(c₆) $ct(\forall x_k \alpha) = ct(\exists x_k \alpha) = ct(\alpha)$

Finally, we define $zz(\alpha)$, the set of bound variables occurring in a formula α, by:

(z₁) $zz(r(t_1, t_2, \ldots, t_i)) = \emptyset$ for $r \in P_i, i \in I_1$;

(z₂) $zz(o\alpha) = zz(\alpha)$ for $o \in L_1$;

(z₃) $zz(\alpha o \beta) = zz(\alpha) \cup zz(\beta)$ for $o \in L_2$;

(z₄) $zz(\forall x_k \alpha) = zz(\exists x_k \alpha) = zz(\alpha) \cup \{x_k\}$

6.3 The Rule of Substitution for Individual Variables

The adoption of the rule of substituting terms for individual variables inevitably involves difficulties connected with its limited applicability. There are two ways to overcome this obstacle. The first one, the most commonly used, is to accept the rule via its intuitive characterization, stating constraints as to its applicability in concrete situations, each time the need occurs. We think it is more convenient to define the rule precisely and to formulate the general applicability conditions in an explicit way.

The rule of substitution of a term t_m for a variable x_k in a term t or in a formula α will be denoted by $(t)x_k/t_m$ or $(\alpha)x_k/t_m$, respectively. It is defined as follows:

(p₁) $(x_i)x_k/t_m = \begin{cases} t_m & \text{if } k = i \\ x_i & \text{if } k \neq i \end{cases}$

for $m \in N$, $i, k \in K$;

(p₂) $(c_k)x_k/t_m = c_k$ for $m \in N$, $k \in K$;

(p₃) $(f(t_1, t_2, \ldots, t_i))x_k/t_m = f((t_1)x_k/t_m, (t_2)x_k/t_m, \ldots, (t_i)x_k/t_m)$

for $i \in I_0$, $i \geq 1$, k, m as above;

(p₄) $(r(t_1, t_2, \ldots, t_i))x_k/t_m = r((t_1)x_k/t_m, (t_2)x_k/t_m, \ldots, (t_i)x_k/t_m)$

for $i \in I_1$, k, m as above;

(p₅) $(o\alpha)x_k/t_m = o(\alpha)x_k/t_m$ for $o \in L_1$, $k, m \in N$;

(p₆) $(\alpha o \beta)x_k/t_m = (\alpha)x_k/t_m o(\beta)x_k/t_m$

for $o \in L_2$, k, m as above;

(p₇) $(\forall x_s \alpha)x_k/t_m = \begin{cases} \forall x_s \alpha, & \text{if } s = k \text{ or } x_s \in zw(t_m) \\ \forall x_s(\alpha)x_k/t_m, & \text{if } s \neq k \text{ and } x_s \notin zw(t_m) \end{cases}$

$(\exists x_s \alpha)x_k/t_m = \begin{cases} \exists x_s \alpha, & \text{if } s = k \text{ or } x_s \in zw(t_m) \\ \exists x_s(\alpha)x_k/t_m, & \text{if } s \neq k \text{ and } x_s \notin zw(t_m) \end{cases}$

Thus, according to this definition, the set of terms and the set of formulas are closed under the rule of substitution. Note that the rule does not apply to bound variables; the set of bound variables occurring in a formula α remains unchanged after substitution.

Let us finally introduce a certain three-element relation (x_k, t_m, α) which says that the variable x_k is free for the term t_m in the formula α. This relation is defined by the conditions:

(d_1) if $x_k \notin zw(\alpha)$ then (x_k, t_m, α);

(d_2) suppose $x_k \in zw(\alpha)$; then:

(1) if α is an elementary formula then (x_k, t_m, α);
(2) if $\alpha = o\beta$, $o \in L_1$, then (x_k, t_m, α)
if and only if (x_k, t_m, β);
(3) if $\alpha = \beta o \gamma$, $o \in L_2$, then (x_k, t_m, α)
if and only if (x_k, t_m, β) and (x_k, t_m, γ);
(4) if $\alpha = \forall x_i \beta$ or $\alpha = \exists x_i \beta$,
then (x_k, t_m, α) if and only if $x_i \notin zw(t_m)$ and (x_k, t_m, β).

We will exhibit the properties of this relation as and when we need them.

6.4 Fundamental Semantic Notions

It is assumed in this section that we are given an arbitrarily defined n-valued propositional calculus L, having a strongly adequate matrix $\mathfrak{M}L$ with universe $\bar{n} = \{0, 1, 2, \ldots, n-1\}$ and the set of designated variables $\{r, r+1, \ldots, n-1\}$, where $n > 2$ and $0 < r \leq n-1$. All propositional connectives of this calculus are characterized (or interpreted) through the functions exhibited in $\mathfrak{M}L$, defined on the set $\bar{n}$ and taking values in the same set.

Suppose

$$A = (V, \{F_i : i \in I_0\}, \{P_i : i \in I_1\}, L_0, L_1, L_2, Q, U)$$

is the alphabet of a first-order predicate calculus, T is the set of terms over A, and S is the set of formulas. In the sequel, we shall for brevity omit the symbols L_0, L_1, L_2, Q when defining an alphabet; these are invariant notions of all languages. In most cases, to identify a language we will confine ourselves to giving its alphabet alone, which will occasionally be also called the language; in fact, the alphabet determines the sets of terms and formulas. For convenience, zero-argument functions and function symbols will be called individual constants.

By an interpretation of a language $A = (V, \{F_i : i \in I_0\}, \{P_i : i \in I_1\})$ we mean a triple

$$\mathfrak{A} = (D, \{F_i^* : i \in I_0\}, \{P_i^* : i \in I_1\})$$

in which D is any nonempty set called the interpretation universe; each F_i^* is a set of i-argument functions with domain D and range in D; and each P_i^* is a set of i-argument functions with domain D and range in the set of logical values $\bar{n} = \{0, 1, \ldots, n-1\}$. Thus, if $f^* \in F_i^*$ then $f^* : D^i \to D$, and if $r^* \in P_i^*$ then $r^* : D^i \to \bar{n}$; the asterisk establishes bijective correspondence:

$$F_i \ni f \leftrightarrow f^* \in F_i^*, \quad P_i \ni r \leftrightarrow r^* \in P_i^*.$$

Let $\sum$ be the set of infinite sequences of entries belonging to D:

$$\sum = \{b : b = (b_1, b_2, \ldots), b_i \in D\}.$$

The symbol $w(t, b)$ will stand for the value of a term t on a sequence $b = (b_1, b_2, \ldots)$.
We define:

(wt1) $w(x_i, b) = b_i$ for $x_i \in V$;

(wt2) $w(f_j, b) = f_j^*$ for $f_j \in F_0$;

(wt3) $w(f_k(t_1, t_2, \ldots, t_m), b) = f_k^*(w(t_1, b), w(t_2, b), \ldots, w(t_m, b))$.

Lemma 6.1

If all variables that occur in a term t belong in $\{x_1, x_2, \ldots, x_m\}$ and if two sequences b, b' do not differ on the first m positions, then $w(t, b) = w(t, b')$.

Proof

Induction with respect to the complexity of t. Let $t = x_i$. Then

$$w(x_i, b) = b_i = w(x_i, b')$$

because $1 \leq i \leq m$. Similarly, if $f_j \in F_0$ then

$$w(f_j, b) = f_j^* = w(f_j, b').$$

Now let $t = f_i(t_1, t_2, \ldots, t_k)$ and suppose

$$\begin{aligned} &w(t_1, b) = w(t_1, b'),\\ &w(t_2, b) = w(t_2, b'),\\ &\ldots\\ &w(t_k, b) = w(t_k, b'). \end{aligned}$$

Then

$$\begin{aligned} &w(f_i(t_1, t_2, \ldots, t_k), b)\\ &= f_i^*(w(t_1, b), w(t_2, b), \ldots, w(t_k, b))\\ &= f_i^*(w(t_1, b'), w(t_2, b'), \ldots, w(t_k, b'))\\ &= w(f_i(t_1, t_2, \ldots, t_k), b'), \end{aligned}$$

ending the proof. ∎

Lemma 6.1 asserts that the value of a term t depends only on an initial segment of a sequence b.

Let $\alpha \in S$ be any formula and let $\mathfrak{A}$ be an interpretation of the language. The symbol $w(\alpha, s)$ will denote the value of α on a sequence $s \in \sum$. By definition, (wf1) $w(r_i(t_1, t_2, \ldots, t_m), s) = k$ iff $r_i^*(w(t_1, s), w(t_2, s), \ldots, w(t_m, s)) = k$;

(wf2) $w(e, s) = e_{\mathfrak{A}}$ for any logical constant e;

(wf3) $w(o\alpha, s) = \odot(w(\alpha, s))$ for $o \in L_1$;

(wf4) $w(\alpha o \beta, s) = w(\alpha, s) \odot w(\beta, s)$ for $o \in L_2$;

(wf5) $w(\forall x_i \alpha, s) = \min\{w(\alpha, s') : s' \in \sum,$
s' coincides with s except at i-th position$\}$;

(wf6) $w(\exists x_i \alpha, s) = \max\{w(\alpha, s') : s' \in \sum,$
s' coincides with s except at i-th position$\}$;

in (wf2), $e_{\mathfrak{A}}$ denotes the interpretation of the constant e in the matrix $\mathfrak{M}L$; in (wf3) and (wf4), the symbol $\odot$ stands for the corresponding functions in $\mathfrak{M}L$, interpreting the respective connectives $o \in L_1 \cup L_2$.

Recall the example of Section 2.5.2. In that case,

$$L_0 = \emptyset, \quad L_1 = \{\sim\} \cup \{J_i : 0 \le i \le n-1\}, \quad L_2 = \{\vee, \wedge, \Rightarrow\};$$

the definition of $w(\alpha, s)$ will take the form

(wf1′) $w(r_i(t_1, t_2, \ldots, t_m), s) = k$ iff $r_i^*(w(t_1, s), w(t_2, s), \ldots, w(t_m, s)) = k$;

(wf2′) $w(\sim \alpha, s) = (w(\alpha, s) + 1) \text{ mod n}$;

(wf2i′) $w(J_i\alpha, s) = \begin{cases} n-1 & \text{if } w(\alpha, s) = i, \\ 0 & \text{if } w(\alpha, s) \neq i; \end{cases}$

(wf3′) $w(\alpha \vee \beta, s) = \max\{w(\alpha, s), w(\beta, s)\}$;

(wf4′) $w(\alpha \wedge \beta, s) = \min\{w(\alpha, s), w(\beta, s)\}$;

(wf5′) $w(\alpha \Rightarrow \beta, s) =$

$$= \begin{cases} n-1 & \text{if } w(\alpha, s) \le w(\beta, s), \\ w(\beta, s) & \text{if } w(\alpha, s) > w(\beta, s) \text{ and } r \le w(\alpha, s), \\ n - w(\alpha, s) + w(\beta, s) - 1 & \text{if } w(\alpha, s) > w(\beta, s) \text{ and } r > w(\alpha, s). \end{cases}$$

Items (wf6′) and (wf7′) are identical with (wf5) and (wf6) from the definition in the general case.

In the sequel we restrict our considerations to the situation of this example. They carry over to the general case without any substantial difficulty; this is just a technical task. Given any propositional calculus, the specification of the general definition has to be relativized to its minimal matrix.

We adopt the following terminology. Let $\bar{n} = \{0, 1, 2, \ldots, n-1\}$ be the set of logical values, $n > 2$, and let $\{r, r+1, \ldots, n-1\}$ be the set of designated values. Suppose the value of a formula α for a sequence s is k; i.e., $w(\alpha, s) = k$, with $0 \leq k \leq n-1$. If $r \leq k$, we say that s satisfies the formula α in degree k; if $k < r$, we say that s dissatisfies α in degree k (in a given interpretation $\mathfrak{A}$). A sequence $s \in \sum$ is said to satisfy a formula α if $w(\alpha, s) = k$ and $r \leq k \leq n-1$; it is said to dissatisfy α if $w(\alpha, s) = k$ and $0 \leq k < r$.

We say that a formula α is true in the model (interpretation) $\mathfrak{A}$ in degree k if and only if $r \leq k \leq w(\alpha, s)$ holds for every sequence $s \in \sum$ and there exists a sequence s' such that $w(\alpha, s') = k$. Dually, we say that a formula α is false in model $\mathfrak{A}$ in degree k if and only if $w(\alpha, s) \leq k < r$ holds for every $s \in \sum$, with equality $w(\alpha, s') = k$ holding for some $s' \in \sum$. A formula α is said to be true in the model $\mathfrak{A}$ if it is satisfied by each sequence $s \in \sum$; it is false in $\mathfrak{A}$ if it is satisfied by no sequence $s \in \sum$.

A formula $\alpha \in S$ is tautologous of order k if and only if it is true in degree k in every model of the language S.

A formula $\alpha \in S$ is a tautology of the n-valued first-order predicate calculus if and only if it is true in every model of the language S.

Lemma 6.2

If all variables that occur in a formula α belong to $\{x_1, x_2, \ldots, x_m\}$ and if two sequences s, s' do not differ at the first m positions, then $w(\alpha, s) = w(\alpha, s')$.

Proof

Again we use induction on the construction of α. Let $\alpha = r_i(t_1, t_2, \ldots, t_k)$. Then

$$\begin{aligned} w(\alpha, s) &= w(r_i(t_1, t_2, \ldots, t_k), s) \\ &= r_i^*(w(t_1, s), w(t_2, s), \ldots, w(t_k, s)) \\ &= r_i^*(w(t_1, s'), w(t_2, s'), \ldots, w(t_k, s')) \\ &= w(r_i(t_1, t_2, \ldots, t_k), s') = w(\alpha, s'), \end{aligned}$$

in view of Lemma 6.1.

If $\alpha = e$ then

$$w(\alpha, s) = w(e, s) = e_{\mathfrak{A}} = w(e, s') = w(\alpha, s').$$

Assume inductively that the lemma holds for some formulas α and β. Then

$$w(o\alpha, s) = \odot(w(\alpha, s)) = \odot(w(\alpha, s')) = w(o\alpha, s')$$

and similarly

$$w(\alpha o \beta, s) = w(\alpha, s) \odot w(\beta, s) = w(\alpha, s') \odot w(\beta, s') = w(\alpha o \beta, s').$$

Further,

$$\begin{aligned} w(\forall x_i \alpha, s) = & \min\{w(\alpha, s'') : s'' \in \textstyle\sum, \\ & \quad s'' \text{ coincides with } s \text{ except at } i\text{-th position}\} \\ = & \min\{w(\alpha, s'') : s'' \in \textstyle\sum, \\ & \quad s'' \text{ coincides with } s' \text{ except at } i\text{-th position}\} \\ = & \; w(\forall x_i \alpha, s'), \end{aligned}$$

since $zw(\alpha) \subseteq \{x_1, x_2, \ldots, x_n\}$.

Similarly we show that $w(\exists x_i \alpha, s) = w(\exists x_i \alpha, s')$. The lemma is proved. ∎

It is plain in view of Lemma 6.2 that the value of a formula α on a sequence s depends only on some initial values of s. The same concerns all the concepts derived from the value: being satisfied/dissatisfied, truth/falsity, and tautologousness. The use of infinite sequences $s \in \sum$ is no more than a technical device.

Corollary 6.1

Suppose a formula α is closed (i.e., $zw(\alpha) = \emptyset$) and there exists in a model $\mathfrak{A}$ a sequence $s \in \sum$ satisfying α (in degree k). Then α is true (in degree k) in $\mathfrak{A}$.

Proof

This is immediate by Lemma 6.2, since α has no free variables. ∎

By the n-valued first-order predicate calculus over a propositional calculus L we will understand the set of all tautologies of L. We will denote it by PL.

6.5 The Many-Valued First-Order Predicate Calculus of Post

Our presentation of the Post predicate calculus is patterned on Rasiowa's book [Rasiowa 1974]. Our approach is slightly more general, in that we consider more than one designated value. The calculus is built over the Post many-valued propositional calculus, in a manner analogous to that used in classical logic, by defining the class of first-order languages over the propositional calculus. In the formalized first-order language we define the consequence operation by giving the axiom system and rules of inference.

Let S_n denote the set of well formed formulas over the alphabet of the predicate calculus A, in which $L_0 = \{e_0, e_1, \ldots, e_{n-1}\}$ is the set of propositional constants, $L_1 = \{\sim, d_1, d_2, \ldots, d_{n-1}\}$ is the set of one-argument connectives, $L_2 = \{\vee, \wedge, \Rightarrow\}$ is the set of two-argument connectives, and $Q = \{\forall, \exists\}$ is the set of quantifiers. We also introduce the symbol $\alpha \Leftrightarrow \beta$ to denote the conjunction $(\alpha \Rightarrow \beta) \wedge (\beta \Rightarrow \alpha)$. As in Section 3.4, we denote the axiom system for the Post predicate calculus by A_n.

The system A_n consists of the following schemes, for $\alpha, \beta, \gamma \in S_n$:

(p$_1$) $\alpha \Rightarrow (\beta \Rightarrow \alpha)$

(p$_2$) $(\alpha \Rightarrow (\beta \Rightarrow \gamma)) \Rightarrow ((\alpha \Rightarrow \beta) \Rightarrow (\alpha \Rightarrow \gamma))$

(p$_3$) $\alpha \Rightarrow \alpha \vee \beta$

(p$_4$) $\beta \Rightarrow \alpha \vee \beta$

(p$_5$) $(\alpha \Rightarrow \gamma) \Rightarrow ((\beta \Rightarrow \gamma) \Rightarrow (\alpha \vee \beta \Rightarrow \gamma))$

(p$_6$) $\alpha \wedge \beta \Rightarrow \alpha$

(p$_7$) $\alpha \wedge \beta \Rightarrow \beta$

(p$_8$) $(\alpha \Rightarrow \beta) \Rightarrow ((\alpha \Rightarrow \gamma) \Rightarrow (\alpha \Rightarrow \beta \wedge \gamma))$

(p$_9$) $(\alpha \Rightarrow \sim \beta) \Rightarrow (\beta \Rightarrow \sim \alpha)$

(p$_{10}$) $\sim (\alpha \Rightarrow \alpha) \Rightarrow \beta$

(p$_{11}$) $d_i(\alpha \vee \beta) \Leftrightarrow d_i\alpha \vee d_i\beta$ for $i = 1, 2, \ldots, n-1$;

(p$_{12}$) $d_i(\alpha \wedge \beta) \Leftrightarrow d_i\alpha \wedge d_i\beta$ for $i = 1, 2, \ldots, n-1$;

(p$_{13}$) $d_i(\alpha \Rightarrow \beta) \Leftrightarrow \bigwedge_{j=1}^{i} (d_j\alpha \Rightarrow d_j\beta)$ for $i = 1, 2, \ldots, n-1$,

where $\bigwedge_{j=1}^{i} \alpha_j = \alpha_1 \wedge \alpha_2 \wedge \ldots \wedge \alpha_i$;

(p$_{14}$) $d_i \sim \alpha \Leftrightarrow \sim d_i\alpha$ for $i = 1, 2, \ldots, n-1$;

(p$_{15}$) $d_i d_j \alpha \Leftrightarrow d_j \alpha$ for $i = j = 1, 2, \ldots, n-1$;

(p$_{16}$) if $i \leq j$ then $d_i e_j$; if $i > j$ then $\sim d_i e_j$, for $i = 1, 2, \ldots, n-1$, $j = 0, 1, 2, \ldots, n-1$;

(p$_{17}$) $\alpha \Leftrightarrow \bigvee_{j=1}^{n-1} (d_j\alpha \wedge e_j)$ where $\bigvee_{j=1}^{s} \alpha_j = \alpha_1 \vee \alpha_2 \vee \ldots \vee \alpha_s$;

(p$_{18}$) $d_1\alpha \vee \sim d_1\alpha$.

These axioms are accompanied by the following rules of inference:
– Rule of detachment:

(rd) $$\frac{\alpha, \alpha \Rightarrow \beta}{\beta};$$

– Rule of introduction of the connective d_{n-1}:

(rn) $$\frac{\alpha}{d_{n-1}\alpha};$$

– Rule of substitution for individual variables:

(rs) $$\frac{\alpha(x_1, x_2, \ldots, x_m)}{\alpha(t_1, t_2, \ldots, t_m)},$$

the expression below the line arising by simultaneous replacement of each x_i by the respective term t_i;

– Rule of introduction of the existential quantifier to the implicative premise:

(rie) $$\frac{\alpha(x) \Rightarrow \beta}{\exists y\alpha(y) \Rightarrow \beta},$$

provided the variable x does not occur in β and there are no quantifiers binding y in $\alpha(x)$;

– Rule of introduction of the universal quantifier to the implicative conclusion:

(riu) $$\frac{\alpha \Rightarrow \beta(x)}{\alpha \Rightarrow \forall y\beta(y)},$$

provided x does not occur in α as a free variable and there are no quantifiers binding y in $\beta(x)$;

– Rule of elimination of the existential quantifier from the implicative premise:

(ree) $$\frac{\exists y\alpha(y) \Rightarrow \beta}{\alpha(x) \Rightarrow \beta},$$

provided variable y does not occur in $\alpha(x)$;

– Rule of elimination of the universal quantifier from the implicative conclusion:

(reu) $$\frac{\alpha \Rightarrow \forall y\beta(y)}{\alpha \Rightarrow \beta(x)},$$

provided variable y does not occur in $\beta(x)$.

The consequence operation Cn is defined in the standard way.

The resulting system, called the n-valued Post predicate calculus, will be denoted by L_n.

Lemma 6.3

For $\alpha, \beta, \gamma, \delta \in S_n$, $X \subseteq S_n$,

(a) $(\alpha \Rightarrow \alpha) \in Cn\emptyset$;

(b) if $\alpha, (\alpha \Rightarrow \beta) \in CnX$ then $\beta \in CnX$;

(c) if $(\alpha \Rightarrow \beta), (\beta \Rightarrow \gamma) \in CnX$ then $(\alpha \Rightarrow \gamma) \in CnX$;

(d) if $\alpha \in CnX$ then $\beta \Rightarrow \alpha \in CnX$;

(e) if $(\alpha \Rightarrow \beta), (\beta \Rightarrow \alpha) \in CnX$ then $(\sim \alpha \Rightarrow \sim \beta) \in CnX$ and $(d_i\alpha \Rightarrow d_i\beta) \in CnX$, for $i = 1, 2, \ldots, n-1$;

(f) if $(\alpha \Rightarrow \beta), (\beta \Rightarrow \alpha), (\gamma \Rightarrow \delta), (\delta \Rightarrow \gamma) \in CnX$,
then $(\alpha \vee \gamma \Rightarrow \beta \vee \delta) \in CnX$,
$(\alpha \wedge \gamma \Rightarrow \beta \wedge \delta) \in CnX$ and
$((\alpha \Rightarrow \gamma) \Rightarrow (\beta \Rightarrow \delta)) \in CnX$.

Proof

The formula

$$(\alpha \Rightarrow ((\beta \Rightarrow \alpha) \Rightarrow \alpha) \Rightarrow ((\alpha \Rightarrow (\beta \Rightarrow \alpha)) \Rightarrow (\alpha \Rightarrow \alpha))$$

obviously belongs in $Cn\emptyset$, as a particular case of axiom (p_2). The formula

$$\alpha \Rightarrow ((\beta \Rightarrow \alpha) \Rightarrow \alpha)$$

is in $Cn\emptyset$, as a specific case of axiom (p_1). Hence by detachment

$$((\alpha \Rightarrow (\beta \Rightarrow \alpha)) \Rightarrow (\alpha \Rightarrow \alpha)) \in Cn\emptyset.$$

Applying once more the rule of detachment, to the last formula and axiom (p_1), we infer

$$(\alpha \Rightarrow \alpha) \in Cn\emptyset,$$

proving (a).

The proof of (b) is immediate from the definition of CnX. The proof of (c) results by observing that

$$((\alpha \Rightarrow \beta) \Rightarrow ((\beta \Rightarrow \gamma) \Rightarrow (\alpha \Rightarrow \gamma))) \in Cn\emptyset.$$

Statement (d) follows at once from axiom (p_1) and the rule of detachment.

To prove the first claim in (e) it is enough to note that the formulas

$$(\alpha \Rightarrow \beta) \Rightarrow (\sim \beta \Rightarrow \sim \alpha), (\beta \Rightarrow \alpha) \Rightarrow (\sim \alpha \Rightarrow \sim \beta)$$

belong to $Cn\emptyset$. The second claim of (e) follows from axioms (p_{14}), (p_{15}) and (p_{18}).

The proof of (f) is achieved by appealing to axioms (p_5), (p_8) and (p_2). ∎

Lemma 6.4

Let $\alpha(x)$ be a formula of the Post predicate calculus and suppose y is a variable not occurring in $\alpha(x)$. For any term t, the formulas

$$(\forall y \alpha(y) \Rightarrow \alpha(t)) \quad \text{and} \quad (\alpha(t) \Rightarrow \exists y \alpha(y))$$

belong to $Cn\emptyset$.

Proof

By virtue of Lemma 6.3 (a),

$$(\forall y \alpha(y) \Rightarrow \forall y \alpha(y)) \in Cn\emptyset$$

and

$$(\exists y \alpha(y) \Rightarrow \exists y \alpha(y)) \in Cn\emptyset.$$

Applying the rules (reu), (ree) and (rs) we obtain

$$(\forall y \alpha(y) \Rightarrow \alpha(t)) \in Cn\emptyset$$

and

$$(\alpha(t) \Rightarrow \exists y \alpha(y)) \in Cn\emptyset,$$

as claimed. ∎

We return now to general considerations, involving an arbitrary language, not necessarily a Post language.

The set of terms of a given language together with the set of its function symbols constitutes a free algebra in the class of similar algebras, provided that any term $f_j(t_1, t_2, \ldots, t_k)$ is viewed as the result of operation f_j applied to terms $t_1, t_2, \ldots, t_k$. We consider this algebra as the pair $(T, \{f_j : j \in J\})$ and denote it by $\mathcal{T}$ (J is some index set). Evidently, $\mathcal{T}$ is freely generated by V, the set of individual variables.

Now suppose $\mathcal{B} = (K, \{g_j : j \in J\})$ is an algebra similar to $\mathcal{T}$. Every mopping $v : V \to K$ can be extended to a homomorphism h of $\mathcal{T}$ into $\mathcal{B}$ in the standard way, by

$$h(x_i) = v(x_i) \quad \text{for } x_i \in V,$$
$$h(f_j(t_1, t_2, \ldots, t_k)) = g_j(h(t_1), h(t_2), \ldots, h(t_k)).$$

In particular, if s is a substitution, i.e., a map $s : V \to T$, then s extends uniquely to a homomorphism of $\mathcal{T}$, the equality

$$s(f_j(t_1, t_2, \ldots, t_k)) = f_j(s(t_1), s(t_2), \ldots, s(t_k))$$

holding for every function symbol f_j and any terms $t_1, t_2, \ldots, t_k$.

Given a first-order predicate calculus, by a realization of the terms of its language in a nonempty set U we mean any mapping

$$r : \{f_j : j \in J\} \to U^{U^{arg(f_j)}}$$

where $arg(f_j)$ is the number of arguments of the function symbol f_j. That is to say, to an i-argument symbol f_j there is assigned a function

$$\varphi_{f_j} : U^i \to U.$$

Obviously, any realization of the terms in a nonempty set U assigns to the algebra $\mathcal{T} = (T, \{f_j : j \in J\})$ the algebra $\mathcal{T}_r = (U, \{\varphi f_j : j \in J\})$, free in the class of similar algebras.

A valuation of the n-valued Post predicate calculus in a set $U \neq \emptyset$ is defined as an arbitrary mapping

$$v : V \to U$$

of the set of individual variables V into U.

Let r be a realization of the terms in a set $U \neq \emptyset$. Every valuation v can be regarded as a mapping from the set of generators of $\mathcal{T}$ into the algebra $\mathcal{T}_r$. It extends uniquely to a homomorphism v^* of $\mathcal{T}$ into $\mathcal{T}_r$ by

(a) $v^*(x) = v(x) \quad \text{for} \quad x \in V$,

(b) $v_r(F_j(t_1, t_2, \ldots, t_m)) = \varphi r_j(v_r(t_1), v_r(t_2), \ldots, v_r(t_m))$
for any function symbol F_j and any terms $t_1, t_2, \ldots, t_m$.

Now suppose we fix a term t and let valuation v vary. Then $v^*(t)$ becomes a function of the variable $v \in U^V$. Denoting this function by t^*, we have of course

(c) $t^* : U^V \to U$,

(d) $t^*(v) = v^*(t)$.

Hence

$$x_r(v) = v(x) \quad \text{for} \quad x \in V$$

and

$$(F_j(t_1, t_2, \ldots, t_m))_r(v) = \varphi r_j((t_1)_r(v), (t_2)_r(v), \ldots, (t_m)_r(v))$$

for any m-argument function symbol F_j and any terms $t_1, t_2, \ldots, t_m$.

If s is a substitution, then

$$s(t^*(v)) = t^*(s^*(v)),$$

where

$$s^*(v(x)) = s(x^*(v)).$$

Taking for U the set T of terms and setting

$$\varphi r_j(t_1, t_2, \ldots, t_m) = F_j(t_1, t_2, \ldots, t_m)$$

we obtain a realization r in T, which we call the canonical realization of the terms. Similarly we define the canonical valuation in the set T, denoted by v_0, letting

$$v_0(x) = x \quad \text{for every } x \in V.$$

It is not hard to see that if r is the canonical realization, then $t^*(v) = v(t)$ for every term $t \in T$ and every valuation $v : V \to T$; similarly, for the canonical valuation v_0 we have $t^*(v_0) = t$.

Now, let

$$\mathfrak{A}_n = (A_n, \vee, \wedge, \Rightarrow, \sim, d_1, d_2, \ldots, d_{n-1}, e_0, e_1, \ldots, e_{n-1})$$

be a Post algebra. We define a binary relation in A_n by

$$a \leq b \quad \text{if and only if} \quad a \Rightarrow b = e_{n-1}.$$

This is an order on A_n, having e_{n-1} for the greatest element.

Note that we use the same characters to denote the elements of A_n and algebraic operations, on the one hand, and logical formulas and connectives, on the other. This should not lead to misunderstanding. The actual meaning of the symbol is determined by the context.

Let $\{a_s : s \in S\}$ be an arbitrary indexed set of elements of A_n. The symbols

$$\bigcup_{s \in S} a_s \quad \text{and} \quad \bigcap_{s \in S} a_s$$

stand respectively for the least upper bound and the greatest lower bound of $\{a_s : s \in S\}$, provided they exist. (They are often called the generalized union and intersection, or the generalized join and meet.)

A Post algebra $\mathfrak{A}_n$ is said to be complete if these bounds exist for every nonempty subset of A_n.

The following statements are obviously valid for any set $\{a_s : s \in S\}$, in every complete Post algebra:

(a) $a_s \leq \bigcup_{s \in S} a_s, \quad \bigcap_{s \in S} a_s \leq a_s;$

(b) if, for a certain $c \in A_n$, $a_s \leq c$ holds for all s, then $\bigcup_{s \in S} a_s \leq c$;

dually, if $c \leq a_s$ for all s, then $c \leq \bigcap_{s \in S} a_s$.

Let $\mathcal{B}_n$ be a complete subalgebra of $\mathfrak{A}_n$. Denoting by $\bigcup'$ and $\bigcap'$ the upper and lower bounds relative to $\mathcal{B}_n$, we have

(c) $\bigcup_{s \in S} a_s \leq \bigcup_{s \in S}' a_s,$

(d) $\bigcap_{s \in S}' a_s \leq \bigcap_{s \in S} a_s.$

Now suppose $\mathfrak{A}_n$ and $\mathcal{B}_n$ are any two complete Post algebras. The bounds, upper and lower, relative to these algebras will be denoted by

$$\bigcup_{s \in S}^{\mathfrak{A}} a_s, \bigcap_{s \in S}^{\mathfrak{A}} a_s \quad \text{and} \quad \bigcup_{s \in S}^{\mathcal{B}} a_s, \bigcap_{s \in S}^{\mathcal{B}} a_s.$$

Let h be a homomorphism from $\mathfrak{A}_n$ to $\mathcal{B}_n$. Plainly, if $a \leq b$ then $h(a) \leq h(b)$, for any $a, b \in A_n$. Also, the following inequalities hold:

$$\bigcup_{s \in S}^{\mathcal{B}} h(a_s) \leq h\left(\bigcup_{s \in S}^{\mathfrak{A}} a_s\right),$$

$$h\left(\bigcap_{s \in S}^{\mathfrak{A}} a_s\right) \leq \bigcap_{s \in S}^{\mathcal{B}} h(a_s).$$

In the case when the inequality sign in the last two relations can be replaced by equality, we say that h preserves the least upper bound and the greatest lower bound.

Let us now consider the algebra of formulas of the n-valued first-order predicate calculus defined by E. Post,

$$\wp_n = (S_n, \wedge, \vee, \Rightarrow, \sim, d_1, d_2, \ldots, d_{n-1}, e_0, e_1, \ldots, e_{n-1}).$$

Let X be a fixed set of formulas (for instance, the set of axioms). Define a relation $\leq$ by

$$\alpha \leq \beta \quad \text{iff} \quad \alpha \Rightarrow \beta \in CnX.$$

In view of Lemma 6.3(a, c), this relation is a quasi-order in S_n.

The binary relation $\approx$ defined in S_n by

$$\alpha \approx \beta \quad \text{iff} \quad \alpha \Rightarrow \beta,\ \beta \Rightarrow \alpha \in CnX$$

is a congruence in the algebra φ_n.

The relation $\leq$ defined on $S_n/\approx$ by

$$[\alpha] \leq [\beta] \quad \text{iff} \quad \alpha \Rightarrow \beta \in CnX$$

is an ordering of $S_n/\approx$.

Fix an integer r with $0 < r \leq n-1$ (the least designated value) and consider the element e_r in the algebra φ_n. For any formulas $\alpha, \beta \in S_n$ such that $\alpha, \beta \in CnX$, their cosets $[\alpha], [\beta]$ are not less than $[e_r]$ (modulo the order just introduced). Hence, clearly,

$$[e_r] \leq [\alpha] \quad \text{iff} \quad \alpha \in CnX.$$

Consider the quotient algebra

$$\varphi_n/\approx\, = (S_n/\approx, \vee^*, \wedge^*, \Rightarrow^*, \sim^*, d_1^*, d_2^*, \ldots, d_{n-1}^*, [e_0], [e_1], \ldots, [e_{n-1}])$$

with the induced operations

$$\begin{aligned}
&[\alpha] \wedge^* [\beta] = [\alpha \wedge \beta],\\
&[\alpha] \vee^* [\beta] = [\alpha \vee \beta],\\
&[\alpha] \Rightarrow^* [\beta] = [\alpha \Rightarrow \beta],\\
&\sim^* [\alpha] = [\sim \alpha]\\
&d_i{}^*[\alpha] = [d_i \alpha] \quad \text{for } i = 1, 2, \ldots, n-1.
\end{aligned}$$

We call $\varphi_n/\approx$ the algebra of the elementary theory of axiomatics X.

When X is the empty set, this algebra is called the algebra of the n-valued predicate calculus.

It is not hard to see that generalized joins and meets exist in $S_n/\approx$. Moreover, the substitution $sb(x) = x$ for each $x \in V$ yields

$$\bigcup_{t \in T} [\alpha(t)] = [\exists x \alpha(x)],$$

$$\bigcap_{t \in T} [\alpha(t)] = [\forall x \alpha(x)],$$

where as before T denotes the set of terms of the language S_n.

It is not hard to see that if ϱ is a realization in $\bar{n}$, then we have for any formula α and any valuation v

$$(\alpha(v))_\varrho = ([v(\alpha)])_{\varrho/\approx}$$

where $\varrho/\approx$ denotes the induced realization of $S_n/\approx$.

As in the classical calculus, a valuation $v : V \to U$ is said to satisfy a formula α in degree $0, k \leq n-1$, in a realization ϱ, if and only if $(\alpha(v))_\varrho = k$. A valuation v is said to satisfy α if v satisfies α in degree not less than the smallest designated value.

A formula is true in realization ϱ if and only if it is satisfied by every valuation v in realization ϱ.

A formula is a tautology of the n-valued Post predicate calculus if and only if it is true in every realization ϱ in a set of cardinality not less than $n-1$.

It is a matter of straightforward verification that the axioms of the n-valued Post predicate calculus are tautologies. Also, the rules of inference lead from tautologies to tautologies, which implies that if the premises are tautologous then so is the conclusion. Hence it follows easily that a formula which is derivable in the n-valued Post predicate calculus is a tautology of the calculus. The converse statement that all tautologies of the calculus are derivable (the completeness theorem) is valid, as well. In a somewhat more general setting, it is formulated and proved in the monograph by H. Rasiowa [Rasiowa 1974], which we warmly recommend to the reader.

7 The Method of Finitely Generated Trees in n-valued Logical Calculi

7.1 Introductory Remarks

The method referred to in the title is a generalization of a well known technique of proofs, employed in the classical two-valued logic and known under the names of Hintikka's trees or Beth tableaux. The generalization to n-valued propositional calculi is due to Suchoń and Surma [Surma 1977]; its extension to n-valued predicate calculi is due to Carnielli [Carnielli 1987]. Our presentation of the topic follows Carnielli, whose particularly valuable contribution was to devote attention to the so-called n-valued distributive quantifiers.

We will be working within the framework of a maximally generalized n-valued logical calculus, adopting Carnielli's version of the first-order predicate calculus. We give the most general treatment possible of propositional connectives and quantifiers, but we confine ourselves to the predicate calculus free of function symbols.

It has to be emphasized that the method of finitely generated trees is algorithmic; it provides a way of automatic proving theorems of the propositional calculus. (Its use for the predicate calculus requires certain refinements.) For technical reasons, we define the set of well formed formulas anew, for propositions as well as for predicates. This will facilitate a concise formulation of various facts and concepts, for example that of the level of a formula.

7.2 Finitely Generated Trees for n-valued Propositional Calculi

Let $\Omega = \{\omega_i : i \in I\}$ be a finite set of propositional connectives; ω_i is assumed to be an m_i-argument connective. The set of formulas, denoted by S, is constructed inductively:

$$\begin{aligned} S_0 &= \{p_i : i \in N\}, \\ S_{k+1} &= S_k \cup \{\omega_i(\alpha_1, \alpha_2, \ldots, \alpha_{m_i}) : \alpha_1, \alpha_2, \ldots, \alpha_{m_i} \in S_k,\ i \in I\}, \\ S &= \bigcup_{k \in N} S_k. \end{aligned}$$

The elements of S_0 are called atomic formulas or propositional variables; the formulas that belong to $S_k - S_{k-1}$ are said to be of degree k, or of level k, or of k-th degree of complexity.

Write $\bar{n} = \{0, 1, \ldots, n-1\}$ and let $h_1 : \bar{n}^{m_i} \to \bar{n}$ be functions such that the algebras

$$\bar{S} = (S, \Omega) \quad \text{and} \quad \mathfrak{A} = (\bar{n}, \{h_i : i \in I\})$$

are similar. Fix an integer r with $0 < r \leq n-1$. The set

$$\mathfrak{M} = (\mathfrak{A}, \{r, \ldots, n-1\})$$

is a model for the language S. Then $\{r, \ldots, n-1\}$ is called the set of designated values. The propositional calculus $\mathfrak{M}L$ can be defined as the pair

$$\mathfrak{M}L = (S, \mathfrak{M}),$$

with $\mathfrak{M}$ regarded as a minimal strongly adequate matrix for $\mathfrak{M}L$.

In the classical two-valued logic we can write $T\alpha$ or $F\alpha$, to indicate that a formula α is true or false, respectively. The symbols T and F act something like metaoperators; the resulting expressions $T\alpha$, $F\alpha$ are often called designated formulas.

In analogy to this, we equip L with a certain set $\{s_i : i \in \bar{n}\}$ whose elements are metaoperators for the designation of formulas. Given a formula $\alpha \in S$, we can form n designated formulas $s_i(\alpha)$ $(i = 0, 1, \ldots, n-1)$. The set of all designated formulas will be denoted by S'. Thus

$$S' = \{s_i(\alpha) : \alpha \in S, \, 0 \leq i \leq n-1\}.$$

Now, let $\omega(\alpha_1, \alpha_2, \ldots, \alpha_m) \in S$.

The rule of elimination of connective ω from the designated formula $s_i(\omega(\alpha_1, \alpha_2, \ldots, \alpha_m))$ is described by the following scheme:

$$(\Pi\omega, i) \qquad \frac{s_i(\omega(\alpha_1, \alpha_2, \ldots, \alpha_m))}{\bigvee\{\bigwedge_{k=1}^{t} s_{j_k}(\alpha_{i_k}) : j_k \in \bar{n}, \, 1 \leq k \leq t, \, H_i(\omega, j_1, j_2, \ldots, j_t)\}}$$

where $H_i(\omega, j_1, j_2, \ldots, j_t)$ says that there exists a homomorphism $f : S \to \mathfrak{M}$ with the properties:

(a) $f(\alpha_{i_k}) = j_k \quad \text{for} \quad 1 \leq k \leq t$;

(b) if h is the interpretation of the connective ω in $\mathfrak{M}$ then
$$h(j_1, j_2, \ldots, j_t) = i;$$

(c) there is no t_1 smaller than t and satisfying conditions (a) and (b).

The rule $(\Pi\omega, i)$ is also said to decompose $s_i(\omega(\alpha_1, \alpha_2, \ldots, \alpha_m))$. The designated formula $s_i(\omega(\alpha_1, \alpha_2, \ldots, \alpha_m))$ is called the premise of $(\Pi\omega, i)$, and $s_{j_k}(\alpha_{i_k})$ are the conclusions.

Example

Consider $\bar{n} = \{0, 1, 2\}$ and let h be defined by the table of values

h	0	1	2
0	2	2	2
1	1	2	2
2	0	1	2

Then:

$$\frac{s_0(\alpha \Rightarrow \beta)}{s_2\alpha \wedge s_0\beta},$$

$$\frac{s_1(\alpha \Rightarrow \beta)}{(s_1\alpha \wedge s_0\beta) \vee (s_2\alpha \wedge s_1\beta)},$$

$$\frac{s_2(\alpha \Rightarrow \beta)}{\bigvee\limits_{i,j=0, i\leq j}^{2} (s_i\alpha \wedge s_j\beta)} =$$

$$\frac{s_2(\alpha \Rightarrow \beta)}{(s_0\alpha \wedge s_0\beta) \vee (s_0\alpha \wedge s_1\beta) \vee \ldots \vee (s_1\alpha \wedge s_2\beta) \vee (s_2\alpha \wedge s_2\beta)}.$$

Rule $(\Pi\omega, i)$ is a certain relation holding between $s_i(\alpha)$ and its immediate designated subformulas; by an immediate subformula of a formula of degree $k > 0$ we mean a subformula of degree $k - 1$. A formula of degree $k = 0$ is not decomposable and cannot become a premise of scheme $(\Pi\omega, i)$. Moreover, if the condition $H_i(\omega, j_1, j_2, \ldots, j_t)$ fails to hold, the rule $(\Pi\omega, i)$ is undefined. By $(\Pi\omega)$ we will denote the set of all decomposing rules of the connective ω.

With a designated formula $s_i(\omega(\alpha_1, \alpha_2, \ldots, \alpha_m))$ we associate a tree defined as follows. The formula $s_i(\omega(\alpha_1, \alpha_2, \ldots, \alpha_m))$ itself is taken for the root (origin); it constitutes the top level. The construction continues as follows:

(p_1) The immediate successors of the root are the conclusions of rule $(\Pi\omega, i)$ applied to the root formula; they constitute the next-to-top level.

(p_2) At each stage of the construction:

If all endpoints of the tree are atomic formulas, the construction is finished. Otherwise we choose any one of the highest-level nonatomic points. Assume it represents a designated formula $s_j(\alpha)$. To the endpoint of each branch passing through that point we attach the conclusions of $(\Pi\omega, j)$ applied to $s_j(\alpha)$.

At each instance when rule $(\Pi\omega, i)$ is applied, expressions $\bigwedge_{k=1}^{t} s_{j_k}(\alpha_{i_k})$ enter as successive points of the same branch.

The tree thus constructed is called the tree associated with formula $s_i(\omega(\alpha_1, \alpha_2, \ldots, \alpha_m))$, or for short, the tree of $s_i(\omega(\alpha_1, \alpha_2, \ldots, \alpha_m))$.

A branch of the tree is considered to be closed if one of the following situations occurs: either it contains two points $s_i(\beta), s_j(\beta)$ with $i \neq j$ or there is on it a nonatomic designated formula $s_i(\omega(\beta))$ with the rule $(\Pi\omega, i)$ undefined.

The tree is closed when all its branches are closed. Otherwise the tree is said to be open. Clearly, for any atomic formula p_j, the trees of formulas $s_i(p_j)$ are all open.

It can be noted that the tree of a designated formula $s_i(\alpha)$ is closed when the assumption of the existence of a valuation v (into the set $\bar{n}$), such that $v(\alpha) = i$, results in a contradiction, or when no valuation with $v(\alpha) = i$ can be a priori given.

Theorems of the tree calculus are those formulas which never take a nondesignated value. Thus, a formula α is a theorem if and only if the formulas $s_i(\alpha)$ for $1 \leq i < r$ are all closed. To prove that a formula α is a theorem, one has to consider the decompositions of $s_i(\alpha)$ for all nondesignated i and verify that the corresponding trees are closed.

Example

We show that the formula $\alpha \Rightarrow (\beta \Rightarrow \alpha)$ is a theorem of the three-valued propositional calculus of Lukasiewicz; the calculus is given by the matrix $\mathfrak{M}\mathrm{L} = (\{0,1,2\}, \{2\}, \Rightarrow, \sim)$ with connectives characterized by

$\Rightarrow$	0	1	2
0	2	2	2
1	1	2	2
2	0	1	2

$\sim$	$\sim x$
0	2
1	1
2	0

The rules of decomposition for the formulas $s_i(\alpha \Rightarrow \beta)$, $i = 0, 1, 2$, are:

$$(\Pi \Rightarrow 0) \quad \frac{s_0(\alpha \Rightarrow \beta)}{s_2(\alpha) \wedge s_0(\beta)}$$

$$(\Pi \Rightarrow 1) \quad \frac{s_1(\alpha \Rightarrow \beta)}{(s_1(\alpha) \wedge s_0(\beta)) \vee (s_2(\alpha) \wedge s_1(\beta))}$$

$$(\Pi \Rightarrow 2) \quad \frac{s_2(\alpha \Rightarrow \beta)}{\bigvee\{(s_i(\alpha) \wedge s_j(\beta)) : i \leq j;\ i, j = 0, 1, 2\}}$$

The rules of decomposition of $s_i(\sim\alpha)$, $i = 0, 1, 2$:

$(\Pi \sim 0)$ $\dfrac{s_0(\sim\alpha)}{s_2(\alpha)}$

$(\Pi \sim 1)$ $\dfrac{s_1(\sim\alpha)}{s_1(\alpha)}$

$(\Pi \sim 2)$ $\dfrac{s_2(\sim)}{s_0(\alpha)}$

The tree of the formula $s_0(\alpha \Rightarrow (\beta \Rightarrow \alpha))$ has the form:

$$s_0(\alpha \Rightarrow (\beta \Rightarrow \gamma))$$
$$s_2(\alpha)$$
$$s_0(\beta \Rightarrow \alpha)$$
$$s_2(\beta)$$
$$s_0(\alpha)$$
$$*$$

The asterisk indicates the termination of a closed branch. The tree above consists of a single branch and is closed.

Now we create the tree of $s_1(\alpha \Rightarrow (\beta \Rightarrow \alpha))$:

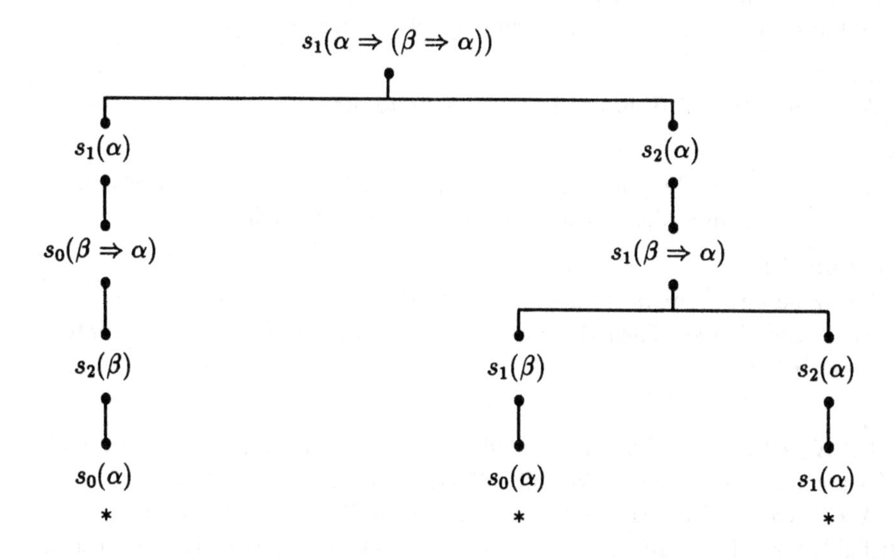

All branches are closed, hence the tree is closed. Consequently $\alpha \Rightarrow (\beta \Rightarrow \alpha)$ is a theorem of the calculus under consideration; the proof has been achieved by the tree method.

Now suppose that $B = \{\alpha_1, \alpha_2, \ldots\}$ is a finite or countable set of formulas. Consider the set

$$B' = \{s_{i_1}(\alpha_1), s_{i_2}(\alpha_2), \ldots\}$$

where $s_{i_1}, s_{i_2}, \ldots$ are arbitrary designation metaoperators. We now define the tree of B'.

(t_1) Design the tree for the formula $s_{i_1}(\alpha_1)$; call it the tree of the first degree.

(t_2) Assume the tree of degree k has been constructed. To each open branch (if there is any) attach the tree of $s_{i_{k+1}}(\alpha_{k+1})$ to obtain the tree of degree $k+1$.

(t_3) Stop when there is no open branch any more.

Every tree that arises in this way (i.e., by an arbitrary enumeration of the elements of B and arbitrary assignment of designation operators s_i to the formulas of B) is called a tree of B.

The notions of a closed tree and an open tree for a set B' of designated formulas are analogous to the case of a single formula. In order that there exist an open tree for B', it is necessary (though not sufficient) that there exist an open tree for each formula in B'. The arrangement (enumeration) of the elements of B' is irrelevant to the question of closedness or openness of the tree. If the set B is infinite, then there exists an infinite open tree for B'.

Lemma 7.1

An infinite tree of B' has at least one infinite branch.

Proof

The tree is finitely generated; hence, each of its points has finitely many successors. By König's lemma there exists an infinite branch. ∎

Theorem 7.1

Let α be any formula from S. Suppose that the designated formula $s_i(\alpha)$ induces a closed tree. Then there exists no valuation v of the variables into $\bar{n}$ such that $v^*(\alpha) = 1$.

Proof

Let X be the set of those designated formulas $s_i(\alpha)$ for which $v^*(\alpha) = i$ holds under a certain valuation v. We will call these formulas i-satisfiable.

A branch of a tree is said to be i-satisfiable if each of its points is an i-satisfiable formula. A tree is i-satisfiable if it has at least one i-satisfiable branch.

Assume there exists a valuation v with $v^*(\alpha) = i$ such that the tree of the set $\{s_i(\alpha)\}$ is i-satisfiable. Then so is every extension of it; a contradiction. ∎

A set H of designated formulas is called a Hintikka set if the following conditions are fulfilled:

(h_1) If $s_j(p) \in H$ then $s_k(p) \notin H$, for any $j, k \in \bar{n}$ with $j \neq k$ and any propositional variable p.

(h_2) If $\alpha = \omega(\alpha_1, \alpha_2, \ldots, \alpha_m)$ is such that $s_i(\alpha) \in H$, then at least one of the conclusions of rule $(\Pi\omega, i)$ applied to $s_i(\alpha)$ belongs to H.

A Hintikka set H^* is said to be saturated when it satisfies two additional conditions:

(h_3) For every propositional variable p there exists $j \in \bar{n}$ such that $s_j(p) \in H^*$.

(h_4) If $\alpha = \omega(\alpha_1, \alpha_2, \ldots, \alpha_m)$ then: $s_i(\alpha) \in H^*$ iff at least one of the conclusions of rule $(\Pi\omega, i)$ applied to $s_i(\alpha)$ belongs to H^*.

Lemma 7.2

Every Hintikka set H can be extended to a saturated set H^*.

Proof

Obvious. ∎

Theorem 7.2

For every saturated Hintikka set H^* there exists a valuation $v : S_0 \to \bar{n}$ such that $v^*(\beta) = i$ holds if and only if $s_i(\beta) \in H^*$; in particular, if there exists an open tree for $s_i(\alpha)$, then there is a valuation v with $v^*(\alpha) = i$.

Proof

For each propositional variable $p \in S_0$ set $v^*(p) = v(p) = i$ iff $s_i(p) \in H^*$. Since H^* is saturated, v is well defined and extends in a standard way to a valuation on S. Easy induction with respect to the complexity of α shows that $v^*(\alpha) = i$ holds iff $s_i(\alpha) \in H^*$.

Suppose that a designated formula $s_{i_0}(\alpha)$ admits an open tree. It has at least one open branch B. Obviously, B is a Hintikka set and can be extended to a saturated set B^* with $s_{i_0}(\alpha) \in B^*$. The proof is complete. ∎

We see from the above that, given an arbitrary formula $\alpha \in S$ and any $i \in \bar{n} = \{0, 1, \ldots, n-1\}$, we can find a valuation $v : S_0 \to \bar{n}$ such that $v^*(\alpha) = i$. Moreover, there exist only a finite number of trees for $s_i(\alpha)$, for a given operator s_i.

7.3 The Existence of Models for the Propositional Calculus

Following [Carnielli 1987], we now present the consistency, completeness and compactness theorems for the finitely generated tree method for n-valued propositional calculi. The notion of a tautology will be used here in its usual meaning (the same as hitherto). The concepts of a theorem and consequence will be defined anew for the tree calculus.

As before, we assume the logic to be a pair $L = (S, \mathfrak{M})$, where

$$\mathfrak{M} = (\bar{n}, \{r, r+1, \ldots, n-1\}, \{\omega_i : i \in J\}),$$

$\bar{n} = \{0, 1, \ldots, n-1\}$, $0 < r \leq n-1$, $n \geq 3$; r is the least of the designated values.

A formula $\alpha \in S$ is a theorem if and only if, for each i with $0 \leq i < r$, there exists a closed tree for the designated formula $s_i(\alpha)$. The set of all closed trees for the formulas $s_i(\alpha)$, $0 \leq i < r$, will be called the proof of α. To indicate that α is a theorem, we write $\overset{L}{\vdash} \alpha$; if L is fixed and no confusion is likely to arise, we write simply $\vdash \alpha$.

Suppose Y is any set of formulas. Let $K \subseteq \bar{n}$ and let f be a function from Y into K. Define

$$S_{K,f}(Y) = \{s_i(\alpha) : f(\alpha) = i, i \in K, \alpha \in Y\}.$$

When f is fixed, the subscript f can be omitted and we can write $S_K(Y)$.

Let $Y \subseteq S$ and $\alpha \in S$. The formula α is said to belong to $Cn(Y)$ if and only if the following is valid: if there exists no closed tree for the set $S_{\bar{n}^*}(Y)$, where $\bar{n}^* = \{r, r+1, \ldots, n-1\}$, then there exist closed trees for the sets $S_{\bar{n}^*}(Y) \cup \{s_i(\alpha)\}$, for all i, $0 \leq i < r$.

It is obvious, in view of this definition, that $\vdash \alpha$ if and only if $\alpha \in Cn(\emptyset)$.

Now we recall the notion of semantic consequence. Let $Y \subseteq S$ be any set of formulas and let $\alpha \in S$. We say that α is a semantic consequence of Y, in symbols $Y \models \alpha$, iff every valuation which satisfies (that is, assumes designated values on) the set Y, satisfies α as well.

Let R be a finite family of sets of definite formulas. We say that R has the analytic consistency property (for short, R has acp) iff, for every $X \in R$, the following conditions are fulfilled:

(a_1) If $p \in S_0$ and $i \neq j$, then the formulas $s_i(p)$ and $s_j(p)$ do not belong to X simultaneously.

(a_2) If $\beta = \omega(\alpha_1, \alpha_2, \ldots, \alpha_m)$ and $s_i(\beta) \in X$ then, denoting by W the set of conclusions of $s_i(\beta)$ via rule $(\Pi\omega, i)$, we have $X \cup W \in R$.

A formula α that belongs to a set $X \in R$ is called R-consistent. The conditions imposed on R are of finite character, and therefore, if $Y \subseteq X$ and $X \in R$, then $Y \in R$.

Let Y be any set of formulas, $Y \subseteq S$, and let R be a family with acp. The set Y is said to be R-consistent iff there exists a set $X \in R$ such that, for every formula $\alpha \in Y$, there is an operator s_i with $s_i(\alpha) \in X$. The set Y is satisfiable if every formula in it is satisfiable.

Theorem 7.3 (Carnielli)

Suppose R has acp and let $Y \subseteq S$. If Y is R-consistent then there exists a valuation $v : S_0 \to \bar{n}$ such that, for any $\alpha \in Y$, equality $v^*(\alpha) = i$ holds iff $s_i(\alpha)$ belongs to a certain $X \in R$. In particular, if $S_{\bar{n}^*,f}(Y) \in R$ for a certain F, then the set Y is satisfiable.

Proof

Let S^* denote the set of all definite formulas of the language S. Arrange S^* into a sequence $\gamma_1, \gamma_2, \ldots$. Consider the sequence of sets $\{Z_k : k \in N\}$ defined inductively by

$$Z_1 = X,$$
$$Z_{k+1} = \begin{cases} Z_k \cup \{\gamma_k\} & \text{if} \quad Z_k \cup \{\gamma_k\} \in R, \\ Z_k & \text{otherwise} \end{cases}$$

The sequence $\{Z_k : k \in N\}$ is increasing with respect to $\subseteq$ and of course each of its members belongs to R.

Let $Z = \bigcup\{Z_k : k \in N\}$. We are going to show that Z can be extended to a saturated Hintikka set. Firstly, Z is in R because every finite subset of Z is contained in some Z_k; every finite subset of Z belonging to R, the set Z belongs there too.

Suppose that, for a certain k, the set $Z \cup \{\gamma_k\}$ is in R. Then $\gamma_k \in Z$ by the definition of Z; in fact, $\gamma_k \in Z_{k+1} \subseteq Z$. By Lemma 7.2, Z can be extended to a saturated Hintikka set Z^*. Define a valuation $v : S_0 \to \bar{n}$ by setting (for each variable $p \in S_0$)

$$v(p) = i \quad \text{iff} \quad s_i(p) \in Z^*.$$

This definition is correct, by the definition of a saturated set. Since S is freely generated by S_0, the valuation v uniquely extends to the whole of S. This ends the proof. ∎

Lemma 7.3

Let $Y \subseteq S$ and let $i : S \to N$ be any function. Suppose there exists an open tree for the set $\{s_{i(\alpha)}(\alpha) : \alpha \in Y\}$. Then there exists a valuation v such that $v^*(\alpha) = i(\alpha)$ for all $\alpha \in Y$.

Proof

The open tree of the assumption of the lemma has at least one open branch, either finite or countably infinite. This branch is a Hintikka set and can be extended to a saturated set. The valuation is defined according to Theorem 7.2.

■

Corollary 7.1

Let $Y \cup \{\alpha\} \subseteq S$. If $\alpha \in Cn(Y)$ then $Y \models \alpha$.

Proof

Immediate from the definition of Cn, Lemma 7.3 and Theorem 7.3. ■

Lemma 7.4

Let Y be a finite or countably infinite set of designated formulas. An open tree for Y exists if and only if there exists an open tree for every finite subset $X \subseteq Y$.

Proof

The "only if" part is trivial. So is the "if" part when Y is finite.

Assume Y is infinite and arrange its formula into a sequence $\{\gamma_k : k \in N\}$. Consider then finite sets $Y_k = \{\gamma_1, \gamma_2, \ldots, \gamma_k\}$. The open tree of Y is constructed by means of the following algorithm:

(a) Form an open tree for Y_1.

(b) Suppose the tree for Y_k has been constructed.

To each open branch attach a tree of γ_{k+1} so as to obtain an open tree; this is possible on account of the assumption of the lemma.

The repeated application of (b) produces an open tree for the entire set Y.

■

Theorem 7.4 (Completeness Theorem)

Let $Y \subseteq S$ and $\alpha \in S$. If $Y \models \alpha$ then $\alpha \in Cn(Y)$.

Proof

Assume $\alpha \notin Cn(Y)$. Then there exists an open tree for the set $S_{\bar{n}^*}(Y) \cup \{s_i(\alpha)\}$ for some i, a nondesignated value. Let R be the class of all open trees for sets of definite formulas. By Lemma 7.4, R has acp. In virtue of Theorem 7.3, the set $Y \cup \{\alpha\}$ is R-consistent. So there exists a valuation $v : S_0 \to \bar{n}$ such that $v^*(\beta) \in \bar{n}^*$ for $\beta \in Y$ and $v^*(\alpha) \notin \bar{n}^*$. However, this conflicts with $Y \models \alpha$. Hence $\alpha \in Cn(Y)$. ■

Corollary 7.2 (Compactness Theorem)

Let $Y \subseteq S$. The set Y is satisfiable if and only if every finite subset of S is satisfiable.

Proof

Immediate from Lemma 7.4 and Theorem 7.4. ∎

7.4 Finitely Generated Trees for n-valued First-Order Predicate Calculi

A slight modification of the definition of an n-valued predicate calculus will be practical. As in the preceding chapter, the predicate calculus will be defined over a propositional calculus $L = (S, \mathfrak{M})$ with a minimal strongly adequate n-valued matrix $\mathfrak{M}$. The symbol S will also stand for the language of the predicate calculus, i.e., for the set of its formulas.

The definition of S and the semantics of the calculus will be adopted from [Carnielli 1987]; this approach seems to be most convenient for our purposes. Thus:

The alphabet of the n-valued predicate calculus of the first order is defined as the union of all entries of the ordered system

$A = (V, \{p_i : i \in I\}, \{\omega_j : j \in J\}, \{c_k : k \in K\}, \{Q_r : r \in R\}, M)$,

in which

V is the set of individual variables;

$\{p_i : i \in I\}$ is the set of relation symbols (predicates);

$\{w_j : j \in J\}$ is the set of propositional connectives; it is assumed that each p_i involves a positive, finite number of arguments, whereas each ω_j involves a nonnegative, finite number of arguments, and that both these sets are nonempty; by zero-argument connectives we mean the logical constants acceptable in the given calculus;

$\{c_k : k \in K\}$ is the set of symbols representing individual constants, sometimes called the parameters;

$\{Q_n : h \in H\}$ is the set of quantifiers; M is the set of auxiliary symbols (parentheses, commas, dots).

The union $V \cup \{c_k : k \in K\}$ is referred to as the set of terms, denoted by T.

If p_i is an m-argument predicate and $t_1, t_2, \ldots, t_m \in T$ then $p_i(t_1, t_2, \ldots, t_m)$ is called an atomic formula or an elementary formula. The set of elementary formulas will be denoted, as in propositional calculus, by S_0.

We define S, the set of all formulas of the first-order predicate calculus, as follows:

(a_1) S_0 is the set of atomic formulas;

(a_2) $S_{k+1} = S_k \cup \{\omega_j(\alpha_1, \alpha_2, \ldots, \alpha_{m_j}) : j \in J, \alpha_l \in S_n \text{ for } 1 \leq l \leq m_j\}$
$\cup \{Q_h x_i \alpha : x_i \in V, \alpha \in S_k, h \in H\}$;

(a_3) $S = \bigcup_{k \in N} S_k$.

A formula $\alpha \in S_k \setminus S_{k-1}$ is said to have level (order, degree) k. Atomic formulas have level 0. Formulas which do not contain symbols representing individual constants are called pure. The notions of a closed formula and a subformula have their usual meaning. Now we introduce the concepts of universe and relativization.

Any nonempty set U can be considered as a universe. Suppose U is a universe and $\alpha \in S$ is any formula. When we replace all individual constants that occur in α by elements of U, we obtain what we shall call a U-formula. Every U-formula arising in this way from α will be denoted by α^U. The set of all U-formulas will be denoted by $S(U)$. Clearly, if α is a pure formula then $\alpha^U = \alpha$, for any universe U; the sets of pure formulas and pure U-formulas coincide.

Our further considerations concern the so called distributive quantifiers; as mentioned earlier, we follow the method of [Carnielli 1987]. A semantic characterization of those quantifiers will be needed.

Suppose U is a certain fixed universe. By a distributive function we will mean any function

$$f : S(U) \rightarrow \{0,1\}^n \setminus \{(0,0,\ldots,0)\}.$$

Thus f assigns to each U-formula α^U an ordered n-tuple of zeros and ones. This n-tuple is to provide information as to which values i, $i \in \bar{n}$, occur among all values assumed by subformulas of α^U. If a value i occurs there, the i-th coordinate equals 1; otherwise it is 0.

Let Q be an admissible n-valued quantifier. Let

$$\sigma_Q : \{0,1\}^n \setminus \{(0,0,\ldots,0)\} \rightarrow \bar{n}$$

be the interpretation function of Q. The composition $\sigma_Q \circ f$ defines the interpretation of U-formulas involving the quantifier Q.

With this characterization of n-valued quantifiers, we define the n-valued predicate calculus of the first-order over the propositional calculus L as the triple

$$L^H = (S, \mathfrak{M}, \{\sigma_Q : Q \in \{Q_h : h \in H\}\})$$

with $\mathfrak{M}$ the n-valued matrix strongly adequate for L.

Let $\overline{S(U)}$ denote the set of all free formulas from $S(U)$ that do not contain free variables. Those formulas in $\overline{S(U)}$ which do not contain an initial quantifier, i.e., are not of the from $Qx_i\alpha$, for any quantifier Q, constitute a subset of $\overline{S(U)}$, which we denote by S'. By a valuation we mean a function $v : S' \rightarrow \bar{n}$, extended to a function v^* on the whole of $\overline{S(U)}$ as follows:

(v_1) On S', v is a propositional valuation.

(v_2) For every formula α and any variable x_j,

$$v^*(Qx_j\alpha) = \sigma_Q(D_{x_j,v}(\alpha))$$

where $D_{x_j,v} : S(U) \to \{0,1\}^n \setminus \{(0,0,\dots,0)\}$

is defined by

$$D_{x_j,v}(\alpha) = (d(0), d(1), \dots, d(n-1))$$

with

$$d(i) = \begin{cases} 1 & \text{iff} \quad \| (\alpha)x_j/k \| (v) = i \quad \text{for} \quad \text{some } k \in U, \\ 0 & \text{otherwise.} \end{cases}$$

We say that Q is a universal quantifier when its interpretation function σ_Q satisfies the condition:

$$\sigma_Q(j_0, j_1, \dots, j_{n-1}) \in \{r, r+1, \dots, n-1\}$$
$$\text{iff} \quad j_i = 0 \quad \text{for all} \quad i \in \{0, 1, \dots, r-1\}.$$

Similarly, Q is said to be an existential quantifier when

$$\sigma_Q(j_0, j_1, \dots, j_{n-1}) \in \{r, r+1, \dots, n+1\}$$
$$\text{iff} \quad j_i = 1 \quad \text{for some} \quad i \in \{r, r+1, \dots, n-1\}.$$

To denote universal and existential quantifiers we will use the symbols $\forall$ and $\exists$, respectively.

A predicate calculus is called regular if it contains at least one universal quantifier and at least one existential quantifier.

A valuation defined on atomic formulas will be called an atomic valuation.

Lemma 7.5

An atomic valuation extends uniquely to a valuation on the whole of $\overline{S(U)}$.

Proof

Use induction with respect to the order of a formula. When a quantifier occurs, notice that it has an interpretation function. ∎

By a semantic interpretation of the first-order predicate calculus in a nonvoid universe U we mean any function I which

(i_1) to each symbol representing an individual constant assigns an element of U (the assignment need not be one-to-one);

(i_2) to each m-argument predicate assigns a partition of the Cartesian product U^m into n classes $p_0^*, p_1^*, \dots, p_{n-1}^*$.

The classes occurring in (i_2) are meant as counterparts of the n values; the partition induced by p is called the p-partition of U^m.

The pair $\mathfrak{U} = (U, I)$ will be called a structure, or a model, for the n-valued predicate calculus.

It follows from the above that, in a given structure $\mathfrak{U} = (U, I)$, a closed atomic U-formula $p(t_1, t_2, \ldots, t_m)$ assumes value i if and only if $(t_1, t_2, \ldots, t_m) \in p_i^*$. If, moreover, $i \in \{r, r+1, \ldots, n-1\}$, we say that $p(t_1, t_2, \ldots, t_m)$ is true in $\mathfrak{U}$, or that $\mathfrak{U}$ is a model for $p(t_1, t_2, \ldots, t_m)$.

Lemma 7.6

A structure $\mathfrak{U} = (U, I)$ determines the valuation of $S(U)$ uniquely; and conversely, a valuation v on $S(U)$ determines the structure.

Proof

Given a structure, the valuation is defined immediately by Lemma 7.5. Now suppose we are given the valuation. With each relation symbol p we associate the partition $p_0^*, p_1^*, \ldots, p_{n-1}^*$ by assigning an m-tuple $(t_1, t_2, \ldots, t_m)$ to class P_i^* iff $v(p_i(t_1, t_2, \ldots, t_m)) = i$. This defines the desired I. To individual constants assign elements of U, according to the definition of $S(U)$. ∎

Now, let S_p denote the set of all pure closed formulas in S (i.e., propositions). Evidently, S_p is contained in every $\overline{S(U)}$, for any universe U. A proposition α is said to be S_p-satisfiable if there exists a structure $\mathfrak{U}$ in which α assumes a designated value i; or, in other words (see Lemma 7.6): α is S_p-satisfiable if there exists a valuation v such that $v^*(\alpha) \in \{r, \ldots, n-1\}$. A proposition α is a tautology of the n-valued predicate calculus if α is satisfiable relative to every structure.

Suppose $\alpha(c_1, c_2, \ldots, c_l) \in S \setminus S_p$ contains individual constants $c_1, c_2, \ldots, c_l$ only. We say that $\alpha(c_1, c_2, \ldots, c_l)$ is satisfiable (or is a tautology) if there exists a structure $\mathfrak{U} = (U, I)$ such that the proposition $\alpha' = \alpha(I(c_1), I(c_2), \ldots, I(c_l))$ is true in $\mathfrak{U}$ (or if α' is true in every structure $\mathfrak{U}$, respectively).

Suppose the n-valued predicate calculus under consideration is regular. Consider any formula of the type $\alpha(c_1, c_2, \ldots, c_l)$. Let $x_1, x_2, \ldots, x_l$ be distinct individual variables, not occurring in $\alpha(c_1, c_2, \ldots, c_l)$, and let

$$\alpha(x_1, x_2, \ldots, x_l) = (\alpha(c_1, c_2, \ldots, c_l))c_1/x_1\, c_2/x_2 \ldots c_l/x_l.$$

It is not very difficult to ascertain that the formula $\alpha(c_1, c_2, \ldots, c_l)$ is satisfiable if and only if the pure proposition $\exists x_1 \exists x_2 \ldots \exists x_l\, \alpha(x_1, x_2, \ldots, x_l)$ is satisfiable; similarly, $\alpha(c_1, c_2, \ldots, c_l)$ is a tautology iff the pure proposition $\forall x_1 \forall x_2 \ldots \forall x_l\, \alpha(x_1, x_2, \ldots, x_l)$ is a tautology.

This fact (not obvious, perhaps) allows us to restrict the investigation of truth and satisfiability in the n-valued predicate calculus to pure formulas alone.

7.5 Finitely Generated Trees for n-valued Quantifiers

Let $\bar{S}$ be the set of all constant-free formulas in S and let $\bar{S}^*$ denote the set of designated formulas arising from formulas $\alpha \in \bar{S}$; i.e.,

$$\bar{S}^* = \{s_i(\alpha) : \alpha \in \bar{S}, 0 \leq i \leq n-1\},$$

s_i denoting the designating metaoperators, as in the tree method for the propositional calculus.

Suppose Q is a quantifier with an interpretation function σ_Q. Let $m < n$. We say that an $(m+1)$-tuple $(i, j_1, j_2, \ldots, j_m)$ satisfies condition $D_i(Q; j_1, j_2, \ldots, j_m)$, or that $D_i(Q; j_1, j_2, \ldots, j_m)$ holds, if the n-tuple $(d(0), \ldots, d(n-1))$ belongs to $\sigma^Q_{-1}(i)$, where

$$d(t) = \begin{cases} 1 & \text{iff} \quad t \in \{j_1, j_2, \ldots, j_m\}, \\ 0 & \text{otherwise.} \end{cases}$$

In other words, $D_i(Q; j_1, j_2, \ldots, j_m)$ holds if σ_Q assumes value i on the n-tuple in which ones appear at positions $j_1, j_2, \ldots, j_m$ and zeros elsewhere.

Let $\alpha = Q \times \beta$, where Q is any n-valued quantifier. We introduce the rule $(\lambda_{Q,i})$ which assigns to a designated formula $s_i(\alpha)$ the tree

$$(\lambda_{Q,i}) \qquad \frac{s_i(Qx\,\beta)}{\bigvee\{\bigwedge_{l=1}^{m} s_{j_l}((\beta)x/c_l) : D_i(Q : j_1, j_2, \ldots, j_m),\, j_1, j_2, \ldots, j_m < n\}}$$

provided that no symbol representing any individual constant c_l, with $1 \leq l \leq m$, appears in the branch containing $s_{j_l}((\beta)x/c_l)$ (for $m \geq 2$).

If $D_i(Q; j_1, j_2, \ldots, j_m)$ does not hold, the rule is undefined. We denote by λ_i the set of all rules $\lambda_{Q,i}$, for all quantifiers Q.

Let $\alpha \in S$ and consider the definite formula $s_i(\alpha)$. We now define the quantifier systematic tree (for short, the s-tree) for $s_i(\alpha)$. The construction is as follows:

(q_1) Formula $s_i(\alpha)$ is the origin (root) of the tree.

(q_2) Assume a branch terminates in a point c of the form $\omega(\alpha_1, \alpha_2, \ldots, \alpha_m)$; the attachment procedure is the same as in propositional calculus, with propositional variables replaced by elementary formulas and with rule $(\Pi\omega, i)$ modified accordingly.

(q_3) When a branch ends with a point c of the form $s_i(Qx\,\beta)$, we attach at this endpoint the corresponding tree defined by rule $(\lambda_{Q,i})$.

(q_4) The procedure continues until all endpoints are atomic formulas; at each step choose the leftmost of the shortest branches ending in a nonatomic point and follow instruction (q_2) or (q_3).

Open branches and closed branches are defined in exactly the some way as in the propositional calculus. The same concerns the terms open tree and closed tree. The definition of a proof of a formula is adapted accordingly.

To indicate that a formula $\alpha \in S$ is a theorem in the n-valued predicate calculus (in the sense of the tree method), we write $\vdash \alpha$ or $\alpha \in PL$.

Now, let $Y^* = \{s_{i_1}(\alpha_1), s_{i_2}(\alpha_2), \ldots\}$ be a finite or countable set of designated formulas (enumerated arbitrarily). The tree for Y^* is constructed in the following fashion:

(q_1^*) Formula s_{i_1} is the root.

(q_2^*) Suppose we have already constructed the tree up to its n-th point. We then proceed as in the case of a tree generated by a single formula; after performing step (q_2) or (q_3) we attach $s_{i_{n+1}}(\alpha_{n+1})$ at the end of each open branch.

(q_3^*) Instruction (q_4) applies.

The concepts of syntactic and semantic consequence are defined analogously to how they were defined in the propositional calculus. We use the notation: $\alpha \in CnY$ and $Y \models \alpha$.

This general construction is neatly exemplified in the case of the three-valued predicate calculus with two designated values, an existential quantifier and a universal one; for details, we refer to [Carnielli 1987].

Let now a universe U be fixed. A designated formula in which all the constants that occur are elements of U is said to be U-designated. The set of all such formulas will be denoted by H. We call H a quantifier Hintikka set over U, or for short, a QU-Hintikka set, if the following conditions are satisfied:

(c_1) For every atomic formula $\alpha \in \overline{S(U)}$ and every $k, j < n$, $k \neq j$, if $s_k(\alpha) \in H$ then $s_j(\alpha) \notin H$.

(c_2) If $\beta = s_i(\omega(\beta_1, \beta_2, \ldots, \beta_m)) \in H$, then H contains the set of conclusions of rule $(\Pi\omega, i)$ applied to β.

(c_3) If $s_i(Qx\,\beta) \in H$, then there are $j_1, j_2, \ldots, j_m < n$ such that:

(i) $D_i(Q; j_1, j_2, \ldots, j_m)$ is valid;

(ii) for every formula of the type $(\beta)x/c$, where $c \in U$, there exists l with $1 \leq l \leq m$ such that $s_{j_l}((\beta)x/c) \in H$;

(iii) for each l, $1 \leq l \leq m$, there exists $c \in U$ such that $s_{j_l}((\beta)x/c) \in H$.

Lemma 7.7

Let H be a QU-Hintikka set. There exists a valuation v^* such that, for any $\alpha \in \overline{S(U)}$,

$$v^*(\alpha) = i \quad \text{iff} \quad s_i(\alpha) \in H.$$

Proof

For each atomic formula α, set

$$v(\alpha) = \begin{cases} i & \text{if } s_i(\alpha) \in H, \\ \text{arbitrary value} & \text{if } s_i(\alpha) \notin H \text{ for any } i. \end{cases}$$

(This definition is correct, in view of (c_1).) Then V extends to a valuation v^* on the whole of $\overline{S(U)}$. It remains to show that $s_i(\alpha) \in H$ forces $v^*(\alpha) = i$. For atomic formulas, this is ensured by the definition. Now consider $\alpha = \omega(\alpha_1, \alpha_2, \ldots, \alpha_m)$ and suppose $s_i(\alpha) \in H$; the claim follows by condition (c_2). Finally, consider $\alpha = Qx\,\beta$ and suppose $s_i(\alpha) \in H$. By condition (c_3), $D_i(Q; j_1, j_2, \ldots, j_m)$ holds for some $j_1, j_2, \ldots, j_m < n$ and, for each l with $1 \leq l \leq m$, there is $c \in U$ such that $s_{j_l}((\beta)x/c) \in H$. Then, according to the definition of $D_i(Q; j_1, j_2, \ldots, j_m)$ and the definition of v, we get

$$v^*(Qx\,\beta) = \sigma_Q(D_{x,v}(\beta)) = i,$$

which ends the proof. ∎

We introduce the notions of consistency as we did in propositional calculus. Consider a finite family R of subsets of $\overline{S(U)}^*$. Let $Y \subseteq \overline{S(U)}$; i.e., let Y be a set of indefinite U-propositions. We say that Y is R-consistent if the set

$$Z = \{s_i(\alpha) : \alpha \in Y, r \leq i \leq n-1\}$$

belongs to R.

We say that R has the analytic consistency property (for short, acp) iff, for every $Z \in R$, the following conditions are fulfilled:

(a_1) For every atomic formula $\alpha \in \overline{S(U)}$ and every i, j with $i \neq j$, the formulas $s_i(\alpha)$ and $s_j(\alpha)$ do not belong to Z simultaneously.

(a_2) If $s_i(\omega(\alpha_1, \alpha_2, \ldots, \alpha_m)) \in Z$ and if C denotes the set of conclusions of rule $(\Pi\omega, i)$ applied to $s_i(\omega(\alpha_1, \alpha_2, \ldots, \alpha_m))$, then $C \cup Z \in R$.

(a_3) If $s_i(Qx\,\alpha) \in Z$, then there exist $j_1, j_2, \ldots, j_m < n$ with $D_i(Q : j_1, \ldots, j_m)$ and there exists a set Z' such that

(i) for every formula $(\alpha)x/c$ (with $c \in U$), $s_{j_l}(\alpha)x/c \in Z'$ holds for a certain l, $1 \leq l \leq m$;

(ii) for each l, $1 \leq l \leq m$, there is a constant c such that $s_{j_l}(\alpha)x/c \in Z'$;

(iii) $Z \cup Z' \in R$.

Theorem 7.5 [Carnielli 1987]

Let Y be a set of pure propositions. Suppose Y is R-consistent, for a certain family R with acp. Then Y is satisfiable in a countable universe.

Proof

Since Y is R-consistent, there exists a set $Z \in R$ composed of definite formulas $s_i(\alpha)$, α ranging over Y and i ranging over designated values.

Arrange all formulas in Z into a sequence $\gamma_1, \gamma_2, \gamma_3, \ldots$. Define a sequence m_k as follows:

Let $m_1 = \gamma_1$. Assume m_k has been already defined. If m_k has the form $s_i\omega(\alpha_1, \alpha_2, \ldots, \alpha_l)$, we take all $s_{j_z}(\alpha_z)$ (conclusions of rule $(\Pi_{\omega,i})$ applied to $s_i\omega(\alpha_1, \alpha_2, \ldots, \alpha_l)$) to be the subsequent members of the sequence. If m_k has the form $s_iQx\,\alpha$, we continue the construction by adjoining the expressions $s_{j_1}(\alpha)x/d_1, s_{j_2}(\alpha)x/d_2, \ldots, s_{j_l}(\alpha)x/d_l$, where $j_1, j_2, \ldots, j_l$ are as in condition (a_3) and $d_1, d_2, \ldots, d_l$ are symbols not occurring in the hitherto constructed portion of the sequence and representing individual constants. Then we continue the m-sequence with the earliest non-utilized γ_j.

Evidently, the set of all elements of this sequence belongs to R and is a QU-Hintikka set, where U is the set of symbols representing the individual constants. In view of Lemma 7.7, this set is satisfiable. ∎

Corollary 7.3 (completeness)

Suppose Y is a set of pure propositions. Then Y is satisfiable if and only if there exists an open tree for the set $S_D(Y)$. That is, if $Y = Z \cup \{\alpha\}$, then $\alpha \in Cn(Z)$ if and only if $Z \models \alpha$.

Proof

Immediate in view of the fact that the family of all sets of designated formulas giving rise to open trees has the property of analytic consistency. ∎

Corollary 7.4 (compactness)

Let Y be a set of pure propositions. Y is satisfiable iff every finite subset of Y is so. ∎

Corollary 7.5 (Löwenheim-Skolem Theorem)

If Y is a satisfiable set of pure propositions, then Y is satisfiable in a countable universe.

Proof

If Y is satisfiable, then by Corollary 7.3 there exists an open tree for a certain set $S_D(Y)$ with at least one open branch composed entirely of formulas $s_i(\alpha)$ with

$\alpha \in Y$ and $i \geq r$. This branch constitutes a QU-Hintikka set. So the assertion follows in virtue of Lemma 7.7. ∎

For more subtle interpretations and for numerous examples of application of the tree method, the reader should consult the papers of Carnielli [Carnielli 1987] and Surma [Surma 1977]. It has to be emphasized that the tree method for finite n-valued propositional calculi is algorithmic. To a question formulated in terms of the n-valued predicate calculus, it may or may not provide an answer; failure occurs in undecidable cases.

8 Fuzzy Propositional Calculi

8.1 Introductory Remarks

Further generalization of many-valued logics is achieved by considering the so-called fuzzy logics. In fact, every finite-valued logic, even the classical two-valued one, can be viewed as a specific case of a fuzzy logic. Let us once more express the opinion (which may be disputable) that algebra is the framework best suited to the definition and investigation of all logical structures. The adequate semantics of each logic constitutes an algebra. Certain elements of its universe are then specified as designated values. It is the type of this algebra, together with the choice of designated elements, that decide on the kind of logic in question.

Much the same is also true of fuzzy logics. From the semantic viewpoint, their definition is somewhat unusual. The approach which we present below is based entirely on Pavelka's papers [Pavelka 1979], which seem to be of most interest, containing many important results. We discuss those results following the order (slightly unusual, we feel) in which they appeared in Pavelka's works.

8.2 Fuzzy Sets

Let X be any nonempty set (viewed as the "space") and let $(L, \leq)$ be some ordered set; in particular, $(L, \leq)$ can be a lattice.

By an L-fuzzy set in X we understand an arbitrary mapping $A : X \to L$. When $(L, \leq) = (\{0,1\}, \leq)$ is the two-element Boolean algebra, L-fuzzy sets in any space X can be identified with the usual subsets of X.

Formally, an L-fuzzy set is a set of pairs $(x, A(x))$, hence a subset of the Cartesian product $X \times L$.

When the lattice is fixed and a misunderstanding cannot arise, we omit the prefix and speak of fuzzy sets rather than L-fuzzy.

It is plain from the definition that the concept of a fuzzy set generalizes that of the characteristic function of a set. Therefore one often writes μ_A (rather than A itself) to denote a fuzzy set A.

Now suppose that U is some other nonempty set and $m : U \to L(X)$ is a mapping from U into $L(X)$, the family of all L-fuzzy sets in X. We regard the elements of U as fuzzy objects. For $\mathrm{B} \in U$, the value $m(\mathrm{B}) = B$ is a fuzzy set, hence a certain function; we call it the adherence (incidence) function of the object B; the value $B(x)$ at a point $x \in X$ is called the degree of adherence (incidence) of x to B.

A fuzzy set A is said to be contained in a fuzzy set B, in symbols $A \lesssim B$, if $A(x) \leq B(x)$ holds for every $x \in X$. Similarly, a fuzzy object A is contained in a fuzzy object B if $m(\mathrm{A}) \lesssim m(\mathrm{B})$. Two fuzzy sets A and B are considered as identical if $A \lesssim B$ and $B \lesssim A$.

In the sequel we will be mainly concerned with the cases in which L is either a finite chain or the interval $[0, 1]$ with the usual real-line order.

8.3 Syntactic Introduction

Let L be a complete lattice and let S be a set of propositional formulas, supposed to constitute an algebra similar to lattice L. By L-valuation we shall mean the operation

$$v : L^S \to L^S$$

which assigns to each fuzzy set $X : S \to L$ the fuzzy set

$$v(X) : S \to L$$

defined by

$$v(X)(\alpha) = \bigwedge\{T(\alpha) : T \in L^S \quad \text{and} \quad \forall \beta \in S \quad (T(\beta) \geq X(\beta))\}$$

for every α in S; here $\bigwedge$ denotes the generalized lattice meet in L. In the terminology of fuzzy set theory, we call $v(X)(\alpha)$ the degree of incidence of α to X.

Let $A : S \to L$ be a certain fuzzy set, called the axiom system, and suppose R is a set of L-valued rules of inference in S; the rules will be specified later on (in which we will follow an idea due to Goguen).

Let S and L be fixed. The relation $\leq$ defined in L^S by

$$Y \leq Z \quad \text{iff} \quad \forall \alpha \in S \quad (Y(\alpha) \leq Z(\alpha))$$

is a partial ordering; the resulting poset (partially ordered set) L^S is a complete lattice; in fact, it is just a direct Cartesian product of copies of the lattice L. For instance, we have

$$\bigwedge X(\alpha) = \bigwedge\{Y(\alpha) : Y \leq X\}$$

for any $\alpha \in S, X \subseteq L^S$. The greatest element and the least element in any of the lattices under discussion will be denoted by the usual symbols 1 and 0. The elements of L^S will be referred to as L-subsets of S, or simply L-sets if the context leaves no doubt as to what is S.

We now introduce the notion of L-consequence in S. It is defined as any closure operation in L^S, i.e., a mapping

$$C : L^S \to L^S$$

satisfying the conditions:

(c_1) $X \leq C(X)$;

(c_2) if $X \leq Y$ then $C(X) \leq C(Y)$;

(c_3) $C(C(X)) \leq C(X)$.

An L-set X is consistent with respect to an L-consequence C when $C(X) \neq 1$.

Lemma 8.1

Let C be an L-consequence in S. The following statements are valid for any sets $X, Y \in L^S$:

(c_4) if $X \leq C(Y)$ and $Y \leq C(Z)$ then $X \leq C(Z)$;

(c_5) $C(X \vee Y) = C(X \vee C(Y)) = C(C(X) \vee C(Y))$.

Proof

These are immediate consequences of (c_1), (c_2), (c_3); see also Tarski [1956].

■

Theorem 8.1

Let $F \subseteq L^S$ be any set. Consider the mapping

$$C_F : L^S \to L^S$$

which assigns to an L-set X the L-set

$$C_F(X) = \bigwedge\{T \in F : T \geq X\}.$$

The operation C_F is an L-consequence on S.

Proof

Condition (c_1) follows from the definition of C_F; clearly $X \leq C_F(X)$.

If $X \leq Y$, then

$$\{T \in F : T \geq Y\} \subseteq \{T \in F : T \geq X\}.$$

Hence $C_F(X) \leq C_F(Y)$, so (c_2) holds.

To get (c_3) it suffices to note that, for any $T \in F$ and $X : S \to L$, the inequality $T \geq X$ holds if and only if $T \geq C_F(X)$. ■

Given $F \subseteq L^S$, $X \subseteq L^S$, $\alpha \in S$ and $x \in L$, the fact that

$$C_F(X)(\alpha) \geq x$$

will be stated as follows: in the semantic F, the formula α is a logical consequence of the set X in degree at least x; in symbols,

$$X \overset{F,x}{\models} \alpha.$$

Obviously, this is equivalent to:

$$\text{for each } T \in F, \quad \text{if } T \geq X \quad \text{then } T(\alpha) \geq x.$$

Also the following identify has to be noticed

$$C_F(X)(\alpha) = \bigvee\{x \in L : X \overset{F,x}{\models} \alpha\},$$

for all $\alpha \in S$ and $X : S \to L$.

Now we introduce the concept of an n-ary rule of inference in S. According to the ideas of Goguen [Goguen 1967] and Pavelka [Pavelka 1979], a rule will be defined as an ordered pair composed of two operations, the first acting in S and the second in L.

Thus, by an n-argument L-rule of inference in S we mean a pair $r = (p, q)$ such that

(r_1) p is an n-argument partial function with domain contained in S^n and values in S;

(r_2) q is an n-argument function in L satisfying the condition
$q(x_1, \ldots, x_{k-1}, \bigvee\{x_{k_m} : m \in M\}, x_{k+1}, \ldots, x_n) =$
$\bigvee\{q(x_1, \ldots, x_{k-1}, x_{k_m}, x_{k+1}, \ldots, x_n) : m \in M\}$
for $M \neq \emptyset$, $k = 1, 2, \ldots, n$ and $x_j \in L$.

In other words, q is assumed to preserve (nonempty) lattice joins, in each variable separately. We say that q is semi-continuous.

An L-set X is closed with respect to (or under) an L-rule $r = (p, q)$ if the inequality

$$X(p(\alpha_1, \alpha_2, \ldots, \alpha_n)) \geq q(X(\alpha_1), X(\alpha_2), \ldots, X(\alpha_n))$$

holds for every n-tuple $(\alpha_1, \alpha_2, \ldots, \alpha_n)$ in the domain of p. A set is closed with respect to a set of L-rules R when it is closed under each rule $r \in R$.

Lemma 8.2

Suppose R is a set of L-rules of inference in S and let $\mathfrak{A}$ be the family of all L-subsets of S closed with respect to R. Then $\mathfrak{A}$ is closed under the formation of meets in L^S.

Proof

Plainly, $1 = \bigwedge\emptyset$ is closed with respect to R.

Take a set $\mathfrak{B} \subseteq \mathfrak{A}$, $\mathfrak{B} \neq \emptyset$, and let $r = (p, q)$ be an n-argument rule, $r \in R$.

For any $\alpha_1, \alpha_2, \ldots, \alpha_n \in S$ we then have

$$(\bigwedge \mathfrak{B})(p(\alpha_1, \alpha_2, \ldots, \alpha_n)) = \bigwedge\{T(p(\alpha_1, \alpha_2, \ldots, \alpha_n)) : T \in \mathfrak{B}\} \geq$$
$$\bigwedge\{q(T(\alpha_1), T(\alpha_2), \ldots, T(\alpha_n)) : T \in \mathfrak{B}\} \geq$$
$$q(\bigwedge\{T(\alpha_1) : T \in \mathfrak{B}\}, \bigwedge\{T(\alpha_2) : T \in \mathfrak{B}\}, \ldots, \bigwedge\{T(\alpha_n) : T \in \mathfrak{B}\}) =$$
$$q((\bigwedge \mathfrak{B})(\alpha_1), (\bigwedge \mathfrak{B})(\alpha_2), \ldots, (\bigwedge \mathfrak{B})(\alpha_n)),$$

which proves the lemma. ∎

A fuzzy propositional calculus over L is defined as a pair (A, R) with $A \in L^S$ an axiom system and R a set of L-rules of inference in S. Such a pair will also be called an L-syntax (in S).

Let (A, R) be a fuzzy calculus over L in S. For any L-set $X : S \to L$ there exists the least L-set $Cn_{A,R}(X) : S \to L$ with the following properties:

(a_1) $A \leq Cn_{A,R}(X)$;

(a_2) $X \leq Cn_{A,R}(X)$;

(a_3) $Cn_{A,R}(X)$ is closed under R.

Instead of $Cn_{A,R}(X)(\alpha) \geq x$, we will also write

$$X \overset{A,R}{\underset{x}{\vdash}} \alpha,$$

for $X \in L^S$, $\alpha \in S$, $x \in L$.

For each $r = (p, q) \in R$ and any $X \in L^S$, if $(\alpha_1, \alpha_2, \ldots, \alpha_n)$ is in the domain of p and if $X \underset{x_i}{\vdash} \alpha_i$ for $i = 1, 2, \ldots, n$, then

$$X \underset{q(x_1, x_2, \ldots, x_n)}{\vdash} p(\alpha_1, \alpha_2, \ldots, \alpha_n);$$

this is obvious from the definition of $Cn_{A,R}$.

Let R be a set of L-rules of inference in S. Consider

$$\sum = S \cup (S \times \{0\}) \cup (S \times R \times N^*)$$

where $N^* = \bigcup_{m=1}^{\infty} N^m$ and N is the set of positive integers. By an R-proof in S we mean an arbitrary nonvoid sequence $w = (w_1, w_2, \ldots, w_m)$, $w_i \in \sum$, with the following property:

For each k with $1 \leq k \leq m$, if the k-th element of w has the form

$$w_k = (\alpha, r, (i_1, i_2, \ldots, i_n)),$$

then $r = (p, q)$ is an n-argument L-rule belonging to R, each i_j is an integer with $1 \leq i_j < k$, and we have

$$\alpha = p({}^*w_{i_1}, {}^*w_{i_2}, \ldots, {}^*w_{i_n}),$$

${}^*w_{i_j}$ denoting the first component of w_{i_j}. The integer m is called the length of the proof and is denoted by $l(w)$.

The set of all R-proofs in S will be denoted by $Pr(S, R)$. Suppose $w = (w_1, w_2, \ldots, w_m) \in Pr(S, R)$. Each w_k, $1 \leq k \leq m$, is an element of $\sum$; hence, it is either a single formula α or a pair $(\alpha, 0)$, or even a certain triple. In every case, the first item occurring in w_k is a formula; we will denote it by $pr_1(w_k)$. (Thus pr_1 is the projection to the first component of an element of $\sum$.) In other words, $pr_1(w_k)$ indicates which formula constitutes the k-th link of the proof.

Following Pavelka, we introduce the notation ${}^+w = pr_1(w_1)$ and $w^+ = pr_1(w_m)$, for $w = (w_1, \ldots, w_m)$.

Evidently, if $w = (w_1, w_2, \ldots, w_m)$ is an R-proof in S, then every initial segment $w^k = (w_1, w_2, \ldots, w_k)$, $1 \leq k \leq m$, is an R-proof, as well.

Let (A, R) be an L-syntax in S. We now assign to each proof $w \in Pr(S, R)$ a certain function $\bar{w} : L^S \to L$. The definition is by induction on $l(w)$.

For $X : S \to L$:

(a_1) If $l(w) = 1$ then

$$\bar{w}(X) = \begin{cases} X(\alpha) & \text{if} \quad w = ((\alpha)), \\ A(\alpha) & \text{if} \quad w = ((\alpha, 0)). \end{cases}$$

(a_2) Suppose the bars have been defined for proofs w' of length $l(w') < m$.

If $w = (w_1, w_2, \ldots, w_m) \in Pr(S, R)$, then

$$\bar{w}(X) = \begin{cases} X(\alpha) & \text{if} \quad w_m = (\alpha), \\ A(\alpha) & \text{if} \quad w_m = (\alpha, 0), \\ q(\bar{w}^{i_1}(X), \bar{w}^{i_2}(X), \ldots, \bar{w}^{i_n}(X)) & \text{if} \quad w_m = (\alpha, r, (i_1, \ldots, i_n)). \end{cases}$$

For a proof $w \in Pr(S, R)$ and an L-set $X : S \to L$, the element $\bar{w}(X) \in L$ will be called the value of w with respect to X.

Theorem 8.2 [Pavelka 1979]

Let (A, R) be an L-syntax in S. The equality

$$Cn_{A,R}(X)(\alpha) = \bigvee\{\bar{w}(X) : w \in Pr(S, R), w^+ = \alpha\}$$

holds in L for every $X \in L^S$ and every formula $\alpha \in S$.

Proof

Consider the mapping $\bar{X} : S \to L$ defined by

$$\bar{X}(\alpha) = \bigvee\{\bar{w}(X) : w \in Pr(S,R), w^{+} = \alpha\},$$

where X is a fixed L-set in S.

First we show that $\bar{X} \leq Cn_{A,R}(X)$. This will be done as soon as we show that $\bar{w}(X) \leq U(w^{+})$ holds in L wherever $w \in Pr(S,R)$ and $U \geq X \vee A$ is an L-set closed under R. If $l(w) = 1$, $\bar{w}(X)$ equals either $X(w^{+})$ or $A(w^{+})$, and we have $X(w^{+}) \vee A(w^{+}) \leq U(w^{+})$. Assume $\bar{w'}(X) \leq U((w')^{+})$ holds for all $w' \in Pr(S,R)$ with $l(w') < m$. Let $l(w) = m$. If $w_m \in S \cup (S \times \{0\})$, we argue as above. When $w_m = (\alpha, r, (i_1, i_2, \ldots, i_n))$, we have

$$\begin{aligned}\bar{w}(X) &= q(\bar{w}^{i_1}(X), \bar{w}^{i_2}(X), \ldots, \bar{w}^{i_n}(X)) \\ &\leq q(U(w^{i_1+}), U(w^{i_2+}), \ldots, U(w^{i_n+})) \\ &= q(U(^{+}w_{i_1}), U(^{+}w_{i_2}), \ldots, U(^{+}w_{i_n})) \\ &\leq U(p(^{+}w_{i_1}, \ldots, ^{+}w_{i_n})) = U(^{+}w_m) = U(w^{+})\end{aligned}$$

To establish the opposite inequality we note that $((\alpha))$ and $((\alpha, 0))$ are R-proofs, whatever $\alpha \in S$ may be, and hence $A \vee X \leq \bar{X}$. It remains to show that $\bar{X}$ is closed under R.

Choose $r = (p,q) \in R$ and $(\alpha_1, \alpha_2, \ldots, \alpha_n)$ in the domain of p. By the definition of $\bar{X}$, we can find for each k, $1 \leq k \leq n$, a family of R-proofs

$$\left\{w^{(k,j_k)} = \left(w_1^{(k,j_k)}, w_2^{(k,j_k)}, \ldots, w_{m(k,j_k)}^{(k,j_k)}\right) : j_k \in J_k\right\},$$

with J_k an index set, such that

$$^{+}w_{m(k,j_k)}^{(k,j_k)} = \alpha_k \quad \text{for all} \quad j_k \in J_k$$

and

$$\bar{X}(\alpha_k) = \bigvee\left\{\overline{w^{(k,j_k)}}(X) : j_k \in J_k\right\}.$$

Write

$$i_k = m(1,j_1) + m(2,j_2) + \ldots + m(k,j_k)$$

for $k = 1, 2, \ldots, n$. For every n-tuple $(j_1, j_2, \ldots, j_n) \in \prod_{i=1}^{n} J_i$ the word $w(j_1, j_2, \ldots, j_n)$ written below is an R-proof of $p(\alpha_1, \alpha_2, \ldots, \alpha_n)$ in S:

$$\begin{aligned} w(j_1, j_2, \ldots, j_n) = (&w_1^{(1,j_1)}, w_2^{(1,j_1)}, \ldots, w_{m(1,j_1)}^{(1,j_1)},\\ &w_1^{(2,j_2)}, w_2^{(2,j_2)}, \ldots, w_{m(2,j_2)}^{(2,j_2)},\\ &\ldots\\ &w_1^{(n,j_n)}, w_2^{(n,j_n)}, \ldots, w_{m(n,j_n)}^{(n,j_n)},\\ &(p(\alpha_1, \alpha_2, \ldots, \alpha_n), r, (i_1, i_2, \ldots, i_n))). \end{aligned}$$

By the definition of $r = (p, q)$,

$$\begin{aligned} &q(\bar{X}(\alpha_1), \bar{X}(\alpha_2), \ldots, \bar{X}(\alpha_n))\\ &= q(\bigvee\{\overline{w^{(1,j_1)}}(X) : j_1 \in J_1\}, \ldots, \bigvee\{\overline{w^{(n,j_n)}}(X) : j_n \in J_n\})\\ &= \bigvee\{q(\overline{w^{(1,j_1)}}(X), \ldots, \overline{w^{(n,j_n)}}(X) : (j_1, j_2, \ldots, j_n) \in \prod_{i=1}^{n} J_i\}\\ &= \bigvee\{\overline{w(j_1, j_2, \ldots, j_n)}(X) : (j_1, j_2, \ldots, j_n) \in \prod_{i=1}^{n} J_i\}\\ &\leq \bar{X}(\alpha_1, \alpha_2, \ldots, \alpha_n)), \end{aligned}$$

which completes the proof. ∎

Corollary 8.1

Suppose the lattice L is a complete chain. Let (A, R) be an L-syntax in S and let $X \in L^S$, $\alpha \in S$. Then:

(a$_1$) If $X \vdash^{A,R}_{x} \alpha$ and $y < x$, then there exists $w \in Pr(S, R)$ such that ${}^{+}w = \alpha$ and $\bar{w}(X) > y$.

(a$_2$) If L is a descending well-ordering (in particular, if L is finite), then there exists an R-proof of α, whose value with respect to X equals $Cn_{A,R}(X)(\alpha)$. ∎

Corollary 8.2

Let (A, R) be an L-syntax in S and suppose that the rule $r_0 = (p_0, q_0)$ given by the inference schemes

$$p_0 : \frac{\alpha, \alpha}{\alpha} \qquad q_0 : \frac{x, y}{x \vee y}$$

belongs to R. Further suppose that L is an ascending chain. Let $x \in L^S$, $\alpha \in S$. The set

$$\{\bar{w}(X) : w \in Pr(S, R),\ w^{+} = \alpha\} \subseteq L$$

is closed under finite joins and has a maximal element. Hence, there exists in S an R-proof w such that $w^{+} = \alpha$ and $\bar{w}(X) = Cn_{A,R}(X)(\alpha)$. ∎

Let $X \in L^S$ and $P \subseteq S$. By $X \upharpoonright P$ we denote the fuzzy set

$$(X \upharpoonright P)(\beta) = \begin{cases} X(\beta) & \text{if } \beta \in P, \\ 0 & \text{otherwise.} \end{cases}$$

Let Cn be an L-consequence operation in S. We say that Cn is compact if, for any $X \in L^S$ and any $\alpha \in S$, there is a finite set $P \subseteq S$ such that

$$(Cn(X))(\alpha) = (Cn(X \upharpoonright P))(\alpha).$$

Evidently, the operations $Cn_{A,R}$ considered in Corollaries 8.1 and 8.2 are compact, provided L is an ascending chain.

On the other hand, if L fails to be an ascending chain then there need not exist an R-proof with $w^+ = \alpha$ and $\bar{w}(X) = (Cn_{A,R}(X))(\alpha)$. In the sequel we introduce a certain class $\mathfrak{A}$ of propositional calculi over complete lattices of logical values; the calculi will be characterized in terms of their semantic systems $(S, \mathfrak{B})$, $\mathfrak{B} \subseteq L^S$. If $(S, \mathfrak{B})$ is an arbitrary system, the L-consequence $Cn_{\mathfrak{B}}$ need not be compact, as it is not required that L be an ascending or descending chain. Now assume L is not an ascending chain and suppose that an L-semantic system $(S, \mathfrak{B}) \in \mathfrak{A}$ admits substitution in an L-syntactic system (S, A, R); that is to say, $Cn_{\mathfrak{B}} = Cn_{A,R}$ for a certain L-syntax (A, R) in S. We will see below that this can indeed be the case. Theorem 8.2 only asserts the existence of a family $\{w^{(j)} : j \in J\}$ of R-proofs of α in S such that $(Cn_{A,R}(X))(\alpha) = \bigvee\{\bar{w}^{(j)}(X) : j \in J\}$. These properties seem to be a reasonable compromise between the finiteness of proofs and the locally infinite character of the lattice L.

Now we inspect how these properties are influenced by deleting the condition of semicontinuity from the definition of an L-rule and replacing it by a weaker one. The modified condition requires that

(i) $q : L^n \to L$ is an isotonic map.

A rule satisfying this modified definition will be called an L-quasirule; a pair (A, R) with $A \in L^S$ and R a set of L-quasirules in S will be called an L-quasisyntax in S. Lemma 8.2 and Theorem 8.2, accordingly modified, remain valid. Setting

$$\bar{X}(\alpha) = \bigvee\{\bar{w}(X) : w \in Pr(S, R), w^+ = \alpha\}$$

for $X \in L^S$, $\alpha \in S$, we define a consequence operation $C : L^S \to L^S$ which satisfies conditions (c_1) and (c_2), and also

$$\bar{X} \leq Cn_{A,R} \quad \text{for every} \quad X \in L^S.$$

Using transfinite induction we define a family of operations, indexed by ordinals. To every set X and each $\xi \in On$ (the class of ordinal numbers) we assign a set $\bar{X}^{\xi}$ according to the following rules:

(O_1) $\bar{X}^O = X$;

(O_2) $\bar{X}^{\xi+1} = \overline{(\bar{X}^\xi)}$;

(O_3) $\bar{X}^\lambda = \bigvee\left\{\bar{X}^\xi : \xi < \lambda\right\}$ for limit ordinals λ.

Clearly enough, for every given L-quasisyntax (A, R) there exists an ordinal number α such that

$$\bar{X}^\alpha = Cn_{A,R}(X)$$

holds for every fuzzy set $X \in L^S$.

Theorem 8.3

Suppose L contains a well-ordered ascending chain of ordinal type γ, a limit ordinal number. For every ordinal α with $1 < \alpha \leq \gamma$ there exist a set A, an L-quasisyntax (A, R) and an L-set $X : S \to L$ such that

(a_1) $\bar{X}^\alpha = Cn_{A,R}(X)$,

(a_2) $\bar{X}^\xi < Cn_{A,R}(X)$ if $\xi < \alpha$.

Proof

Let H be the chain of type γ, whose existence is assumed, and let $\beta : \gamma \to H$ be the ordering map, $\beta(\xi) = h_\xi \in H$. For any limit ordinal $\lambda \leq \gamma$ let h^*_λ denote $\bigvee\{h_\xi : \xi < \lambda\}$. Obviously $h_\lambda \geq h^*_\lambda > h_\xi$ for $\xi < \lambda$. Take a set P of cardinality equal to the cardinality of γ and choose α with $1 < \alpha \leq \gamma$. Fix a bijection between the sets $(\alpha - \{0\})X(\gamma + 2)$ and P, assigning to any pair (ξ, n) an element $y_{\xi,n} \in P$. Let $(S(P), \Rightarrow)$ be a free binary groupoid over P and let $r_1 = (p_1, q_1)$ be the L-quasirule given by the schemes of inference

$$p_1 : \frac{\alpha, \alpha \Rightarrow \beta}{\beta}, \qquad q_1 : \frac{x, y}{x \wedge y}.$$

(L need not be infinitely distributive.)

The required L-quasisyntax (A, R) will consist of:

(i_1) $A = \emptyset$, the empty fuzzy set;

(i_2) the set

$$R = \{r_1, r_\gamma\} \cup \{r_\lambda : \lambda \text{ a limit ordinal} < \alpha\},$$

where $r_\gamma = (p_\gamma, q_\gamma)$ is unary,

$$Dp_\gamma = \{y_{\xi,\gamma} : 1 \le \xi \le \alpha\},$$
$$p_\gamma(y_{\xi,\gamma}) = y_{\xi,\gamma+1},$$
$$q_\gamma(x) = \begin{cases} 1 & \text{if } x \ge h^*_\gamma, \\ 0 & \text{otherwise,} \end{cases}$$

and, for a limit ordinal $\lambda < \alpha$, $r_\lambda = (p_\lambda, q_\lambda)$ is the unary L-quasirule with

$$Dp_\lambda = \{y_{\xi,\gamma}\},$$
$$p_\lambda(y_{\lambda,\gamma}) = y_{\lambda,\gamma+1},$$
$$q_\lambda(x) = \begin{cases} 1 & \text{if } x \ge h^*_\gamma, \\ 0 & \text{otherwise.} \end{cases}$$

We define X, an L-subset of $S(P)$, by

$$X(\beta) = \begin{cases} h_\eta & \text{if } \beta = y_{1,\eta},\ \eta < \gamma, \\ & \text{or } \beta = y_{\xi,\gamma+1} \Rightarrow y_{\xi+1,\eta}, \\ & 1 \le \xi,\ \xi+1 < \alpha,\ \eta < \gamma; \\ 1 & \text{if } \beta = y_{\xi,\eta} \Rightarrow y_{\xi,\eta}, \\ & 1 \le \xi < \alpha,\ \eta < \gamma, \\ & \text{or } \beta = y_{\xi,\xi} \Rightarrow y_{\xi,\gamma}, \\ & \xi < \lambda < \alpha,\ \lambda \text{ a limit number}; \\ 0 & \text{in any other case.} \end{cases}$$

Then, by transfinite induction, $\bar{X}^\alpha = C_{A,R}(X)$, and we have

$$\bar{X}^\alpha(y_{\xi,\gamma+1}) = 1$$

for every ξ with $1 \le \xi < \alpha$, and

$$\bar{X}^\alpha(y_{\xi,\gamma+1}) = 0.$$

for other ξ. ∎

We say that an L-rule of inference $r = (p, q)$ is consistent with respect to an L-semantic $\mathfrak{L} \subseteq L^S$ if every set $T \in \mathfrak{L}$ is closed under r.

For instance, the rule $r_0 = (p_0, q_0)$ of patterns

$$p_0 : \frac{\alpha, \alpha}{\alpha}, \qquad q_0 : \frac{x, y}{x \vee y}$$

is consistent with respect to L^S.

To see this, assume $T : S \to L$ and $\alpha \in S$. Then

$$T(p_0(\alpha, \alpha)) = T(\alpha) = T(\alpha) \vee T(\alpha) = q_0(T(\alpha), T(\alpha)).$$

Further, we say that an L-syntax (A, R) on S is consistent with respect to a semantic $\mathfrak{L} \subseteq L^S$ is the following conditions are fulfilled:

(s_1) $A \leq C_{\mathfrak{L}}(\emptyset)$, with $\emptyset$ denoting the least element of the lattice L^S;

(s_2) every rule $r \in R$ is consistent with respect to $\mathfrak{L}$.

Theorem 8.4
An L-syntax (A, R) in S is consistent with respect to an L-semantic $\mathfrak{L} \subseteq L^S$ if and only if

$$Cn_{A,R}(X) \leq C_{\mathfrak{L}}(X)$$

holds for every $X : S \to L$.

Proof
Assume (A, R) is $\mathfrak{L}$-consistent and let $X \in L^S$. By the definition of $Cn_{A,R}$ we then have $Cn_{A,R} \leq T$ for every $T \in \mathfrak{L}$ such that $X \leq T$. Obviously,

$$Cn_{A,R}(X) \leq \bigwedge\{T \in \mathfrak{L} : X \leq T\} = C_{\mathfrak{L}}(X).$$

Conversely, assume $Cn_{A,R} \leq C_{\mathfrak{L}}$ in $(L^S)^{L^S}$. Then $A \leq Cn_{A,R}(\emptyset) \leq C_{\mathfrak{L}}(\emptyset)$, proving ($s_1$).
To show (s_2), take any rule $r = (p, q)$ belonging to R. Let $T \in \mathfrak{L}$ and let $(\alpha_1, \alpha_2, \ldots, \alpha_n) \in Dp$, the domain of p.
Then

$$\begin{aligned} & q(T(\alpha_1), T(\alpha_2), \ldots, T(\alpha_n)) \\ & \quad \leq (Cn_{A,R}(T))(p(\alpha_1, \alpha_2, \ldots, \alpha_n)) \\ & \quad \leq (C_{\mathfrak{L}}(T))(p(\alpha_1, \alpha_2, \ldots, \alpha_n)) \\ & \quad = T(p(\alpha_1, \alpha_2, \ldots, \alpha_n)), \end{aligned}$$

ending the proof. ■

An L-syntax (A, R) in S is said to be complete relative to an L-semantic $\mathfrak{L} \subseteq L^S$ if the L-consequence operations $Cn_{A,R}$ and $C_{\mathfrak{L}}$ coincide.
Given an L-semantic $\mathfrak{L} \subseteq L^S$, the L-semantic system $(S, \mathfrak{L})$ is axiomatizable if there exists an L-syntax (A, R) in S which is complete relative to $\mathfrak{L}$; i.e., such that $Cn_{A,R} = C_{\mathfrak{L}}$.

8.4 Semantic Basis for Fuzzy Propositional Logics

As mentioned several times already, a logic is in fact determined by its semantics. Our standpoint is that algebra is the appropriate framework for the construction of a logical system (except, maybe, the classical two-valued logic). This algebraic background need not be an ultimately formalized system; it can be created and "furnished" as the need arises. Such a standpoint is of course disputable; one can criticize it for example by pointing out the cases of modal calculi or the intuitionistic calculus. The latter has arisen rather incidentally, as a result of attempts toward formalization of a logical system admitting a third truth status besides the usual truth-falsity extremes. The discovery of adequate models for the intuitionistic calculus has become an inspiration for the creation of a variety of nonclassical logics. These have a rather uniform nature; however, it is difficult to classify them.

Now, in some measure, fuzzy calculi provide a most general class of many-valued logics, containing in particular finitely-valued logics, whose admissible values constitute a finite chain. It is largely a matter of agreement whether or not a given logic should be considered as a fuzzy logic.

We adopt the ideas of J. A. Goguen [Goguen 1967] for the basis of our presentation, as we did in the preceding section. In the discussion of complete residual lattices (considered well suited as semantics for fuzzy propositional calculi) we will closely follow the approach of J. Pavelka [Pavelka 1979]. For further motivation and justification of our choice, see [Pavelka 1979] and [Goguen 1967]. Taking the class of residual lattices (of a certain specific type) as the semantic basis for logics which we call "fuzzy" is essentially optional. It is largely a matter of convention, to be accepted not, according to one's own taste and preferences in the study of fuzzy logics.

We begin with a brief presentation of the notion and properties of residual lattices and their extensions. These will be illustrated by examples due to Pavelka. At the end of the section we define the consequence operation for the class of lattices in question.

Let P be a partially ordered set.

Let $(\otimes, \rightarrow)$ be a pair of binary operations in P with the following properties:

(a_1) operation $\otimes$ is isotonic;

(a_2) operation $\rightarrow$ is antitonic in the first argument and isotonic in the second argument;

(a_3) for any $x, y, z \in P$, $x \otimes y \leq z$ holds if and only if $x \leq y \rightarrow z$.

(Recall that an operation is isotonic if it preserves the lattice operations; it is antitonic if it carries joins to meets and vice versa.)

A pair $(\otimes, \rightarrow)$ with these properties will be called an adjoint pair.

Lemma 8.3

Suppose $\otimes$ and $\rightarrow$ are binary operations in a partially ordered set P, satisfying conditions (a_1) and (a_2). Then $(\otimes, \rightarrow)$ is an adjoint pair if and only if

(s_1) $x \leq y \to (x \otimes y)$,

(s_2) $(x \to y) \otimes x \leq y$

hold for every $x, y \in P$. ■

Lemma 8.4

Suppose $(\otimes, \to)$ is an adjoint pair in P. Then:

(m_1) for each $x \in P$, the function $y \mapsto x \otimes y$ preserves all lattice joins within P; in particular, if $0 = \bigvee\emptyset$ then $0 \otimes x = 0$;

(m_2) for each $y \in P$, the function $x \mapsto (y \to x)$ preserves all lattice meets within P; in particular, if $1 = \bigwedge\emptyset$ then $a \to 1 = 1$. ■

Lemma 8.5

Let L be a complete lattice. Then:

(b_1) If a binary operation $\otimes$ in L fulfills (a_1) and (m_1) then there exists a unique binary operation $\to$ in L such that $(\otimes, \to)$ is an adjoint pair; this operation is given by

$$x \to y = \bigvee\{z : z \otimes x \leq y\}.$$

(b_2) If a binary operation $\to$ in L fulfills (a_2) and (m_2) then there exists a unique binary operation $\otimes$ in L such that $(\otimes, \to)$ is an adjoint pair; this operation is given by

$$x \otimes y = \bigwedge\{z : x \leq y \to z\}. \quad ■$$

Lemma 8.6

Let P be a partially ordered set with a greatest element 1, and let $(\otimes, \to)$ be an adjoint pair in P. The following equivalence statements are valid:

(r_1) For every $y \in P$, the function $x \mapsto (y \otimes x)$ preserves lattice joins within P if and only if the function $x \mapsto (x \to y)$ carries all joins (existing within P) into meets.

(r_2) For every $x, y, z \in P$,

$$(x \otimes y) \otimes z \leq x \otimes (y \otimes z)$$

holds if and only if

$$(y \to z) \leq (x \to y) \to (x \to z).$$

(r_3) For every $x \in P$, $x \otimes 1 = x$ holds if and only if $x = 1 \to x$.

(r_4) For every $x, y \in P$,

$$1 \otimes x = x$$

holds if and only if

$$x \leq y \text{ is equivalent to } 1 = x \to y.$$

(r_5) For every $x, y, x \in P$,

$$x \otimes y = y \otimes x$$

holds if and only if

$$x \leq y \to z \text{ is equivalent to } y \leq z \to z.$$

(r_6) For every $x, y, z \in P$,

$$(x \otimes y) \otimes z \leq x \otimes (y \otimes z)$$

holds if and only if

$$x \to y \leq (x \otimes z) \to (y \otimes z).$$

(r_7) For every $x, y, z \in P$,

$$(x \otimes y) \otimes z = x \otimes (y \otimes z)$$

holds if and only if

$$(x \otimes y) \to z = x \to (y \to z). \quad \blacksquare$$

By a residual lattice we mean a triple $\mathrm{L} = (L, \otimes, \to)$ in which

(a_1) L is a bounded lattice with zero 0 and unit 1;

(a_2) $(\otimes, \to)$ is an adjoint pair in L;

(a_3) $(L, \otimes, 1)$ is a commutative monoid.

Then $\otimes$ will be called multiplication and $\to$ the residuum operation. The symbol $x \to y$ is read as "x residues y".

An adjoint pair $(\otimes, \to)$ is referred to as an mr-pair on L if L is a bounded lattice and condition (a_3) is satisfied.

A residual lattice $\mathrm{L} = (L, \otimes, \to)$ is complete if L is a complete lattice.

When card $L = 1$ (i.e., $1 \leq 0$), the lattice is said to be degenerated.

Conditions (s_1) and (s_2) are automatically satisfied in every residual lattice. Moreover,

(b_3) $\qquad x \otimes y \leq (x \otimes 1) \wedge (1 \otimes y) = x \wedge y$

is valid in a residual lattice.

For any $y \in L$ and any positive integer n we can consider the power y^n relative to the monoid structure of $(L, \otimes, 1)$. If $y \in L$ is fixed, the assignment $n \mapsto y^n$ is an antitonic map from N into L.

Examples

1. Let L be a Brouwer lattice (see [Birkhoff 1948]). Consider the mr-pair $(\wedge, \rightarrow)$ with $\wedge$ the lattice meet and $\rightarrow$ the Heyting implication. L is a residual lattice.

If L is a Boolean algebra then

$$x \rightarrow y = -x \vee y$$

where $x \mapsto -x$ denotes complementation.

If L is a bounded chain then L is a Brouwer lattice and we have

$$x \rightarrow y = \begin{cases} 1 & \text{if } x \leq y, \\ 0 & \text{otherwise.} \end{cases}$$

2. Let R be a commutative ring with unit (1). Let $\mathrm{F}(R)$ denote the complete lattice of all ideals in R. For $F_1, F_2 \in \mathrm{F}(R)$ define

$$F_1 \otimes F_2 = \left\{ \sum_{i=1}^{n} x_i y_i : n \in N,\ x_i \in F_1,\ y_i \in F_2 \right\},$$

$$F_1 \rightarrow F_2 = \{x \in R : \forall y \in F_1\ (x \cdot y \in F_2)\}.$$

Then $(\otimes, \rightarrow)$ is an mr-pair in $\mathrm{F}(R)$. In particular, if m is not a prime then, taking for R the ring Z_m (the integers modulo m) and considering $\mathrm{F}(Z_m)$, we obtain an example of a finite residual lattice which is not a chain and whose multiplication is not idempotent.

3. Let $m \geq 1$ be an integer. Consider an $(m+1)$-element chain

$$C_{m+1} = \{x_0, x_1, \ldots, x_m\},$$

$$0 = x_0 < x_1 < \ldots < x_m = 1.$$

For $i,\ j \in \{0, 1, \ldots, m\}$ define

$$x_i \otimes x_j = x_{\max(0, i+j-m)},$$

$$x_i \rightarrow x_j = x_{\min(m, m-i+j)}.$$

The triple $(C_{m+1}, \otimes, \rightarrow)$ is a residual lattice. It will be referred to as the $(m+1)$-element Łukasiewicz chain.

4. Let $I = [0,1]$ be the unit segment of the real line. Every binary operation, continnous and associative on I, constant on 0 and 1, is isotonic and commutative. Therefore I, viewed as a complete chain, can be equipped with an mr-pair. Obviously, the continuity of $\otimes$ is a stronger condition than the mere preservation of lattice joins. So there exist residual lattices (formed from I) whose multiplication is semicontinuous and not continuous. It is also known that there exist $2^{\aleph_0}$ mutually nonisomorphic continuous $\otimes$-type operations. Two of them play a special role in the theory of topological semigroups on I; these are:

(a₁) the usual multiplication of real numbers, with the right dual defined by

$$x \to y = \begin{cases} 1 & \text{if } x \leq y, \\ \frac{y}{x} & \text{otherwise;} \end{cases}$$

(a₂) the multiplication operation

$$x \otimes y = 0 \vee (x + y - 1),$$

accompanied by the residuum operation

$$x \to y = 1 \wedge (1 - x + y),$$

with $\wedge, \vee$ denoting the usual bounds (min, max) in I. The resulting residual lattice L $= (I, \otimes, \to)$ is a continuous analogue of the $(m+1)$-element chain $(C_{m+1}, \otimes, \to)$. We call $(I, \otimes, \to)$ the Łukasiewicz interval. It is worth noticing that (denoting by x^n the n-th power $x \otimes \cdots \otimes x$) we have for every $x \in (0,1)$ and each n with $\frac{1}{1-x} \leq n$

$$x^n = 0 \vee (1 - n(1-x)) = 0.$$

Thus every x in the open interval $(0,1)$ is a nilpotent element of the semigroup $(I, \otimes)$.

Let L $= (L, \otimes, \to)$ be an arbitrary residual lattice. The binary operation in L given by

$$x \leftrightarrow y = (x \to y) \wedge (y \to x)$$

will be called the biresiduum.

Theorem 8.5

The operation of biresiduum in a residual lattice has the following properties:

(b₁) $x \leftrightarrow 1 = x$;

(b₂) $x = y$ if and only if $x \leftrightarrow y = 1$;

(b₃) $x \leftrightarrow y = y \leftrightarrow x$;

(b_4) $(x \leftrightarrow y) \otimes (y \leftrightarrow z) \leq (x \leftrightarrow z)$;

(b_5) $(x_1 \leftrightarrow y_1) \wedge (x_2 \leftrightarrow y_2) \leq (x_1 \wedge x_2) \leftrightarrow (y_1 \wedge y_2)$;

(b_6) $(x_1 \leftrightarrow y_1) \wedge (x_2 \leftrightarrow y_2) \leq (x_1 \vee x_2) \leftrightarrow (y_1 \vee y_2)$;

(b_7) $(x_1 \leftrightarrow y_1) \otimes (x_2 \leftrightarrow y_2) \leq (x_1 \otimes x_2) \leftrightarrow (y_1 \otimes y_2)$;

(b_8) $(x_1 \leftrightarrow y_1) \otimes (x_2 \leftrightarrow y_2) \leq (x_1 \rightarrow x_2) \leftrightarrow (y_1 \rightarrow y_2)$.

Proof

(b_1), (b_2) and (b_3) follow immediately from the definition of $\leftrightarrow$ and Lemma 8.6.

Choose $x, y, z \in L$. Then

$$\begin{aligned}
&(x \leftrightarrow y) \otimes (y \leftrightarrow z) \leq (x \rightarrow y) \otimes (y \rightarrow z),\\
&(x \rightarrow y) \leq (y \rightarrow z) \rightarrow (x \rightarrow z),\\
&(x \rightarrow y) \otimes (y \rightarrow z) \leq (x \rightarrow z),\\
&(x \leftrightarrow y) \otimes (y \leftrightarrow z) \leq (x \rightarrow z),
\end{aligned}$$

so (b_4) follows from (b_3) and the definition of $\leftrightarrow$.

For the proof of (b_5), note that

$$\begin{aligned}
&((x_1 \leftrightarrow y_1) \wedge (x_2 \leftrightarrow y_2) \otimes (x_1 \wedge x_2))\\
&\quad \leq ((x_1 \rightarrow y_1) \wedge (x_2 \rightarrow y_2)) \otimes (x_1 \wedge x_2)\\
&\quad \leq ((x_1 \rightarrow y_1) \otimes x_1) \wedge ((x_2 \rightarrow y_2) \otimes x_2) \leq y_1 \wedge y_2
\end{aligned}$$

holds for all $x_1, x_2, y_1, y_2 \in L$. Hence

$$(x_1 \leftrightarrow y_1) \wedge (x_2 \leftrightarrow y_2) \leq (x_1 \wedge x_2) \rightarrow (y_1 \wedge y_2).$$

By the definition of biresiduum,

$$(x_1 \leftrightarrow y_1) \wedge (x_2 \leftrightarrow y_2) \leq (y_1 \wedge y_2) \rightarrow (x_1 \wedge x_2).$$

The two inequalities jointly result in (b_5).

Inequalities (b_6), (b_7) and (b_8) are proved in a similar fashion; and thus,

$$\begin{aligned}
&((x_1 \leftrightarrow y_1) \wedge (x_2 \leftrightarrow y_2)) \otimes (x_1 \vee x_2)\\
&\quad = (((x_1 \leftrightarrow y_1) \wedge (x_2 \leftrightarrow y_2)) \otimes x_1) \vee (((x_1 \leftrightarrow y_1) \wedge (x_2 \leftrightarrow y_2)) \otimes x_2)\\
&\quad \leq ((x_1 \rightarrow y_1) \otimes x_1) \vee ((x_2 \rightarrow y_2) \otimes x_2) \leq y_1 \vee y_2
\end{aligned}$$

yielding (b_6);

$$\begin{aligned}
&(x_1 \leftrightarrow y_1) \otimes (x_2 \leftrightarrow y_2) \otimes x_1 \otimes x_2\\
&\quad \leq ((x_1 \rightarrow) \otimes x_1) \otimes ((x_2 \rightarrow y_2) \otimes x_2) \leq y_2 \otimes y_2
\end{aligned}$$

gives (b_7); and finally,

$$x_1 \leftrightarrow y_1) \otimes (x_2 \leftrightarrow y_2) \otimes (x_1 \rightarrow x_2)$$
$$\leq (y_1 \rightarrow x_1) \otimes (x_2 \rightarrow y_1) \otimes (x_1 \rightarrow x_2)$$
$$\leq (y_1 \rightarrow x_1) \otimes (x_1 \rightarrow y_2) \leq y_1 \rightarrow y_2,$$

implying (b_8) and ending the proof of the theorem. ■

Let $\mathrm{L} = (L, \otimes \rightarrow)$ be a residual lattice and let o be an n-ary operation in L $(n \geq 1)$. We say that o is admissible in L if there exist positive integers $k_1, k_2, \ldots, k_n$ such that the inequality

$$(x_1 \leftrightarrow y_1)^{k_1} \otimes (x_2 \leftrightarrow y_2)^{k_2} \otimes \ldots \otimes (x_n \leftrightarrow y_n)^{k_n}$$
$$\leq o(x_1, x_2, \ldots, x_n) \leftrightarrow o(y_1, y_2, \ldots, y_n)$$

holds for any $x_1, x_2, \ldots, x_n, y_1, y_2, \ldots, y_n \in L$; the powers are to be understood as powers in semigroup $(L, \otimes)$. The integers $k_1, k_2, \ldots, k_n$ are then called admissible exponents for o, with respect to the residual lattice L.

It is easy to see that if $k_1, k_2, \ldots, k_n$ are admissible exponents for an operation o, then so are any exponents $l_1, l_2, \ldots, l_n$ with $k_1 \leq l_i$ for $i = 1, 2, \ldots, n$.

Examples

1. In the $(m+1)$-element Łukasiewicz chain $\mathrm{L}_{m+1} = (C_{m+1}, \otimes, \rightarrow)$, each element $x \in C_{m+1}$ satisfies either $x = 1$ or $x^m = o$. Hence, if $o : C_{m+1}^n \rightarrow C_{m+1}$ is any operation, we have for any elements $x_1, x_2, \ldots, x_n, y_1, y_2, \ldots, y_n \in C_{m+1}$ either

$$o(x_1, x_2, \ldots, x_n) = o(y_1, y_2, \ldots, y_n)$$

or

$$(x_1 \leftrightarrow y_1)^m \otimes (x_2 \leftrightarrow y_2)^m \otimes \ldots \otimes (x_n \leftrightarrow y_n)^m = o.$$

Consequently, every finite operation is admissible in L_{m+1}.

2. Let $\mathrm{L} = (I, \otimes, \rightarrow)$ be the Łukasiewicz interval. For any $n \geq 1$ consider the metric ρ_n in I^n,

$$\rho_n((x_1, \ldots, x_n), (y_1, \ldots, y_n)) = \sum_{i=1}^{n} | x_i - y_i | .$$

Every Lipschitz mapping (with respect to ρ_n) is admissible in L. In particular, multiplication (a contractive map) is admissible.

3. Every n-argument operation in I, admissible for the Łukasiewicz interval, is continuous on I^n. Note that the right adjoint $\rightarrow$ to the real multiplication fails to be continuous at (o, o); hence, it is not an admissible operation in L ([Pavelka 1979], pp.125–126).

Theorem 8.6

Let $\mathrm{L} = (L, \otimes, \rightarrow)$ be a residual lattice and let O denote the set of finite-argument operations admissible in L. Then all operations definable in the universal algebra

$$\mathfrak{A} = (\mathrm{L}, L, \wedge, \vee, \otimes, \rightarrow, \mathrm{O})$$

are admissible in L.

Proof [Pavelka 1979]

The proof is by induction on the definability structure of an operation. In view of Theorem 8.5, the primitive functions $\otimes, \rightarrow, \wedge, \vee$ are admissible. The set O is admissible by hypothesis.

For the induction step assume:

1. $f : L^n \rightarrow L$ is admissible with exponents $k_1, k_2, \ldots, k_n$.

2. For each $i = 1, 2, \ldots, n$, $g_i : L^{m_i} \rightarrow L$ is admissible with exponents $k_{i_1}, \ldots, k_{i_{m_i}}$, respectively.

3. $s_i = m_1 + m_2 + \ldots + m_i$ for $i = 1, 2, \ldots, n$.

4. $m \geq 1$ is an integer;

$$p : \{1, 2, \ldots, s_n\} \rightarrow \{1, 2, \ldots, m\}$$

is an arbitrary mapping.

5. Define

$$q : \{1, 2, \ldots, s_n\} \rightarrow \{1, 2, \ldots, n\}$$

by

$$q(j) = i \quad \text{iff} \quad s_{i-1} < j \leq s_i.$$

Consider the m-argument operation

$$\begin{aligned} h(x_1, x_2, \ldots, x_m) = f\big(&g_1(x_{p(1)}, x_{p(2)}, \ldots, x_{p(s_1)}),\\ &g_2(x_{p(s_1+1)}, x_{p(s_1+2)}, \ldots, x_{p(s_2)}),\\ &\ldots\\ &g_n(x_{p(s_{n-1}+1)}, x_{p(s_{n-1}+2)}, \ldots, x_{p(s_n)})\big). \end{aligned}$$

We claim that h is admissible with exponents $k^{(1)}, k^{(2)}, \ldots, k^{(m)}$, where

$$k^{(i)} = \sum_{p(j)=i} k_{q(j)} \; k_{q(j), j-s_{q(j)-1}}$$

for $i = 1, 2, \ldots, m$.

This is so because we have for $x_1, x_2, \ldots, x_m, y_1, y_2, \ldots, y_m \in L$:

$$
\begin{aligned}
&(x_1 \leftrightarrow y_1)^{k(1)} \otimes (x_2 \leftrightarrow y_2)^{k(2)} \otimes \ldots \otimes (x_m \leftrightarrow y_m)^{k(m)} \\
= \;& ((x_{p(1)} \leftrightarrow y_{p(1)})^{k_{11}} \otimes (x_{p(2)} \leftrightarrow y_{p(2)})^{k_{12}} \otimes \ldots \otimes (x_{p(s_1)} \leftrightarrow y_{p(s_1)})^{k_{1m1}})^{k_1} \otimes \ldots \\
& \otimes((x_{p(s_{n-1}+1)} \leftrightarrow y_{p(s_{n-1}+1)}^{k_{n1}}) \otimes \ldots \otimes (x_{p(s_n)} \leftrightarrow y_{p(s_n)})^{k_{nmm}})^{k_n} \\
\leq \;& (g_1(x_{p(s_1)}) \leftrightarrow g_1(y_{p(1)}, \ldots, y_{p(s_1)}))^{k_1} \otimes \ldots \\
& \otimes(g_n(x_{p(s_{n-1}+1)}, \ldots, x_{p(s_n)}) \leftrightarrow g_n(y_{p(s_{n-1}+1)}, \ldots, y_{p(s_n)}))^{k_n} \\
\leq \;& h(x_1, x_2, \ldots, x_m) \leftrightarrow h(y_1, y_2, \ldots, y_m).
\end{aligned}
$$

This accomplishes induction and proves the theorem. ∎

Now, let Δ be any set (the empty set is not excluded). Suppose $Ar : \Delta \to N \setminus \{0\}$ and $Ex : \Delta \to N^* = \bigcup_{m=1}^{\infty} N^m$ are two mappings, so that Ex assigns to each $d \in \Delta$ an $Ar(d)$-tuple $(k_1, k_2, \ldots, k_{Ar(\Delta)})$. Let $\mathrm{L} = (L, \otimes, \to)$ be a residual lattice and let $\mathrm{O} = \{o_d : d \in \Delta\}$ be a Δ-indexed set of operations in L; for each $d \in \Delta$, o_d is assumed to be an $Ar(d)$-ary operation, admissible for L, the entries of $Ex(d)$ being admissible exponents. The pair $\mathrm{E} = (\mathrm{L}, \mathrm{O})$ will then be called an extended residual lattice over L, of type (Ar, Ex).

With every extended residual lattice $\mathrm{E} = ((L, \otimes, \to), \mathrm{O})$ of type (Ar, Ex) there is associated the universal algebra

$$a\mathrm{E} = (L, 0, 1, \wedge, \vee, \otimes, \to, \mathrm{O}).$$

The order structure of L is defined in terms of $\wedge$ and $\vee$ (the inequality $x \leq y$ is represented by equalities $x \wedge y = x$ or $x \vee y = y$). Thus, for a given type (Ar, Ex), universal algebras $a\mathrm{E}$ constitute a variety of algebras.

By a congruence in an extended residual lattice $\mathrm{E} = (\mathrm{L}, \mathrm{O})$ we mean any congruence in the associated universal algebra $a\mathrm{E}$. If $\mathrm{E}_1 = ((L_1, \otimes_1, \to_1), \mathrm{O}_1)$ and $\mathrm{E}_2 = ((L_2, \otimes_2, \to_2), \mathrm{O}_2)$ are of the same type, we call a mapping $f : L_1 \to L_2$ a homomorphism between the extended lattices E_1 and E_2 if and only if f, regarded as a map between the universal algebras $a\mathrm{E}_1$ and $a\mathrm{E}_2$, is an algebra homomorphism.

A filter in a residual lattice $\mathrm{L} = (L, \otimes, \to)$ is defined as a subset $\mathrm{F} \subseteq L$ that satisfies the conditions:

(f_1) $1 \in \mathrm{F}$;

(f_2) if $x \in \mathrm{F}$ and $x \leq y$ then $y \in \mathrm{F}$;

(f_3) if $x, y \in \mathrm{F}$ then $x \otimes y \in \mathrm{F}$.

Evidently, if F is a filter in L, it is automatically a filter in the lattice L. This follows from the fact that $x \otimes y \leq x \wedge y$ and hence $x \otimes y \in \mathrm{F}$ yields $x \wedge y \in \mathrm{F}$.

Theorem 8.7

Let $\mathrm{E} = (\mathrm{L}, \mathrm{O})$ be an extended residual lattice over $\mathrm{L} = (L, \otimes, \to)$. Then:

(i_1) For every congruence $\sim$ in E, the set

$$\mathrm{F} = \{x \in L : x \sim 1\}$$

is a filter in L.

(i_2) For every filter F in L, the relation

$$x \sim y \quad \text{iff} \quad x \leftrightarrow y \in \mathrm{F}$$

is a congruence in E.

(i_3) Conditions (i_1) and (i_2) establish a bijective correspondence between the set of all congruences in E and the set of all filters in L.

Proof

If $\sim$ is a congruence on E, then of course $1 \sim 1$ and $1 \in \mathrm{F}$. If $x \leq y$ and $x \in \mathrm{F}$ then $y = x \vee y$, $x \vee y \sim 1$, $y \sim 1$, so that $y \in \mathrm{F}$. Similarly, $x \otimes y \sim 1 \otimes 1 = 1$. Hence, $x, y \in \mathrm{F}$ forces $x \otimes y \in \mathrm{F}$, and so F is a filter in L and (i_1) is proved.

For a proof of (i_2), suppose F is a filter in L; then the relation

$$\{(x, y) : (x \leftrightarrow y) \in \mathrm{F}\}$$

is a congruence in L. It follows from Theorem 8.6 that this relation preserves all operations in the algebra $a\mathrm{E}$, hence is a congruence in E.

To show (i_3), it is enough to observe that if $\sim$ is a congruence in E then, firstly, if $x = y$ then $x \leftrightarrow y \sim x \leftrightarrow y = 1$; and secondly, if $x \leftrightarrow y = 1$ then

$$\begin{aligned} x \sim x \otimes (x \leftrightarrow y) &= (x \otimes (x \leftrightarrow y)) \wedge y \sim x \wedge y \sim x \wedge ((x \leftrightarrow y) \otimes y) \\ &= (x \leftrightarrow y) \otimes y \sim y, \end{aligned}$$

ending the proof. ∎

Corollary 8.3

Let $\mathrm{E} = (\mathrm{L}, \mathrm{O})$ be an extended residual lattice over $\mathrm{L} = (L, \otimes, \to)$. A congruence $\sim$ in L is a congruence in E if and only if it preserves $\otimes$ and $\to$. ∎

Evidently, if $\mathrm{E} = (\mathrm{L}, \mathrm{O})$ is an extended residual lattice, F is a filter in L and $\sim$ is the induced congruence in $a\mathrm{E}$, then the quotient algebra $a\mathrm{E}/\sim$ is similar to $a\mathrm{E}$. It is associated with an extended residual lattice of the same type as E, which will be denoted by E/F. Let f be the canonical epimorphism of E onto E/F. For any $x, y \in L$, the inequality $f(x) \leq f(y)$ holds in E/F if and only if $x \to y \in \mathrm{F}$.

In the following we assume that $\mathrm{E} = ((L, \otimes, \to), \mathrm{O})$ is a complete extended residual lattice of type (Ar, Ex), $Ar : \Delta \to N \setminus \{0\}$, $Ex : \Delta \to N^*$. Suppose V is a countably infinite set of propositional variables. Consider the absolutely free algebra

$$\mathrm{S}(V, \mathrm{E}) = (S(V, L, \Delta); \{\bar{e} : e \in L\}, \{\bar{d} : d \in \Delta\}, \wedge, \vee, \&, \Rightarrow)$$

where

(a_1) $\{\bar{e} : e \in L\}$ is the set of zero-argument operations;

(a_2) $\{\bar{d} : d \in \Delta\}$ is a set of operations defined in V, the set of absolutely free generators; for each $d \in \Delta$, the arity (number of arguments) of $\bar{d}$ equals $Ar(d)$;

(a_3) $\wedge, \vee, \&, \Rightarrow$ are two-argument operations.

We shall call $\mathrm{S}(V, \mathrm{E})$ the algebra of formulas of the E-fuzzy propositional calculus over V; for short, (V, E)-calculus. The elements of $\mathrm{S}(V, \mathrm{E})$ will be called (V, E)-formulas, or simply formulas; V is the set of propositional variables.

Thus, (V, E)-formulas constitute the smallest set with the following properties:

(f_1) If $p \in V$ then p is a (V, E)-formula.

(f_2) If $e \in L$ then $\bar{e}$ is a (V, E)-formula.

(f_3) If α, β are (V, E)-formulas then $\alpha \wedge \beta$, $\alpha \vee \beta$, $\alpha \& \beta$, $\alpha \Rightarrow \beta$ are (V, E)-formulas.

(f_4) If $d \in \Delta$, $Ar(d) = n$ and $\alpha_1, \alpha_2, \ldots, \alpha_n$ are (V, E)-formulas then $\bar{d}(\alpha_1, \alpha_2, \ldots, \alpha_n)$ is a (V, E)-formula.

Following [Pavelka 1979], we will call zero-argument connectives $\bar{e}$ inner truth values. The binary connectives $\wedge, \vee, \&$ and $\Rightarrow$ will be called conjunction, disjunction, context and implication, respectively.

For technical reasons we introduce certain abbreviations:

(s_1) $\alpha \Leftrightarrow \beta = (\alpha \Rightarrow \beta) \wedge (\beta \Rightarrow \alpha)$ for $\alpha, \beta \in S(V, L, \Delta)$;

this connective will be called biimplication;

(s_2) for $\alpha_1, \alpha_2, \ldots, \alpha_n \in S(V, L, \Delta)$, $n \geq 1$:

$\alpha_1 \& \alpha_2 \& \ldots \& \alpha_n = (\ldots(\alpha_1 \& \alpha_2) \& \ldots) \& \alpha_n$;

(s_3) $\alpha^n = \alpha \& \alpha \& \ldots \& \alpha$ (n times).

Let $f\mathrm{E}$ denote the algebra arising by adjoining to $a\mathrm{E}$ the constants $e \in L \setminus \{0,1\}$; i.e., let

$$f\mathrm{E} = (L; \{e : e \in L\}, \wedge, \vee, \otimes, \rightarrow, \{o_d : d \in \Delta\}).$$

By the semantic of the (V,E)-fuzzy calculus we mean the set

$$Sem(V,\mathrm{E}) = \{h \in L^{S(V,L,\Delta)} : h \text{ is a homomorphism from } \mathrm{S}(V,\mathrm{E}) \text{ to } f\mathrm{E}\}.$$

Since the algebra $\mathrm{S}(V,\mathrm{E})$ is absolutely free over V, it follows that for every mapping $t : V \rightarrow L$ there exists a unique $T \in Sem(V,\mathrm{E})$ such that $t(p) = T(p)$ for all $p \in V$. It is defined by

(t_1) $T(\bar{e}) = e$ for $e \in L$;

(t_2) $T(\alpha \wedge \beta) = T(\alpha) \wedge T(\beta)$
$T(\alpha \vee \beta) = T(\alpha) \vee T(\beta)$
$T(\alpha \& \beta) = T(\alpha) \otimes T(\beta)$
$T(\alpha \Rightarrow \beta) = T(\alpha) \rightarrow T(\beta)$
$T(\bar{d}(\alpha_1, \alpha_2, \ldots, \alpha_n)) = o_d(T(\alpha_1), T(\alpha_2), \ldots, T(\alpha_n))$
for $\alpha, \beta, \alpha_1, \alpha_2, \ldots, \alpha_n \in S(V,L,\Delta)$, $d \in \Delta$, $Ar(d) = n$.

Clearly, the biimplication will be interpreted as biresiduum $\leftrightarrow$, while the n-th context will be carried to the n-th power in semigroup $(L, \otimes)$.

As already observed, the semantics defined on the set of formulas determine the consequence operation. Given an L-fuzzy set $X : S(V,L,\Delta) \rightarrow L$, considered as the set of axioms of the L-semantic system $(S(V,L,\Delta), Sem(V,\mathrm{E}))$, we define the degree of incidence of a formula α to $Cn_{Sem(V,\mathrm{E})}X$ by

$$(Cn_{Sem(V,\mathrm{E})}X)(\alpha) = \bigwedge\{T(\alpha) : T \in Sem(V,\mathrm{E}) \text{ with } T(\beta) \geq X(\beta) \text{ for every } \beta \in S(V,L,\Delta)\}.$$

This "consequence operation" will be referred to as $Sem(V,\mathrm{E})$-consequence. In particular, taking for X the empty L-set

$$\emptyset(\alpha) = 0 \quad \text{for all} \quad \alpha \in S(V,L,\Delta),$$

we obtain its $Sem(V,\mathrm{E})$-consequence

$$Cn_{Sem(V,\mathrm{E})}\emptyset = \bigwedge Sem(V,\mathrm{E}),$$

referred to as the L-set of (V,E)-tautologies. The value of

$$(Cn_{Sem(V,\mathrm{E})}\emptyset)(\alpha)$$

will be called the (V,E)-tautologous degree of α. A formula α is said to be a (V,E)-tautology if and only if

$$(Cn_{Sem(V,\mathrm{E})}\emptyset)(\alpha) = 1.$$

For $X : S(V, L, \Delta) \to L$, $\alpha \in S(V, L, \Delta)$ and $e \in L$, the inequality

$$(Cn_{Sem(V,E)}X)(\alpha) \geq e$$

will be written briefly as

$$X \models_e \alpha.$$

In other words, $X \models_e \alpha$ holds if and only if $T(\alpha) \geq e$ for every $T \in Sem(V, E)$ such that $T \geq X$.

Lemma 8.7
Let $X \in L^{S(V,L,\Delta)}$, $\alpha, \beta \in S(V, L, \Delta)$, $e \in L$. Then

$$X \models_e (\alpha \Rightarrow \beta)$$

holds if and only if

$$e \otimes T(\alpha) \leq T(\beta)$$

for every $T \in Sem(V, E)$ such that $X \leq T$.

Proof

$$X \models_e (\alpha \Rightarrow \beta) \text{ is valid if and only if } e \leq T(\alpha \Rightarrow \beta) = T(\alpha) \to T(\beta)$$

holds for every $T \in Sem(V, E)$ with $X \leq T$; and this is equivalent to the statement of the lemma. ∎

The implication $\alpha \Rightarrow \beta$ is a (V, E)-tautology if and only if the inequality $T(\alpha) \leq T(\beta)$ holds for every $T \in Sem(V, E)$.

Following [Pavelka 1979], we introduce some further abbreviated notation:

$\sigma_1(x, y) = (\bar{x} \wedge \bar{y}) \Rightarrow \bar{x} \wedge \bar{y}; \quad x, y \in L$

$\sigma_2(x, y) = (\bar{x} \Rightarrow \bar{y}) \Rightarrow \bar{x} \Rightarrow \bar{y}; \quad x, y \in L$

$\sigma_3^d(x_1, x_2, \ldots, x_n) = \bar{d}(\bar{x}_1, \bar{x}_2, \ldots, \bar{x}_n) \Leftrightarrow o_d(\bar{x}_1, \bar{x}_2, \ldots, \bar{x}_n)$
for $d \in \Delta$, $Ar(d) = n$; $x_1, x_2, \ldots, x_n \in L$

$\sigma_4(\alpha, \beta, \gamma) = \alpha \Rightarrow 1$

$\sigma_5(\alpha, \beta, \gamma) = \alpha \Rightarrow \alpha$

$\sigma_6(\alpha, \beta, \gamma) = (\beta \Rightarrow \gamma) \Rightarrow ((\alpha \Rightarrow \beta) \Rightarrow (\alpha \Rightarrow \gamma))$

$\sigma_7(\alpha, \beta, \gamma) = (\alpha \Rightarrow (\beta \Rightarrow \gamma)) \Rightarrow (\beta \Rightarrow (\alpha \Rightarrow \gamma))$

$\sigma_8(\alpha, \beta, \gamma) = (\alpha \wedge \beta) \Rightarrow \alpha$

$\sigma_9(\alpha, \beta, \gamma) = (\alpha \wedge \beta) \Rightarrow \beta$

$\sigma_{10}(\alpha, \beta, \gamma) = (\gamma \Rightarrow \alpha) \Rightarrow ((\gamma \Rightarrow \beta) \Rightarrow (\gamma \Rightarrow (\alpha \wedge \beta)))$

$\sigma_{11}(\alpha, \beta, \gamma) = \alpha \Rightarrow (\alpha \vee \beta)$

$\sigma_{12}(\alpha, \beta, \gamma) = \beta \Rightarrow (\alpha \vee \beta)$

$\sigma_{13}(\alpha, \beta, \gamma) = (\alpha \Rightarrow \gamma) \Rightarrow ((\beta \Rightarrow \gamma) \Rightarrow ((\alpha \vee \beta) \Rightarrow \gamma))$

$\sigma_{14}(\alpha, \beta, \gamma) = \alpha \Rightarrow (\beta \Rightarrow (\alpha \& \beta))$

$\sigma_{15}(\alpha, \beta, \gamma) = (\alpha \Rightarrow (\beta \Rightarrow \gamma)) \Rightarrow ((\alpha \Rightarrow \beta) \Rightarrow \gamma)$

$\sigma_{16}(\alpha, \beta, \gamma) = (\alpha \& 1) \Leftrightarrow \alpha$

$\sigma_{17}^{d}(\alpha_1, \alpha_2, \ldots, \alpha_n, \beta_1, \beta_2, \ldots, \beta_n) = ((\alpha_1 \Leftrightarrow \beta_1)^{k_1} \& (\alpha_2 \Leftrightarrow \beta_2)^{k_2} \& \ldots$
$\& (\alpha_n \Leftrightarrow \beta_n)) \Rightarrow (\bar{d}(\alpha_1, \alpha_2, \ldots, \alpha_n) \Leftrightarrow \bar{d}(\beta_1, \beta_2, \ldots, \beta_n))$
for $d \in \Delta$, $Ar(d) = n$, $Ex(d) = (k_1, k_2, \ldots, k_n)$, and
$\alpha_1, \alpha_2, \ldots, \alpha_n, \beta_1, \beta_2, \ldots, \beta_n \in S(V, L, \Delta)$.

Theorem 8.8

Given any set V of propositional variables and any extended complete residual lattice E of type (Ar, Ex), the semantically induced L-consequence operation $Cn_{Sem(V,E)}$ has the following properties:

(c_0) For any $e, e_1, e_2 \in L$,

$\emptyset \models_e \bar{e}$,

$\emptyset \models_{e_1 \otimes e_2} \bar{e_1} \& \bar{e_2}$,

$\emptyset \models_{e_1 \to e_2} \bar{e_1} \Rightarrow \bar{e_2}$.

(c_1) For any $e_1, e_2 \in L$, the formulas

$\sigma_1(e_1, e_2)$ and $\sigma_2(e_1, e_2)$ are (V, E)-tautologies.

(c_2) For any $d \in \Delta$ and any $e_1, e_2, \ldots, e_n \in L$, where $n = Ar(d)$, the formula $\sigma_3^d(e_1, e_2, \ldots, e_n)$ is a (V, E)-tautology.

(c_3) For any α, β, γ, the formulas $\sigma_i(\alpha, \beta, \gamma)$, $i = 4, 5, \ldots, 16$, are (V, E)-tautologies.

(c_4) For any $d \in \Delta$ and any $\alpha_1, \alpha_2, \ldots, \alpha_n, \beta_1, \beta_2, \ldots, \beta_n \in S(V, L, \Delta)$, where $n = Ar(d)$, the formula $\sigma_{17}^d(\alpha_1, \alpha_2, \ldots, \alpha_n, \beta_1, \beta_2, \ldots, \beta_n)$ is a (V, E)-tautology.

Proof

Immediate from the definition and properties of an extended complete residual lattice. For instance, consider $\sigma_{10}(\alpha, \beta, \gamma)$. It will be enough to show that

$$T(\gamma \Rightarrow \alpha) \leq T((\gamma \Rightarrow \beta) \Rightarrow (\gamma \Rightarrow (\alpha \wedge \beta))),$$

i.e.,

$$T(\gamma) \to T(\alpha) \leq (T(\gamma) \to T(\beta)) \to (t(\gamma) \to (T(\alpha) \wedge T(\beta))),$$

holds in L for every $T \in Sem(V, E)$ and all (V, E)-formulas α, β, γ.

This is so indeed because the formula

$$(z \to x) \otimes (z \to y) \leq (z \to x) \wedge (z \to y) = z \to (x \wedge y)$$

implies via conjugation

$$z \to x \leq (z \to y) \to (z \to (x \wedge y)).$$

The proof is similar in all other cases. ∎

Theorem 8.9

Let E be a complete residual lattice over a chain L and let, for $\alpha, \beta \in S(V, L, \Delta)$,

$$\lambda_0(\alpha, \beta) = (\alpha \Rightarrow \beta) \vee (\beta \Rightarrow \alpha)$$

and

$$\lambda_n(\alpha, \beta) = (\alpha \vee \beta)^n \Rightarrow (\alpha^n \vee \beta^n)$$

for integers $n \geq 1$.

The formulas $\lambda_0(\alpha, \beta)$ and $\lambda_n(\alpha, \beta)$ for $n \geq 1$ are (V, E)-tautologies, for every $\alpha, \beta \in S(V, L, \Delta)$.

Proof

For $\lambda_0(\alpha, \beta)$, since L is a chain, either $x \leq y$ or $y \leq x$ holds. In the first case we have $x \to y = 1$, and in the second, $y \to x = 1$.
In either case

$$(x \to y) \vee (y \to x) = 1$$

follows; hence

$$\emptyset \models_1 (\alpha \Rightarrow \beta) \vee (\beta \Rightarrow \alpha),$$

for all $\alpha, \beta \in S(V, L, \Delta)$.

As regards formulas $\lambda_n(\alpha, \beta)$, note that for any $x, y \in L$ we have $x \vee y \in \{x, y\}$, because L is a chain. Since the multiplication $\otimes$ is isotonic in L, we get $(x \vee y)^n = x^n \vee y^n$ for all $n \geq 1$. Consequently

$$\emptyset \models_1 (\alpha \vee \beta)^n \Rightarrow (\alpha^n \vee \beta^n)$$

holds for arbitrary $\alpha, \beta \in S(V, L, \Delta)$. ∎

Theorem 8.10

Let E be an extended complete residual lattice over the $(m+1)$-element chain

$$\mathrm{L} = (C_{m+1}, \otimes, \to), C_{m+1} = \{e_0, e_1, \ldots, e_m\},\ 0 = e_0 < e_1 < \ldots < e_m = 1.$$

Then, for any $\alpha \in S(V, C_{m+1}, \Delta)$ and each $k \in \{0, 1, \ldots, m-1\}$, the formula

$$\kappa_k(\alpha) = (\alpha \Rightarrow \bar{e}_k) \vee (\bar{e}_{k+1} \Rightarrow \alpha)$$

is a (V, E)-tautology.

Proof

Fix $e \in C_{m+1}$, $k \in \{0, 1, \ldots, m-1\}$ and $\alpha \in S(V, C_{m+1}, \Delta)$. Then either $e \leq e_k$ or $e_{k+1} \leq e$ holds, implying $e \to e_k = 1$ or $e_{k+1} \to e = 1$ respectively. In either case,

$$(e \to e_k) \vee (e_{k+1} \to c) = 1,$$

so that

$$\emptyset \models_1 (\alpha \Rightarrow \bar{e}_k) \vee (\bar{e}_{k+1} \Rightarrow e),$$

as needed. ∎

Theorem 8.11

Let E be an extended residual lattice over the Lukasiewicz interval L $= (I, \otimes, \rightarrow)$ and let $\alpha \in S(V, I, \Delta)$ be arbitrary.

(a_1) If $x, y, u, v \in I$ and the inequalities $0 \leq u < x, y < v < 1, n \cdot x + y \leq n \cdot u + v$ are satisfied for an integer $n \geq 1$, then the formula

$$((\bar{x} \Rightarrow \alpha)^n \Rightarrow \bar{y}) \Rightarrow ((\bar{u} \Rightarrow \alpha)^n \Rightarrow \bar{v})$$

is a (V, E)-tautology.

(a_2) If $x, y, u, v \in I$ and the inequalities $x < u \leq 1, y < v < 1, n \cdot u - v \leq n \cdot x - y$ are satisfied for an integer $n \geq 1$, then the formula

$$((\alpha \Rightarrow \bar{x})^n \Rightarrow \bar{y}) \Rightarrow ((\alpha \Rightarrow \bar{u}) \Rightarrow \bar{v})$$

is a (V, E)-tautology.

Proof

In a lattice over the Lukasiewicz interval, the inequality

$$(x \rightarrow y)^n \rightarrow u = 1 \wedge (u + (0 \vee n \cdot (x - y)))$$

is valid. On account of the inequalities postulated in the conditions of claims (a_1) and (a_2) we obtain

$$(x \rightarrow e)^n \rightarrow y \leq (u \rightarrow e)^n \rightarrow v$$

in case (a_1), and

$$(e \rightarrow x)^n \rightarrow y \leq (e \rightarrow u)^n \rightarrow v$$

in case (a_2). This ends the proof. ■

Recall that an L-consequence Cn in the set S is said to be compact if, for any L-set $X : S \rightarrow L$ and any $\alpha \in S$, there exists a finite set $Q \subseteq S$ such that

$$(Cn(\alpha))(x) = (Cn(X \restriction Q))(\alpha),$$

where $\restriction$ is the restriction operation (the fuzzy set $X \restriction Q$ coincides with X on Q and is equated to zero off Q).

Theorem 8.12 [Pavelka 1979]

Let E be an extended complete residual lattice over L $= (L, \otimes, \rightarrow)$. If the operation of L-consequence $Cn_{Sem(V,\mathrm{E})}$ on $S(V, L, \Delta)$ is compact, then L does not contain an infinite ascending or descending chain.

Proof

For an indirect proof, assume that L contains an infinite ascending chain $x_1 < x_2 < x_3 < \ldots$. Consider $x = \bigvee\{x_n : n \in N, n \geq 1\}$. Choose an infinite sequence of distinct propositional variables $p_0, p_1, p_2, \ldots$ and define

$$X(\alpha) = \begin{cases} 1 & \text{if} \quad \alpha = p_n \Rightarrow p_0 \quad \text{for some integer} \quad n \geq 1, \\ x_n & \text{if} \quad \alpha = p_n \quad \text{for some integer} \quad n \geq 1, \\ 0 & \text{in other cases.} \end{cases}$$

Then $(Cn_{Sem(V,E)}(X))(p_0) = x$, while $(Cn_{Sem(V,E)}(X \upharpoonright Q))(p_0) \leq x_{cardQ} < x$ for every finite set $Q \subseteq S$, contrary to the compactness of $Cn_{Sem(V,E)}$.

Now assume that L contains an infinite descending chain $x_1 > x_2 > x_3 > \ldots$. Setting $x = \bigwedge\{x_n : n \in N, n \geq 1\}$ and choosing a variable $p \in V$ we can define an L-set $X : S \to L$ by

$$X(\alpha) = \begin{cases} 1 & \text{if } \alpha = p \Rightarrow \bar{x}_n \quad \text{for an integer} \quad n \geq 1, \\ 0 & \text{otherwise.} \end{cases}$$

Obviously, $X \models_1 p \Rightarrow \bar{x}$. On the other hand,

$$(Cn_{Sem(V,E)}(X \upharpoonright Q))(p \Rightarrow \bar{x}) \leq x_{cardQ} \to x < 1$$

for every finite $Q \subseteq S$, contrary to compactness. The proof is complete. ∎

8.5 Remarks on the Incompleteness of Fuzzy Propositional Calculi

We have pointed out several times that a basis for the construction of any logical system has to be constituted by a certain algebraic structure. The same viewpoint has been emphasised also by J. Pavelka in his paper [Pavelka 1979] concerning a broad class of fuzzy logics.

Having accepted an algebraic structure to be taken for the semantics of the logic in question, our next step is to axiomatize the system (axiomatization need not necessarily be understood in the sense of Hilbert's method). The correctness of an axiomatization is then established by a suitable completess theorem.

Now, it is often the case that some fuzzy logics fail to be axiomatizable in that sense. A thorough discussion of this question can be found in the papers by J. Pavelka [Pavelka 1979] and J. Menu and J. Pavelka [Menu and Pavelka 1976]. The results reported below are taken from [Pavelka 1979], as are also the examples that follow. All concepts, notation and assumptions are adopted from the same paper. Of course, the axiomatization of a fuzzy (V, E)-calculus depends on the underlying structure E, that is to say, on the specific complete residual lattice $\mathrm{L} = (L, \otimes, \to)$.

Theorem 8.13

Suppose L is a complete lattice which contains an infinite descending chain and does not contain an infinite ascending chain. No fuzzy propositional (V, E)-calculus is then axiomatizable.

Proof

Assume, conversely, that there exists an L-syntax (A, R) on the set $S(V, L, \Delta)$ such that the two L-consequences $Cn_{A,R}$ and $Cn_{Sem(V,E)}$ coincide. By adjoining to R the rule

$$(\mathrm{r}_0) \qquad \frac{\alpha, \alpha}{\alpha}, \quad \frac{x, y}{x \vee y}$$

we do not change the operation $Cn_{A,R}$. Thus $Cn_{A,R\cup\{r_0\}} = Cn_{Sem(V,E)}$. Hence, for any $X : S(V, L, \Delta) \to L$ and any formula α there exists an $R \cup \{r_0\}$-proof in $S(V, L, \Delta$ which "tends" to α and whose degree of incidence to X equals $(Cn_{Sem(V,E)}(X))(\alpha)$. It follows that $Cn_{Sem(V,E)}$ is compact, which on account of Theorem 8.12 contradicts the assumption. ■

Example

Let N_ω denote the chain of natural numbers with the upper bound ω attached; let $\tilde{N}_\omega$ be the dual set to N_ω, so that 0 is the unit and ω is the zero of $\tilde{N}_\omega$. Define

$$(\mathrm{a}_1) \qquad x \otimes y = \begin{cases} x + y & \text{if } x, y \in N \\ \omega & \text{if } x = \omega \text{ or } y = \omega \end{cases}$$

$$(\mathrm{a}_2) \qquad x \to y = \begin{cases} 0 & \text{if } y \leq x & \text{in } N_\omega \\ y - x & \text{if } x < y < \omega & \text{in } N_\omega \\ \omega & \text{if } x < y = \omega & \text{in } N_\omega \end{cases}$$

$$(\mathrm{a}_3) \qquad x \otimes' y = \begin{cases} xy + x + y & \text{if } x, y \in N \\ \omega & \text{if } x = \omega \text{ or } y = \omega \end{cases}$$

$$(\mathrm{a}_4) \qquad x \to y = \begin{cases} 0 & \text{if } x \geq y & \text{in } N_\omega \\ \left[\frac{y-x}{x+1}\right] & \text{if } x < y < \omega & \text{in } N_\omega \\ \omega & \text{if } x < y = \omega & \text{in } N_\omega, \end{cases}$$

$[x]$ denoting the greatest integer not exceeding a rational number x. Evidently, the triples

$$\mathfrak{N}_\omega = (\tilde{N}_\omega, \otimes, \to)$$

and

$$\mathfrak{N}'_\omega = (\tilde{N}_\omega, \otimes', \to')$$

are residual lattices.

Since $\tilde{N}_\omega$ is a well ordered infinite ascending chain, it follows from Theorem 8.13 that neither $\mathfrak{N}_\omega$ nor $\mathfrak{N}'_\omega$ can provide a semantic for an $\tilde{N}_\omega$-valued propositional calculus.

Now, let C be any complete chain. We assign to C a cardinal number, denoted by $char(C)$ and defined as follows:

Let C be a complete chain. Then $char(C)$ will denote the cardinal number defined as follows:

(c_1) if C is descending and well ordered then $char(C) = 1$;

(c_2) if C fails to be a well ordered descending chain then $char(C)$ is the least cardinal ξ such that for every $x \in C \setminus \{0\}$ which has no predecessor in C there exists an ascending well ordered chain $B \subseteq C_x = \{y \in C \mid y < x\}$ with the following properties:

($c_2$1) B is cofinal with C_x,

($c_2$2) the ordinal type of B is a cardinal number not exceeding ξ.

Theorem 8.14

Let C be a complete chain and let (A, R) be a C-syntax on S (the set of formulas). For every C-subset $X : S \to C$ and each $\alpha \in S$ there exists a family $\mathfrak{M} = \{\omega^{(j)} : j \in J\}$ of R-proofs in S such that

(a_1) card $J \leq char(C)$;

(a_2) each $\omega^{(j)}$ tends to α;

(a_3) $(Cn_{A,R}(X))(\alpha) = \bigvee\{\omega^{(j)}(X) : j \in J\}$.

Proof

Note that for $X \in C^S$ and $\alpha \in S$

$$(Cn_{A,R}(X)) = \bigvee\{\omega(X) : \omega \text{ is an } R\text{-proof in } S, \text{ tending to } \alpha\}.$$

For a fixed C-set X and a fixed formula α consider the value $x = (Cn_{A,R}(X))(\alpha)$. Clearly, $x \in C$. If $x = 0$, then $X(x) = 0$ and $\mathfrak{M} = \{\ll x \gg\}$.

Thus assume $x > 0$. If x has a predecessor in C, then there is an R-proof $\omega^{(0)}$ in S which tends to α, such that $\omega^{(0)}(X) = x$. So we take $\mathfrak{M} = \{\omega^{(0)}\}$.

If $x = \bigvee C_x$, there exists a cardinal $\beta \leq char(C)$ and an embedding $\xi \longmapsto c_\xi$, $\xi < \beta$, into C_x, such that $x = \bigvee\{c_\xi : \xi < \beta\}$. For each $\xi < \beta$ we choose an R-proof $\omega^{(\xi)}$ in S which tends to α and whose value with respect to X satisfies the inequality

$$\omega^{(\xi)}(X) \geq c_\xi.$$

Taking

$$\mathfrak{M} = \{\omega^{(\xi)} : \xi < \beta\}$$

we obtain

$$x \leq \bigvee\{c_\xi : \xi < \beta\} \leq \bigvee\{\omega^{(\xi)}(X) : \xi < \beta\} \leq x,$$

and the proof is complete. ∎

Corollary 8.4

Suppose C is a complete chain and (A, R) is a C-syntax on S. The C-consequence operation $Cn = Cn_{A,R}$ has the following property:

(c) for every C-set $X : S \to C$ and every formula $\alpha \in S$,

$$(Cn(X))(\alpha) = \bigvee\{(Cn(X \upharpoonright G))(\alpha) : G \subseteq S,\ G \text{ finite}\}.$$

Proof

Given $X \in C^S$ and $\alpha \in S$, consider a family $\mathfrak{M} = \{\omega^{(j)} : j \in J\}$ with properties as in Theorem 8.14. Let $\omega^{(j)} = (\omega_2^{(j)}, \ldots, \omega_{m(j)}^{(j)})$ for $j \in J$. Then (writing Cn for $Cn_{A,R}$)

$$\begin{aligned}(Cn(X))(\alpha) &= \bigvee\{\omega^{(j)}(X) : j \in J\} \\ &\leq \bigvee\{(Cn(X \upharpoonright \{{}^*\omega_1^{(j)}, {}^*\omega_2^{(j)}, \ldots, {}^*\omega_{m(j)}^{(j)}\})(\alpha) : j \in J\} \\ &\leq \bigvee\{(Cn(X \upharpoonright G))(\alpha) : G \subseteq S, G \text{ finite}\} \\ &\leq (Cn(X))(\alpha) \quad \blacksquare\end{aligned}$$

Theorem 8.15

Let $\mathrm{L} = (C, \otimes, \to)$ be a complete residual chain. Consider C as topological space, with the order topology τ, having the family

$$B = \{\{z \in C : x < z < y\} : x, y \in C\}$$

for its base. If the residuum operation $\to$ fails to be a continuous mapping of $C \times C$ into C, then every (V, E)-calculus over L is nonaxiomatizable.

Proof

In every residual lattice, the residuum operation preserves lattice meets in the second variable and sends joins into meets in the first variable. We claim that the residuum $\to$ in the lattice $\mathrm{L} = (C, \otimes, \to)$ is a continuous mapping of $C \times C$ into C (with respect to topology τ) if and only if it satisfies, in addition to the lattice structure properties just mentioned, the following two conditions:

(a_1) for each $x \in C$, the function $y \mapsto (x \to y)$ from C into C preserves nonvoid joins;

(a_2) for each $x \in C$, the function $y \mapsto (y \to x)$ from C into C carries nonvoid meets into joins.

On account of Corollary 8.4, it will be enough to show that, wherever the residuum fails to satisfy (a_1) or (a_2), the C-consequence $Cn_{Sem(V,\mathrm{E})}$ does not have property (c).

Assume (a_1) does not hold. Then there are $x \in C$ and $B \neq \emptyset$ such that

$$z = \bigvee\{x \to y : y \in B\} < (x \to \bigvee B).$$

Define

$$X(\alpha) = \begin{cases} 1 & \text{if } \alpha = \overline{\beta} \to p_0,\ \beta \in B, \\ 0 & \text{otherwise,} \end{cases}$$

where p_0 is a selected element of V. Then

$$(Cn_{Sem(V,\mathrm{E})}(X))(\bar{x} \to p_0) = (x \to \bigvee B)$$

and we have

$$(Cn_{Sem(V,\mathrm{E})}(X \restriction G))(\bar{x} \to p_0) \leq z$$

for every finite $G \subseteq S(V, L, \Delta)$.

Hence, $Cn_{Sem(V,\mathrm{E})}$ does not satisfy condition (c) of Corollary 8.4.

In the case where the residuum does not fulfill (a_2), the proof is similar. ∎

In [Menu and Pavelka 1976] it is shown that every residual lattice over universe I, with residuum operation continuous on $I \times I$, is isomorphic to the Łukasiewicz interval $LL = (I, \otimes, \to)$. Consequently, fuzzy logics modelled over those residual lattices are the only instances of axiomatizable fuzzy logics.

In the sequel we present several results due to Pavelka concerning the rules of inference. For detailed proofs we refer to [Pavelka 1979].

A rule of inference (or rather an L-rule, to be precise) in the set $S(V, L, \Delta)$ is defined as a pair $r = (p, q)$ in which p is an n-ary partial function in $S(V, L, \Delta)$ and q is the corresponding n-ary function on L, preserving nonvoid lattice joins in each variable. The infallibility of a rule $r = (p, q)$ with respect to $Sem(V, \mathrm{E})$ is expressed by the following condition:

If $(\alpha_1, \alpha_2, \ldots, \alpha_n) \in Dp$ and $T \in S(V, \mathrm{E})$, then
$T(p(\alpha_1, \alpha_2, \ldots, \alpha_n)) \geq q(T(\alpha_1), T(\alpha_2), \ldots, T(\alpha_n))$.

We now state the basic properties of the notions just introduced:

(r_1) For every set V of propositional variables and every extended residual lattice $\mathrm{E} = (\mathrm{L}, \mathrm{O}), \mathrm{L} = (L, \otimes, \to)$, the L-rule of detachment $r_1 = (p_1, q_1)$ given by the schemes

$$p_1 : \frac{\alpha, \alpha \Rightarrow \beta}{\beta}, \qquad q_1 : \frac{x, y}{x \otimes y}$$

is infallible in $S(V, L, \Delta)$ with respect to $Sem(V, \mathrm{E})$.

(r_2) Let L be a complete lattice and let (A, R) be an L-syntax on S. Suppose that each $r = (p, q) \in R$ satisfies

$$q(x_1, x_2, \ldots, x_n) \leq x_1 \wedge x_2 \wedge \ldots \wedge x_n$$

for $x_1, x_2, \ldots, x_n \in L$. Then we have for every $X : S \to L$ and every $\alpha \in S$ the inequality

$$(Cn_{A,R}(X))(\alpha) \leq y \vee B(x),$$

where $B = Cn_{A,R}(\emptyset)$ and $y = \bigvee\{X(\beta) : \beta \in S\}$.

(r₃) For every set V of propositional variables and every extended residual lattice $\mathrm{E} = (\mathrm{L}, \mathrm{O}), \mathrm{L} = (L, \otimes, \to)$, the following is valid: If C is a complete chain such that, for every $x \in C$, the function $y \mapsto (x \to y)$ (from C to C) preserves nonvoid joins, then for every $x \in L$ the rule $r_2 x = (p_2 x, q_2 x)$ given by

$$p_2 x : \frac{\alpha}{x \Rightarrow \alpha}, \qquad q_2 x : \frac{y}{x \to y}$$

is a unary rule in $S(V, L, \Delta)$, infallible with respect to $Sem(V, \mathrm{E})$.

Now, let L be a bounded lattice and let L^* denote its dual lattice. We say that L^* is Brouwerian iff there exists a binary operation

$$(x, y) \mapsto x \vdash y$$

from $L \times L$ into L such that, for any $x, y, z \in L$:

(a₁) $x \vee y \geq z$ if and only if $x \geq y \vdash z$.

Obviously, a pair $(\vee, \vdash)$ satisfies (a_1) if and only if the inequalities

(a_1') $x \geq y \vdash (x \vee y)$,

(a_2') $(x \vdash y) \vee x \geq y$

hold in the lattice $(L, \vee, \vdash)$.

If L is a bounded chain then L^* is a Brouwerian lattice with $\vdash$ defined by

$$x \vdash y = \begin{cases} 0 & \text{if } x \geq y \text{ in } L, \\ y & \text{otherwise.} \end{cases}$$

(r₄) Let L be a complete lattice and suppose its dual L^* is Brouwerian. Then, for every set V, every E with support L, and every $x \in L$, the rule $r_3 = (p_3, q_3)$ of (L, x)-elimination given by the patterns

$$p_3 : \frac{\alpha \vee \bar{x}}{\alpha}, \qquad q_3 : \frac{y}{x \vdash y}$$

is a unary rule in $S(V, L, \Delta)$, infallible with respect to $Sem(V, \mathrm{E})$.

(r_5) For every set V, every lattice E with support L, and every $x \in L$, the L-rule $r_4 = (p_4, q_4)$ with $Dp_4 = \{\bar{x}\}$, $p_4(\bar{x}) = 0$ and

$$q_4(y) = \begin{cases} 0 & \text{if } y \leq x, \\ 1 & \text{otherwise,} \end{cases}$$

is a unary L-rule of inference in $S(V, L, \Delta)$, infallible with respect to $Sem(V, \mathrm{E})$.

Recall that $X \vdash_{x} \alpha$ is an abbreviated notation for $(Cn_{A,R}(X))(\alpha) \geq x$. When $x = \emptyset$, we write simply $\vdash_{x} \alpha$. Similarly, we write S rather than $S(V, L, \Delta)$ when it is clear what L and Δ are under consideration.

Theorem 8.16 (Completeness Theorem) [Pavelka 1979]

Let V be a set of propositional variables and let E be an extended complete residual lattice of type $(Ar : \Delta \to N \setminus \{0\}, Ex : \Delta \to N^*)$. Suppose (A, R) is an L-syntax on the set $S(V, L, \Delta)$ satisfying the following conditions:

(i_1) The rule $r_1 = (p_1, q_1)$ with

$$p_1 : \frac{\alpha, \alpha \Rightarrow \beta}{\beta}, \qquad q_1 : \frac{x, y}{x \otimes y}$$

is in R, and so are the rules $r_2x = (p_2x, q_2x)$,

$$p_2x : \frac{\alpha}{\bar{x} \Rightarrow \alpha}, \qquad q_2x : \frac{y}{x \to y},$$

for all $x \in L$;

(i_2) for any $x, y \in L$

$\vdash_{x} \bar{x}, \vdash_{x \otimes y} (\bar{x} \& \bar{y}), \vdash_{x \to y} (\bar{x} \Rightarrow \bar{y}),$

$\vdash_{1} (\bar{x} \wedge \bar{y}) \Rightarrow (\overline{x \wedge y}), \vdash_{1} (\bar{x} \Rightarrow \bar{y}) \Rightarrow \overline{x \to y};$

(i_3) if $d \in \Delta$, $Ar(d) = n$ and $x_1, x_2, \ldots, x_n \in L$,

then

$\vdash_{1} \bar{d}(\bar{x}_1, \bar{x}_2, \ldots, \bar{x}_n) \Leftrightarrow o_d(\bar{x}_1, \bar{x}_2, \ldots, \bar{x}_n);$

(i_4) for any $\alpha, \beta, \gamma \in S(V, L, \Delta)$

$\vdash_{1} \alpha \Rightarrow 1$

$\vdash_{1} \alpha \Rightarrow \alpha$

$\vdash_{\bar{1}} (\beta \Rightarrow \gamma) \Rightarrow ((\alpha \Rightarrow \beta) \Rightarrow (\alpha \Rightarrow \gamma))$
$\vdash_{\bar{1}} (\alpha \Rightarrow (\beta \Rightarrow \gamma)) \Rightarrow (\beta \Rightarrow (\alpha \Rightarrow \gamma))$
$\vdash_{\bar{1}} (\alpha \wedge \beta) \Rightarrow \alpha$
$\vdash_{\bar{1}} (\alpha \wedge \beta) \Rightarrow \beta$
$\vdash_{\bar{1}} (\gamma \Rightarrow \alpha) \Rightarrow ((\gamma \Rightarrow \beta) \Rightarrow (\gamma \Rightarrow \alpha \wedge \beta))$
$\vdash_{\bar{1}} \alpha \Rightarrow (\alpha \vee \beta)$
$\vdash_{\bar{1}} \beta \Rightarrow (\alpha \vee \beta)$
$\vdash_{\bar{1}} (\alpha \Rightarrow \gamma) \Rightarrow ((\beta \Rightarrow \gamma) \Rightarrow ((\alpha \vee \beta) \Rightarrow \gamma))$
$\vdash_{\bar{1}} \alpha \Rightarrow (\beta \Rightarrow (\alpha \& \beta))$
$\vdash_{\bar{1}} (\alpha \Rightarrow (\beta \Rightarrow \gamma)) \Rightarrow ((\alpha \Rightarrow \beta) \Rightarrow \gamma)$
$\vdash_{\bar{1}} (\alpha \& \bar{1}) \Leftrightarrow \alpha$

(i_5) if $d \in \Delta$, $Ar(d) = n$ and $\alpha_1, \alpha_2, \ldots, \alpha_n, \beta_1, \beta_2, \ldots, \beta_n \in S(V, L, \Delta)$, then

$$\vdash_{\bar{1}} ((\alpha_1 \Leftrightarrow \beta_1)^{k_1} \& (\alpha_2 \Leftrightarrow \beta_2)^{k_2} \& \ldots \& (\alpha_n \Leftrightarrow \beta_n)^{k_n}) \Rightarrow (\bar{d}(\alpha_1, \alpha_2, \ldots, \alpha_n) \Leftrightarrow \bar{d}(\beta_1, \beta_2, \ldots, \beta_n)).$$

Under these assumptions, the following claims are valid for every $X : S(V, L, \Delta) \to L$:

(s_1) The relation $\leq$ in $S(V, L, \Delta)$ defined by

$$\alpha \leq \beta \quad \text{iff} \quad X \vdash_{\bar{1}} \alpha \Rightarrow \beta$$

is a quasi-ordering with the property:
for any $x \in L$ and $\alpha \in S(V, L, \Delta)$,

$$\bar{x} \leq \alpha \quad \text{iff} \quad X \vdash_{\bar{x}} \alpha.$$

(s_2) The equivalence relation $\approx$ in $S(V, L, \Delta)$ defined by

$$\alpha \approx \beta \quad \text{iff} \quad \alpha \leq \beta \quad \text{and} \quad \beta \leq \alpha$$

is a congruence in the structure $S(V, \mathrm{E})$.

(s_3) For $\alpha \in S(V, L, \Delta)$ denote by $[\alpha]$ the coset modulo $\approx$ represented by α, and let $S(X) = S(V, L, \Delta)/\approx$. Further let

$$\bar{S}(X) = (S(X), \{[\bar{x}] : x \in L\}, \wedge, \vee, \otimes, \to, \{[\bar{d}] : d \in \Delta\})$$

be the quotient structure of $S(V, \mathrm{E})$ modulo congruence $\approx$.

Then the algebra

$$\mathrm{E}(X) = (S(X), [0], [1], \wedge, \vee, \otimes, \rightarrow, \{[\bar{d}\,] : d \in \Delta\})$$

is an extended residual lattice of the same type as E.

(s_4) The mapping j from L to $S(X)$ defined by

$$j(x) = [x]$$

extends uniquely to a lattice homomorphism of E into E(X) preserving all lattice joins.

Proof

Write S for $S(V, L, \Delta)$. Evidently, for every $X \in L^S$ and any $\alpha \in S, x \in L$, if $\vdash_{\bar{x}} \alpha$ then $X \vdash_{\bar{x}} \alpha$. For every $\alpha \in S$ we have $\vdash_{\bar{1}} \alpha \Rightarrow \alpha$, hence $X \vdash_{\bar{1}} \alpha \Rightarrow \alpha$, and so $\alpha \leq \alpha$.

Assuming $\alpha \leq \beta$ and $\beta \leq \gamma$ we have:

1. $X \vdash_{\bar{1}} \beta \Rightarrow \gamma$ (by assumption),
2. $X \vdash_{\bar{1}} (\beta \Rightarrow \gamma) \Rightarrow ((\alpha \Rightarrow \beta) \Rightarrow (\alpha \Rightarrow \gamma))$,
3. $X \vdash_{\overline{q_1(1,1)=1}} (\alpha \Rightarrow \beta) \Rightarrow (\alpha \Rightarrow \gamma)$ (by (r_1), 1, 2),
4. $X \vdash_{\bar{1}} (\alpha \Rightarrow \beta)$ (by assumption),
5. $X \vdash_{\bar{1}} (\alpha \Rightarrow \gamma)$ (by (r_1), 3, 4);

hence $\alpha \leq \gamma$, proving that $\leq$ is a quasi-order in S.

If $\bar{x} \leq \alpha$ then applying rule r_1 to $X \vdash_{\bar{x}} \bar{x}$ and $X \vdash \bar{x} \Rightarrow \alpha$ we obtain

$$X \vdash_{\overline{q_1(x,1)=x}} \alpha.$$

And conversely, if $X \vdash_{\bar{x}} \alpha$ then applying $r_2 x$ we get $X \vdash_{\bar{1}} \bar{x} \Rightarrow \alpha$. This ends the proof of (s_1).

Since $X \vdash_{\bar{0}} \alpha$ holds for any $\alpha \in S$, we get $\bar{0} \leq \alpha$. Similarly, $\alpha \leq \bar{1}$ holds for every $\alpha \in S$, in view of $\vdash_{\bar{1}} \alpha \Rightarrow 1$. So

$$\bar{0} \leq \alpha \leq \bar{1} \quad \text{for} \quad \alpha \in S.$$

By (i_4),

$$\vdash_{1} \alpha \wedge \beta \Rightarrow \alpha,$$
$$\vdash_{1} \alpha \wedge \beta \Rightarrow \beta.$$

Since $\vdash_{x} \alpha$ forces $X \vdash_{x} \alpha$, we get

$$\alpha \wedge \beta \leq \alpha,$$
$$\alpha \wedge \beta \leq \beta.$$

Now assume $\gamma \leq \alpha, \gamma \leq \beta$. Then

$$X \vdash_{1} \gamma \Rightarrow \alpha,$$
$$X \vdash_{1} \gamma \Rightarrow \beta,$$

and by (i_4)

$$\vdash_{1} (\gamma \Rightarrow \alpha) \Rightarrow ((\gamma \Rightarrow \beta) \Rightarrow (\gamma \Rightarrow (\alpha \wedge \beta)).$$

Applying r_1 twice we obtain

$$X \vdash_{1} \gamma \Rightarrow (\alpha \wedge \beta),$$

implying $\gamma \leq \alpha \wedge \beta$, hence showing that $\alpha \wedge \beta$ corresponds to the greatest lower bound of $\{\alpha, \beta\}$ in $(S, \leq)$.

Similar reasoning shows that $\alpha \vee \beta$ corresponds to the least upper bound of $\{\alpha, \beta\}$. So (s_2) is proved.

Consider equivalence classes $[\alpha], [\beta] \in S(X)$ and define

$$[\alpha] \leq [\beta] \quad \text{iff} \quad \alpha \leq \beta.$$

We get an ordering relation in $S(X)$. Since the connectives $\wedge, \vee$ are preserved under $\leq$, they are preserved under $\approx$ as well. So we may write

$$[\alpha] \wedge [\beta] = [\alpha \wedge \beta],$$
$$[\alpha] \vee [\beta] = [\alpha \vee \beta];$$

the structure

$$L(X) = (S(X), [0], [1], \wedge, \vee)$$

is a bounded lattice.

We have already shown that $\beta \leq \gamma$ implies $\alpha \Rightarrow \beta \leq \alpha \Rightarrow \gamma$; implication $\Rightarrow$ is order-preserving in the second argument. Applying the rule of detachment to

$$X \vdash_{1} (\beta \Rightarrow \gamma) \Rightarrow ((\alpha \Rightarrow \beta) \Rightarrow (\alpha \Rightarrow \gamma))$$

and to

$$X \vdash_{1} ((\beta \Rightarrow \gamma) \Rightarrow ((\alpha \Rightarrow \beta) \Rightarrow (\alpha \Rightarrow \gamma)) \Rightarrow ((\alpha \Rightarrow \beta) \Rightarrow ((\beta \Rightarrow \gamma) \Rightarrow (\alpha \Rightarrow \gamma)))$$

we get

$$X \vdash_{1} (\alpha \Rightarrow \beta) \Rightarrow ((\beta \Rightarrow \gamma) \Rightarrow (\alpha \Rightarrow \gamma)).$$

Hence, if $\alpha \leq \beta$ then $\beta \Rightarrow \gamma \leq \alpha \Rightarrow \gamma$ for every $\gamma \in S$. Consequently, $\Rightarrow$ viewed as an operation in the ordered set $(S, \leq)$ is antitonic in the first argument and isotonic in the second.

Let $\alpha, \beta, \gamma \in S$. In view of
$$X \vdash_{\overline{1}} \alpha \Rightarrow (\beta \Rightarrow (\alpha \& \beta))$$
we obtain
$$\alpha \leq (\beta \Rightarrow (\alpha \& \beta)).$$
Further, assuming there exists $\gamma \in S$ with $\alpha \leq \beta \Rightarrow \gamma$, we can apply detachment to
$$X \vdash_{\overline{1}} \alpha \Rightarrow (\beta \Rightarrow \gamma)$$
and
$$X \vdash_{\overline{1}} (\alpha \Rightarrow (\beta \Rightarrow \gamma)) \Rightarrow ((\alpha \Rightarrow \beta) \Rightarrow \gamma)$$
to obtain
$$X \vdash_{\overline{1}} (\alpha \& \beta) \Rightarrow \gamma,$$
i.e., $\alpha \& \beta \leq \gamma$. Thus, given $\alpha, \beta \in S$, the formula $\alpha \& \beta$ is the least (up to $\approx$ equivalence) element in $(S, \leq)$ such that $\alpha \leq \beta \Rightarrow \gamma$. Hence, & is an isotonic operation in $(S, \leq)$, and the pair of operations $(\&, \Rightarrow)$ is an adjoint pair in $(S, \leq)$. It follows that the equivalence relation $\approx$ is preserved under & and $\Rightarrow$; defining
$$[\alpha] \otimes [\beta] = [\alpha \& \beta],$$
$$[\alpha] \rightarrow [\beta] = [\alpha \Rightarrow \beta],$$
for any $\alpha, \beta \in S$, we obtain a pair of operations $(\otimes, \rightarrow)$ which is an adjoint pair in $(S(X), \leq)$. Since
$$X \vdash_{\overline{1}} (\beta \Rightarrow \gamma) \Rightarrow ((\alpha \Rightarrow \beta) \Rightarrow (\alpha \Rightarrow \gamma)),$$
the inequality
$$[\beta] \rightarrow [\gamma] \leq ([\alpha] \rightarrow [\beta]) \rightarrow ([\alpha] \rightarrow [\gamma])$$
holds for every $[\alpha], [\beta], [\gamma] \in S(X)$. Moreover, we have the obvious inequalities
$$([\alpha] \otimes [\beta]) \otimes [\gamma] \leq [\alpha] \otimes ([\beta] \otimes [\gamma])$$
(holding for all $[\alpha], [\beta], [\gamma]$ in $S(X)$, in view of Theorem 8.14), and
$$[\alpha] \leq [\beta] \rightarrow [\gamma] \quad \text{iff} \quad [\beta] \leq [\alpha] \rightarrow [\gamma].$$
It is also not hard to see that the operation $\otimes$ is commutative and associative in $S(X)$.

By the rule of detachment r_1, applied to
$$X \vdash_{\overline{1}} \alpha \& 1 \Rightarrow \alpha,$$
$$X \vdash_{\overline{1}} ((\alpha \Rightarrow \beta) \wedge (\beta \Rightarrow \alpha)) \Rightarrow (\alpha \Rightarrow \beta),$$
$$X \vdash_{\overline{1}} ((\alpha \Rightarrow \beta) \wedge (\beta \Rightarrow \alpha)) \Rightarrow (\beta \Rightarrow \alpha),$$
we get for every $[\alpha] \in S(X)$
$$[\alpha] \otimes [1] = [\alpha].$$
Therefore the structure $(S(X), \otimes, 1)$ is a commutative monoid, and
$$\mathrm{L}(X) = (L(X), \otimes, \rightarrow)$$
is a residual lattice.

We are now going to show, for any formulas $\alpha_1, \alpha_2, \ldots, \alpha_n \in S$ and any $x_1, x_2, \ldots, x_n \in L$, that if $X \vdash_{x_i} \alpha_i$ for $i = 1, 2, \ldots, n$, then

$$X \vdash_{x_1 \otimes x_2 \otimes \ldots \otimes x_n} \alpha_1 \& \alpha_2 \& \ldots \& \alpha_n.$$

We use induction on n. If $n = 1$, there is nothing to prove. Assume the claim is true for an $n = k$ and suppose $X \vdash_{x_i} \alpha_i$ holds for $i = 1, 2, \ldots, k, k+1$. Let β denote the formula $\alpha_1 \& \alpha_2 \& \ldots \& \alpha_k$ and y denote the product $x_1 \otimes x_2 \otimes \ldots \otimes x_k$. According to assumptions,

$$\begin{aligned} &X \vdash_{y} \beta, \\ &X \vdash_{x_{k+1}} \alpha_{k+1}, \\ &X \vdash_{1} \beta \Rightarrow (\alpha_{k+1} \Rightarrow (\beta \& \alpha_{k+1})). \end{aligned}$$

Applying detachment (twice), we get the inductive claim.

As a result we obtain the following n-ary L-rule of inference $r = (p, q)$:

$$p : \frac{\alpha_1, \alpha_2, \ldots, \alpha_n}{\alpha_1 \& \alpha_2 \& \ldots \& \alpha_n}, \qquad q : \frac{x_1, x_2, \ldots, x_n}{x_1 \otimes x_2 \otimes \ldots \otimes x_n},$$

for any $n \geq 1$.

Consider formulas $\alpha, \beta \in S$. By definition, if $\alpha \approx \beta$, then $1 \leq \alpha \Rightarrow \beta$ and $1 \leq \beta \Rightarrow \alpha$, hence $1 \leq (\alpha \Rightarrow \beta) \wedge (\beta \Rightarrow \alpha)$, hence $X \vdash_{1} \alpha \Leftrightarrow \beta$.

Now, let $d \in \Delta$, $Ar(d) = n$, let $\alpha_1, \alpha_2, \ldots, \alpha_n, \beta_1, \beta_2, \ldots, \beta_n \in S$ and suppose $\alpha_i \approx \beta_i$ for $i = 1, 2, \ldots, n$. Then $X \vdash_{1} \alpha_i \Leftrightarrow \beta_i$ and

$$X \vdash_{1} (\alpha_1 \Leftrightarrow \beta_1)^{k_1} \& (\alpha_2 \Leftrightarrow \beta_2)^{k_2} \& \ldots \& (\alpha_n \Leftrightarrow \beta_n)^{k_n}$$

for $(k_1, k_2, \ldots, k_n) \in Ex(d)$. From

$$\begin{aligned} X \vdash_{1} ((\alpha_1 \Leftrightarrow \beta_1)^{k_1} \& (\alpha_2 \Leftrightarrow \beta_2)^{k_2} \& \ldots \& (\alpha_n \Leftrightarrow \beta_n)^{k_n}) \Rightarrow \\ (\bar{d}(\alpha_1, \alpha_2, \ldots, \alpha_n) \Leftrightarrow \bar{d}(\beta_1, \beta_2, \ldots, \beta_n)) \end{aligned}$$

we get by detachment

$$X \vdash_{1} \bar{d}(\alpha_1, \alpha_2, \ldots, \alpha_n) \Leftrightarrow \bar{d}(\beta_1, \beta_2, \ldots, \beta_n),$$

i.e.,

$$d(\alpha_1, \alpha_2, \ldots, \alpha_n) \approx d(\beta_1, \beta_2, \ldots, \beta_n).$$

This means that all n-ary connectives $\bar{d}$ induced by $d \in \Delta$ are $\approx$-preserving. Consider biresiduation $\leftrightarrow$ in $\mathrm{L}(X)$,

$$\begin{aligned} &[\alpha] \leftrightarrow [\beta] = [\alpha \Leftrightarrow \beta]; \\ &[\alpha] \leftrightarrow [\beta] = ([\alpha] \rightarrow [\beta]) \wedge ([\beta] \rightarrow [\alpha]). \end{aligned}$$

We see that the operation $\bar{d}^*$ in $S(X)$ given by

$$\begin{aligned} \bar{d}^*([\alpha_1], [\alpha_2], \ldots, [\alpha_n]) = [\bar{d}(\alpha_1, \alpha_2, \ldots, \alpha_n)] \\ \text{for } [\alpha_1], [\alpha_2], \ldots, [\alpha_n] \in S(X) \end{aligned}$$

is well defined.

So we arrive at the conclusion that the structure

$$\mathrm{E}(X) = (\mathrm{L}(X), \{\bar{d}^* : d \in \Delta\})$$

is an extended residual lattice of type

$$(Ar : \Delta \to N \setminus \{0\},\ Ex : \Delta \to N^*).$$

This settles claim (s_3).

Finally, let $x, y \in L$ and assume $x \leq y$ and $X \vdash_{y} \bar{y}$. Then $X \vdash_{x} \bar{y}$, so $\bar{x} \leq \bar{y}$, showing that j is an isotonic map:

$$j(x) = [\bar{x}] \leq [\bar{y}] = j(y).$$

Hence

$$j(x \wedge y) = j(x) \wedge j(y),$$
$$j(x \to y) = j(x) \to j(y),$$
$$j(x \otimes y) \leq j(x) \otimes j(y)$$

for all $x, y \in L$. To conclude that j is a homomorphism of lattice L into L(X), we need show $j(x) \otimes j(y) \leq j(x \otimes y)$. This inequality is equivalent to $j(x) \leq j(y) \to j(x \otimes y)$, and the latter follows from $x \leq y \to (x \otimes y)$, by the isotonicity of j.

Consider any $d \in \Delta, Ar(d) = n$, and suppose

$$X \vdash_{1} \bar{d}(\bar{x}_1, \bar{x}_2, \ldots, \bar{x}_n) \Leftrightarrow \bar{o}_d(\bar{x}_1, \bar{x}_2, \ldots, \bar{x}_n)$$

holds for all $x_1, x_2, \ldots, x_n \in L$. Then

$$o_d(x_1, x_2, \ldots, x_n)) = \bar{d}^*(j(x_1), j(x_2), \ldots, j(x_n)),$$

showing that j is a homomorphism between the lattices E and E(X). Now assume $K \subseteq L$ and let α be any formula in S. If $\bar{x} \leq \alpha$ for all $x \in K$, then of course $\bigvee \bar{K} \leq \alpha$, where we have denoted $\bar{K} = \{\bar{x} : x \in K\}$. Therefore

$$(Cn_{A,R}(X))(\alpha) = \bigvee\{x \in L : X \vdash_{x} \alpha\},$$

and this completes the proof of the theorem. ∎

Theorem 8.17

Assume conditions (i_1)–(i_5) of Theorem 8.16. Moreover, assume one of the following two hypotheses concerning the lattice L and syntax (A, R):

(i_1) the class of rules $\{rx : x \in L\}$ is contained in R;

(i_2) each element $x \in L \setminus \{0, 1\}$ is nilpotent in the semi-group $(L, \otimes)$.

Then, for every $X : S(V, L, \Delta) \to L$, either $j : \mathrm{E} \to \mathrm{E}(X)$ is an injection or E(X) is degenerate.

Proof

Let $X : S \to L$ and suppose j is not injective. So there are distinct elements $x, y \in L$ with $j(x) = j(y)$, i.e., such that

$$x \leftrightarrow y = 1 \quad \text{yields} \quad j(x \leftrightarrow y) = j(x) \leftrightarrow j(y) = j(1).$$

If R contains all the rules rx, we apply rule $r(x \leftrightarrow y)$ to $X\vdash_1 \bar{x} \Leftrightarrow \bar{y}$, thus obtaining $X\vdash_1 \bar{0}$. This means that $j(1) \leq j(0)$, and so $\mathrm{E}(X)$ is degenerate.

On the other hand, if all elements in $L\setminus\{0,1\}$ are nilpotent, we find a natural $n \geq 1$ such that $(x \leftrightarrow y)^n = 0$ in L. Then

$$j(1) = (j(1))^n = (j(x \leftrightarrow))^n = j((x \leftrightarrow y)^n) = j(0),$$

again showing that $\mathrm{E}(X)$ is degenerate. ∎

Theorem 8.18

Assume all conditions of Theorems 8.16 and 8.17. Let $X : S(V, L, \Delta) \to L$ be such that $\mathrm{E}(X)$ is non-degenerate. Then:

(i_1) Every filter F in $\mathrm{L}(X)$ with the property

$$\mathrm{F} \cap j(L) = \{j(1)\}$$

can be extended to a filter G in $\mathrm{L}(X)$, maximal with respect to inclusion, and such that

$$\mathrm{G} \cap j(L) = \{j(1)\}.$$

(i_2) Suppose G is a maximal filter in $\mathrm{L}(X)$ satisfying $\mathrm{G} \cap j(L) = \{j(1)\}$, and let $\alpha \in S(X)$. Then $\alpha \notin \mathrm{G}$ if and only if there exist $u \in \mathrm{G}, x \in L$ and $n \in N \setminus \{0\}$ such that $x < 1$ and $\alpha^n \otimes u \leq j(x)$.

Proof

Let Z be a chain of filters in $\mathrm{L}(X)$, each $\mathrm{F} \in Z$ satisfying $\mathrm{F} \cap j(L) = \{j(1)\}$. Obviously, $\bigcup Z$ is a filter in $\mathrm{L}(X)$ and satisfies

$$\bigcup Z \cap j(L) = \bigcup(\mathrm{F} \cap j(L) : \mathrm{F} \in Z) = \{j(1)\}.$$

By the Kuratowski–Zorn Lemma (Theorem 1.1) there exists a maximal filter with the property in question. This settles (i_1).

For a proof of (i_2), let G have properties as stated; we will call G a j-ultrafilter. Clearly, for every $\alpha \notin \mathrm{G}$ there exist u, x and n satisfying the claim of (i_2) because $j(\alpha) \notin \mathrm{G}$. And conversely, if $\alpha \in S(X)$ and there are no u, x, n as needed, then

$$\mathrm{G} = \{y \in S(X) : \exists u \in \mathrm{G}\ \exists n \in N \setminus \{0\}(y \leq \alpha^n \otimes u)\}$$

is a filter in $\mathrm{L}(X)$ with $\mathrm{G} \cap j(L) = \{j(1)\}$ and $\mathrm{G} \cup \{\alpha\} \subseteq \mathrm{G}'$, whence by maximality $\alpha \in \mathrm{G}$, proving (i_2). ∎

Theorem 8.19

Assume all conditions of Theorems 8.15 and 8.16. Suppose L is a chain and suppose

$$\vdash_{\bar{1}} (\alpha \vee \beta)^n \Rightarrow (\alpha^n \vee \beta^n)$$

for all $\alpha, \beta \in S(V, L, \Delta)$ and $n \in N$. Then:

(i_1) For every $X : S(V, L, \Delta) \to L$, every $\alpha, \beta \in S(X)$ and every $n \in N \setminus \{0\}$, the following equalities hold in the residual lattice L(X):

$$(\alpha \to \beta) \vee (\beta \to \alpha) = j(1),$$
$$(\alpha \vee \beta)^n = \alpha^n \vee \beta^n.$$

(i_2) If $X : S(V, L, \Delta) \to L$ is such that E(X) is non-degenerate, then every j-ultrafilter G in E(X) is prime; that is, for every $\alpha, \beta \in S(X)$,

$$\alpha \vee \beta \in \mathrm{G} \quad \text{iff} \quad \alpha \in \mathrm{G} \quad \text{or} \quad \beta \in \mathrm{G}.$$

Proof

Statement (i_1) follows immediately from Theorem 8.18 and from

$$\vdash_{\bar{1}} (\alpha \vee \beta)^n \Rightarrow (\alpha^n \vee \beta^n).$$

For (i_2), let X and G be as stated and suppose that neither α nor β is in G. So there exist $u, v \in$ G, $x, y \in L \setminus \{1\}$ and $m, n \in N \setminus \{0\}$ with

$$\alpha^n \otimes u \leq j(x), \quad \beta^m \otimes v \leq j(y).$$

Write $k = \max(n, m)$. Then

$$\begin{aligned}(\alpha \vee \beta)^k \otimes (u \otimes v) &= (\alpha^k \vee \beta^k) \otimes (u \otimes v)\\ &= (\alpha^n \otimes u) \vee (\beta^m \otimes v)\\ &\leq (\alpha^n \otimes u) \vee (\beta^m \otimes v)\\ &\leq j(x) \vee j(y) = j(x \vee y).\end{aligned}$$

Since G is a filter and L is a chain, $u \otimes v \in$ G and $x \vee y < 1$. Hence, again by Theorem 8.18, $\alpha \vee \beta$ does not belong to G. This ends the proof of (i_2). ∎

In what follows we assume that we are given an extended complete residual lattice E $= ((C, \otimes, \to), \sigma)$, which is a chain, and a C-syntax (A, R) over $S(V, C, \Delta)$, satisfying the following hypotheses:

(s_1) R contains the rule of detachment r_1 and the class of rules $\{rx : x \in C\}$.

(s_2) For every $x, y \in C$,

11. $\vdash_{x} \bar{x}$
12. $\vdash_{x \otimes y} \bar{x} \& \bar{y}$

13. $\vdash_{\overline{x \to y}} \bar{x} \Rightarrow \bar{y}$

14. $\vdash_{\bar{1}} (\bar{x} \wedge \bar{y}) \Rightarrow \overline{(x \wedge y)}$

15. $\vdash_{\bar{1}} (\bar{x} \Rightarrow \bar{y}) \Rightarrow \overline{(x \to y)}$

(s_3) If $d \in \Delta, Ar(d) = n, x_1, x_2, \ldots, x_n \in C$, then

$$\vdash_{\bar{1}} \bar{d}(\bar{x}_1, \bar{x}_2, \ldots, \bar{x}_n) \Leftrightarrow o_d(\bar{x}_1, \bar{x}_2, \ldots, \bar{x}_n).$$

(s_4) For every $\alpha, \beta, \gamma \in S(V, C, \Delta)$,

a1. $a \Rightarrow \bar{1}$

a2. $\alpha \Rightarrow \alpha$

a3. $\vdash_{\bar{1}} (\beta \Rightarrow \gamma) \Rightarrow ((\alpha \Rightarrow \beta) \Rightarrow (\alpha \Rightarrow \gamma))$

a4. $\vdash_{\bar{1}} (\alpha \Rightarrow (\beta \Rightarrow \gamma)) \Rightarrow (\beta \Rightarrow (\alpha \Rightarrow \gamma))$

a5. $\vdash_{\bar{1}} (\alpha \wedge \beta) \Rightarrow \alpha$

a6. $\vdash_{\bar{1}} (\alpha \wedge \beta) \Rightarrow \beta$

a7. $\vdash_{\bar{1}} (\gamma \Rightarrow \alpha) \Rightarrow ((\gamma \Rightarrow \beta) \Rightarrow (\gamma \Rightarrow (\alpha \wedge \beta))$

a8. $\vdash_{\bar{1}} \alpha \Rightarrow (\alpha \vee \beta)$

a9. $\beta \Rightarrow (\alpha \vee \beta)$

a10. $\vdash_{\bar{1}} (\alpha \Rightarrow \gamma) \Rightarrow ((\beta \Rightarrow \gamma) \Rightarrow ((\alpha \vee \beta) \Rightarrow \gamma))$

a11. $\vdash_{\bar{1}} \alpha \Rightarrow (\beta \Rightarrow (\alpha \& \beta))$

a12. $\vdash_{\bar{1}} (\alpha \Rightarrow (\beta \Rightarrow \gamma)) \Rightarrow ((\alpha \Rightarrow \beta) \Rightarrow \gamma)$

a13. $\vdash_{\bar{1}} (\alpha \& \bar{1}) \Leftrightarrow \alpha$

(s_5) If $d \in \Delta, Ar(d) = n, \alpha_1, \alpha_2, \ldots, \alpha_n, \beta_1, \beta_2, \ldots, \beta_n \in S(V, C, \Delta)$, then

d1. $\vdash_{\bar{1}} ((\alpha_1 \Leftrightarrow \beta_1)^{k_1} \& (\alpha_2 \Leftrightarrow \beta_2)^{k_2} \& \ldots \& (\alpha_n \Leftrightarrow \beta_n)^{k_n})$
$\Rightarrow (\bar{d}(\alpha_1, \alpha_2, \ldots, \alpha_n) \Leftrightarrow \bar{d}(\beta_1, \beta_2, \ldots, \beta_n))$
for $(k_1, k_2, \ldots, k_n) \in Ex(d)$.

(s_6) Each $x \in C \setminus \{0, 1\}$ is a nilpotent element of the semi-group $(C, \otimes)$.

(s_7) For every $\alpha, \beta \in S(V, C, \Delta)$ and $n \in N$,

c1. $\vdash_{\bar{1}} (\alpha \vee \beta)^n \Rightarrow (\alpha^n \vee \beta^n)$.

Lemma 8.7

A C-syntax (A, R) satisfying hypotheses (s_1)–(s_7) is consistent with respect to $Sem(V, \mathrm{E})$.

Proof

For any $X : S \to C$ and $\alpha \in S$,

$$(Cn_{A,R}(X))(\alpha) \leq (Cn_{Sem(V,\mathrm{E})}(X))(\alpha).$$

Fix a set $X : S \to C$ and a formula $\alpha \in S$. If $(Cn_{A,R}(X))(\alpha) = 1$, then the inequality above becomes an equality. Thus assume

$$(Cn_{A,R}(X))(\alpha) = x < 1.$$

Since C is a chain,

$$(Cn_{Sem(V,\mathrm{E})}(X))(\alpha) = \bigwedge\{T(\alpha) : T \in Sem(V, \mathrm{E}), T \geq X\}.$$

Define y to be equal to x in the case where x has a successor, and let y be any element greater than x otherwise. It will suffice to show that there exists $T \in Sem(V, \mathrm{E})$ such that $T \geq X$ and $T(x) \leq y$. Since $x < 1, \mathrm{E}(X)$ is nondegenerate; the mapping $j : \mathrm{E} \to \mathrm{E}(X)$ is an injection; and x is the greatest element in C such that $j(x) \leq [\alpha]$ in the lattice $C(X)$. This proves the lemma. ■

Lemma 8.8

Let F be a filter in the lattice $\mathrm{L}(X)$, of the form

$$\mathrm{F} = \{y : y \geq ([\alpha] \to j(\beta))^n \quad \text{for some } n \in N\}.$$

Then

$$\mathrm{F} \cap j(C) = \{j(1)\}.$$

Proof

We extend F to j-ultrafilter G in $\mathrm{L}(X)$. By Theorem 8.18, F is a prime filter. Hence the assertion. ■

Lemma 8.9

For every $\alpha \in S(X)$ there exists $z \in C$ such that $(\alpha \leftrightarrow j(z)) \in \mathrm{G}$.

Proof

Filter G defines a congruence in $\mathrm{E}(X)$. Let f denote the canonical homomorphism $\mathrm{E}(X) \to \mathrm{E}(X)/\mathrm{G}$. Since $\mathrm{G} \cap j(C) = \{j(1)\}$, the composed mapping $f \circ j$ is one-to-one.

Consider $T : S(V, C, \Delta) \to C$ defined as

$$T = (f \circ j)^{-1} \circ f \circ g,$$

where g is the canonical map from $S(V, \mathrm{E})$ to $\mathrm{E}(X)$. Then T preserves all propositional connectives, with the only possible exception of constants $\bar{c}$, for $c \in C, 0 \neq c \neq 1$.

For each $c \in C$,

$$T(c) = ((f \circ j)^{-1} \circ f \circ g)(c) = ((f \circ j)^{-1} \circ (f \circ j))(c) = c.$$

Thus T is a homomorphism of $S(V, \mathrm{E})$ into $f(\mathrm{E})$, and hence belongs to $Sem(V, \mathrm{E})$. For every $\alpha \in S$ we have $X \vdash_{\overline{X(\alpha)}} \alpha$, and so $j(X(\alpha)) \leq [\alpha]$ in $L(X)$. Consequently

$$T(\alpha) = ((f \circ j)^{-1} \circ (f \circ g))(\alpha) = ((f \circ j)^{-1}(f([\alpha])) \geq \\ ((f \circ j)^{-1} \circ (f \circ j))(X(\alpha)) = X(\alpha),$$

showing that $T \leq X$. Finally, if

$$[\alpha] \to j(z) \in \mathrm{F} \subseteq \mathrm{G},$$

then

$$f(\alpha) \leq f(j(z)) \quad \text{in} \quad \mathrm{E}(X)/\mathrm{G}.$$

Hence

$$T(\alpha) = ((f \circ j)^{-1} \circ (f \circ j))(\alpha) = ((f \circ j)^{-1} \circ f)[\alpha] \leq ((f \circ j)^{-1} \circ (f \circ j))(z) = z,$$

ending the proof. ∎

Theorem 8.20

Let $\mathrm{L} = (C_{m+1}, \otimes, \to)$ be an arbitrary $(m+1)$-element residual chain, $m \geq 1$. Given a set V of propositional variables and an extended residual lattice E over L, consider the fuzzy set A defined by

$$A(\alpha) = \begin{cases} x & \text{if} \quad \alpha = \bar{x};\ x \in C_{m+1}, \\ x \otimes y & \text{if} \quad \alpha = \bar{x} \& \bar{y};\ x, y \in C_{m+1}, \\ x \to y & \text{if} \quad \alpha = \bar{x} \Rightarrow \bar{y};\ x, y \in C_{m+1}, \\ 1 & \text{if} \quad \alpha \in \sum \cup \sum' \cup \Lambda \cup K, \\ 0 & \text{in any other case,} \end{cases}$$

where

$$\sum = \{\delta_i(a, b) : i = 1, 2;\ a, b \in C_{m+1}\} \cup \\ \{\delta_i(\alpha, \beta, \gamma) : i = 4, \ldots, 16;\ \alpha, \beta, \gamma \in S\}$$

$$\sum' = \{\delta_3^d(a_1, \ldots, a_{Ar(d)}) \mid d \in \Delta, a_1, \ldots, a_{Ar(d)} \in L\} \cup \\ \{\delta_{17}^d(\alpha_1, \ldots, \alpha_{Ar(d)}, \beta_1, \ldots, \beta_{Ar(d)}) : d \in \Delta, \\ \alpha_1, \alpha_2, \ldots, \alpha_{Ar(d)}, \beta_1, \beta_2, \ldots, \beta_{Ar(d)} \in S\}$$

$$\Lambda = \{\lambda_n(\alpha, \beta) : n \in N;\ \alpha, \beta \in S\} \\ K = \{\kappa_i(\alpha) : 0 \leq i \leq m - 1, \alpha \in S\}.$$

Let

$$R = \{r_1\} \cup \{r_i x : i = 3, 4, 5,\ x \in C_{m+1}\}.$$

Consider the C_{m+1}-syntax (A, R) on the set $S(V, C_{m+1}, \Delta)$. Then (A, R) is complete with respect to the C_{m+1}-semantics $Sem(V, \mathrm{E})$.

Proof

Suppose $0 = y_0 < y_1 < \ldots < y_m = 1$ are all the elements of C_{m+1}. By the previous considerations, the C_{m+1}-syntax (A, R) is consistent with $Sem(V, \mathrm{E})$.

Take an $X : S \to C_{m+1}$ and $\alpha \in S$. Let

$$(Cn_{A,R}(X))(\alpha) = y_k$$

for a certain $k, 0 \leq k \leq m-1$. There exists $z \in C_{m+1}, z < 1$, such that $j(z) \in S$ and

$$([\alpha] \to j(y_k))^n \leq j(z)$$

for a certain $n \in N \setminus \{0\}$. From the definition of A we get

$$X \vdash_{\bar{1}} (\alpha \Rightarrow \bar{y}_k) \vee (\bar{y}_{k+1} \Rightarrow \alpha),$$

and hence

$$([\alpha] \to j(y_k)) \vee (j(y_{k+1}) \to [\alpha]) = j(1)$$

in $\mathrm{L}(X)$.

On the other hand, we have in $\mathrm{L}(X)$

$$j(1) = (j(1))^n = ([\alpha] \to j(y_k))^n \vee (j(y_{k+1}) \to [\alpha])^n \leq j(z) \vee (j(y_{k+1}) \to [\alpha]),$$

and so

$$X \vdash_{\bar{1}} (\bar{y}_{k+1} \Rightarrow \alpha) \vee \bar{z}.$$

Applying the elimination rule $r_3 z$ we obtain

$$X \vdash_{\bar{1}} \bar{y}_{k+1} \Rightarrow \alpha,$$

contrary to the assumption that y_k is the smallest element in C_{m+1} with

$$X \vdash_{\overline{y_k}} \alpha.$$

Now assume that $\alpha \in S(X)$ is any formula and let G be a ultrafilter as in Theorem 8.18. Define

$$D_\alpha = \{z \in C_{m+1} : j(z) \to \alpha \in \mathrm{G}\},$$
$$H_\alpha = \{z \in C_{m+1} : \alpha \to j(z) \in \mathrm{G}\}.$$

Plainly, D_α is an initial segment of C_{m+1}, containing 0, and H_α is a terminal segment of C_{m+1}, containing 1. Moreover,

$$(j(z) \to \alpha) \vee (\alpha \to j(z)) = j(1)$$

for any $z \in C_{m+1}$. Since G is a prime filter, $D_\alpha \cup H_\alpha = C_{m+1}$.

We claim that the intersection $D_\alpha \cap H_\alpha$ is nonempty. Assume the contrary.

Then there is an index $p \leq m-1$ such that y_p is the last element of D_α while y_{p+1} is the first element of H_α. Thus $\alpha = [c]$ for a certain $c \in S$. It follows that

$$X \vdash_{\overline{1}} (c \Rightarrow y_p) \vee (y_{p+1} \Rightarrow c),$$

and consequently either $y_p \in H_\alpha$ or $y_{p+1} \in D_\alpha$, which is absurd. This settles the claim $(D_\alpha \cap H_\alpha \neq \emptyset)$.

Take $z_0 \in D_\alpha \cap H_\alpha$. Then

$$\alpha \leftrightarrow j(z_0) = (\alpha \to j(z_0)) \wedge (j(z_0) \to \alpha)$$

belongs to G. The proof is complete. ∎

Theorem 8.21

Consider the Łukasiewicz interval $(I, \otimes, \to)$, with

$$x \otimes y = 0 \vee (x + y - 1),$$
$$x \to y = 1 \wedge (1 - x + y).$$

Let V be a set of propositional variables and let E be an extended residual lattice over $(I, \otimes, \to)$. Define

$$A(\alpha) = \begin{cases} x & \text{if } \alpha = \bar{x};\ x \in I, \\ x \otimes y & \text{if } \alpha = \bar{x} \& \bar{y};\ x, y \in I, \\ x \to y & \text{if } \alpha = \bar{x} \Rightarrow \bar{y};\ x, y \in I, \\ 1 & \text{if } \alpha \in \sum \cup \sum' \cup \Lambda \cup \Phi_1 \cup \Phi_2, \\ 0 & \text{in any other case,} \end{cases}$$

where

$$\sum = \{\delta_i(a, b) : i = 1, 2;\ a, b \in I\} \cup \{\delta_i(\alpha, \beta, \gamma) : i = 4, \ldots, 16;\ \alpha, \beta, \gamma \in S\},$$

$$\sum' = \{\delta_3^d(a_1, \ldots, a_{Ar(d)}) : d \in \Delta;\ a_1, \ldots, a_{Ar(d)} \in I\} \cup \{\delta_{17}^d(\alpha_1, \ldots, \alpha_{Ar(d)}, \beta_1, \ldots, \beta_{Ar(d)}) : d \in \Delta;\ \alpha_1, \alpha_2, \ldots, \alpha_{Ar(d)}, \beta_1, \beta_2, \ldots, \beta_{Ar(d)} \in S\},$$

$$\Lambda = \{\lambda_n(\alpha, \beta) : n \in N;\ \alpha, \beta \in S\},$$

$$\Phi_1 = \{\iota_1(\alpha, x, y, x', y', n) : 0 \leq y < y' < 1, 0 \leq x' < x \leq 1;\ n \in N, n > 0, nx + y \leq nx' + y';\ \alpha \in S\},$$

$$\Phi_2 = \{\iota_2(\alpha, x, y, x', y', n) : 0 \leq y < y' < 1, 0 \leq x < x' \leq 1;\ n \in N, n > 0, nx - y \geq nx' - y';\ \alpha \in S\}.$$

Let

$$R = \{r_1\} \cup \{rx : x \in I\}.$$

Then (A, R) is an I-syntax over $S(V, I, \Delta)$, complete with respect to the I-semantics $Sem(V, \mathrm{E})$.

Proof

The consistency of (A, R) with respect to $Sem(V, \mathrm{E})$ is obvious.

Given a set $X : S(V, I, \Delta) \to I$ and a formula $\alpha \in S(V, I, \Delta)$; assume

$$(Cn_{A,R}(X))(\alpha) = x,\ x < 1,$$

and consider all elements $y \in I$ such that $x < y$. Fix such a y and assume there exists $z \in I$ with $z < 1, j(z) \in \mathrm{F}$, where F is a ultrafilter in L. Then we have, for a certain $n \in N \setminus \{0\}$,

$$([\alpha] \to j(y))^n \leq j(z)$$

and

$$\begin{aligned} j(1) &= (j(1))^n = ([\alpha] \to j(y))^n \vee (j(y) \to [\alpha])^n \\ &\leq j(z) \vee (j(y) \to [\alpha])^n \leq j(z) \vee (j(y) \to [\alpha]). \end{aligned}$$

The element z is either zero or nilpotent in the semigroup $(I, \otimes)$. Let $m \in N \setminus \{0\}$ be such that $z^m = 0$ in $(I, \otimes)$. Then

$$\begin{aligned} j(1) &= (j(1))^m = (j(z))^m \vee (j(y) \to [\alpha])^m = j(0) \vee (j(y) \to [\alpha])^m \\ &\leq j(y) \to [\alpha], \end{aligned}$$

and hence $j(y) \leq [\alpha]$, contrary to the assumption $y > (Cn_{A,R}(X))(\alpha)$.To finish the proof of completeness, take any formula $\alpha \in S(\alpha)$.

Let G be a ultrafilter as in Theorem 8.18 and define, as in the proof of Theorem 8.20,

$$\begin{aligned} D_\alpha &= \{z \in I : j(z) \to \alpha \in \mathrm{G}\}, \\ H_\alpha &= \{z \in I : \alpha \to j(z) \in \mathrm{G}\}. \end{aligned}$$

Clearly, $0 \in D_\alpha, 1 \in H_\alpha$ and $D_\alpha \cup H_\alpha = I$.

We will show that D_α and H_α are closed subsets of I, in the usual topology of the real line segment. First we show that the set $I \setminus D_\alpha$ is open in I. Fix a point $z \in I \setminus D_\alpha$. By the definition of D_α, if $v \in D_\alpha$ and $u \leq v$, then $u \in D_\alpha$. So it is enough to find a point y such that $y < z$ and $y \in D_\alpha$.

There exist $e \in \mathrm{G}, v \in I$ and $n \in N \setminus \{0\}$ such that $v < 1$ and

$$(j(z) \to \alpha)^n \otimes e \leq j(v).$$

Hence

$$e \leq (j(z) \to \alpha)^n \to j(v) \in \mathrm{G}.$$

Choose u so that $v < u < 1$ and put

$$y = \frac{z - (u - v)}{n}.$$

Then choose $\beta \in S$ with $[\beta] = \alpha$. Then

$$X \vdash_{\bar{1}} ((\bar{z} \Rightarrow \beta) \Rightarrow \bar{v}) \Rightarrow ((\bar{y} \Rightarrow \beta)^n \Rightarrow \bar{u}),$$

so that

$$(j(z) \to \alpha)^n \to j(v) \leq (j(y) \to \alpha)^n \to j(u).$$

Consequently

$$(j(y) \to \alpha)^n \to j(u) \in \mathrm{G}$$

and thus

$$j(y) \to \alpha \in \mathrm{G},$$

proving $y \notin D_\alpha$. Hence, $I \setminus D_\alpha$ is open.

The openness of $I \setminus H_\alpha$ is proved in exactly the same way. So D_α and H_α are closed subsets of I. By the connectedness of I there exists a $z \in D_\alpha \cap H_\alpha$. Thus

$$\alpha \leftrightarrow j(z) \in \mathrm{G},$$

ending the proof. ∎

8.6 First-Order Predicate Calculus for Fuzzy Logics

8.6.1 Introductory Remarks

We now present the first-order predicate calculus over a fuzzy propositional calculus, constructed over the class of extended residual lattices. The approach is entirely due to V. Novák [Novák 1987], who seems to have provided the most extensive treatment of the subject. Following Novák, we resort to Pavelka's results [Pavelka 1979], presented in foregoing sections, and we prove completeness by the classical method. It will be clearly seen that the first-order fuzzy predicate calculus is a direct (though far-reaching) generalization of the classical predicate calculus.

8.6.2 Generalized Residual Lattices

As previously, we assume that the set of logical values constitutes a complete infinitely distributive residual lattice

(i) $\mathrm{L} = (L, \wedge, \vee, \otimes, \to, \bar{0}, \bar{1})$,

in which $\bar{0}$ and $\bar{1}$ are the extreme (least and greatest) elements, and the symbols $\otimes$ and $\to$ denote binary operations, called multiplication and residuum. The properties demanded from these operations are the following:

(a_1) The triple $(L, \otimes, \bar{1})$ is a commutative monoid.

(a_2) Operation $\otimes$ is isotonic in both arguments; operation $\rightarrow$ is antitonic in the first argument and isotonic in the second.

(a_3) For every $x, y, z \in L$,

$x \otimes y \leq z \quad \text{iff} \quad x \leq y \rightarrow z.$

For the case of $L = [0,1]$, it is proved in [Pavelka 1979] that if the residuum operation is not continuous in $[0,1] \times [0,1]$ then there exists no complete fuzzy logic over the lattice in question; moreover, every residual lattice

(i) L with $L = [0,1]$ and with $\rightarrow$ a continuous function on $[0,1] \times [0,1]$ is isomorphic to

(ii) $\mathrm{L} = ([0,1], \wedge, \vee, \otimes, \rightarrow, 0, 1)$,

with

$$x \vee y = \max(x,y),\ x \wedge y = \min(x,y),$$
$$x \otimes y = 0 \vee (x + y - 1),\ x \rightarrow y = 1 \wedge (1 - x + y).$$

One more example worth recalling is

(iii) $\mathrm{L} = (\{a_0, a_1, \ldots, a_m\}, \vee, \wedge, \otimes, \rightarrow, \bar{0}, \bar{1})$

with $a_0 \leq a_1 \leq \ldots \leq a_m$ and operations defined by

$$a_k \vee a_l = a_{\max(k,l)},$$
$$a_k \wedge a_l = a_{\min(k,l)},$$
$$a_k \otimes a_l = a_{\max(0,k+l-m)},$$
$$a_k \rightarrow a_l = a_{\min(m,m-k+l)}.$$

Of course, the symbol $\leq$ denotes the lattice order.

For convenience, we confine attention to lattices of type (ii) or (iii) (cf. [Novák 1987]). But of course the problem can be generalized to other extended residual lattices, not necessarily infinitely distributive.

Lemma 8.10

The following equalities and inequalities hold in every lattice of type (ii) or (iii):

(a_1) $\bigwedge_{i \in J_1} x_i \otimes \bigwedge_{j \in J_2} y_j \leq \bigwedge_{i \in J_2} \bigwedge_{j \in J_2} (x_i \otimes y_j)$,

(a_2) $\bigwedge_{i,j \in J} (x_i \leftrightarrow y_j) \leq \bigwedge_{i \in J} x_i \leftrightarrow \bigwedge_{j \in J} y_j$, (a$_3$) $\bigwedge_{i,j\in J} (x_i \leftrightarrow y_j) \leq \bigvee_{i\in J} x_i \leftrightarrow \bigvee_{j\in J} y_j$,

(a$_4$) $-(-x) = x$,

(a$_5$) $\bigwedge_{i\in J}(x_i \otimes y) = \left(\bigwedge_{i\in J} x_i\right) \otimes y$,

(a$_6$) $x \wedge y = -(-x \vee -y)$,

(a$_7$) $x \to y = -(x \otimes -y)$,

(a$_8$) $x \to y = -y \to -x$;

in (a$_1$)–(a$_8$), x, y, x_i, y_j denote arbitrary elements of L and J, J_1, J_2 are any index sets; the symbol $x \leftrightarrow y$ stands for $(x \to y) \wedge (y \to x)$, and $-x$ denotes $(x \to \bar{0})$.

Proof

All these assertions follow immediately from the definition of a residual lattice and the fact that the underlying set is a chain. ∎

Now we extend L by adjoining a set of operations $\{d_j : j \in J_0\}$,

$$Ar(d_j) = n_j,\ Ex(d_j) = (k_1, k_2, \ldots, k_{n_j});$$

J_0 is a certain index set and n_j are some positive integers. These operations are to be viewed as interpretations of some additional n_j-argument connectives. They are subject to the inequalities

$$(x_1 \leftrightarrow y_1) \otimes (x_2 \leftrightarrow y_2) \otimes \ldots \otimes (x_{n_j} \leftrightarrow y_{n_j}) \leq$$
$$d_j(x_1, x_2, \ldots, x_{n_j} \leftrightarrow d_j(y_1, y_2, \ldots, y_{n_j})$$

for the respective $Ex(d_j)$ [Pavelka 1979].

Moreover, we introduce in L a class of generalized operations

$$Q : P(L) \to L,$$

where $P(L)$ is the power set of L. (In particular, we place $\bigwedge$ and $\bigvee$ among the operations Q.) For a set $K \subseteq L$ we write

$$\mathop{\mathrm{Q}}_{x\in K} x \quad \text{for} \quad Q(\{x : x \in K\})$$

and

$$Qx \quad \text{for} \quad Q(\{x\}).$$

A generalized operation Q is called regular if it satisfies the following conditions:

(o₁) $\mathop{Q}_{x\in K}(x\otimes y)\leq(\mathop{Q}_{x\in K}x)\otimes y,$

(o₂) $\mathop{Q}_{x\in K}((x\otimes y)\to 0)\to 0\leq(\mathop{Q}_{x\in K}(x\to 0)\to 0)\otimes y,$

(o₃) $\bigwedge_{x\in K}x\leq\mathop{Q}_{x\in K}x\leq\bigvee_{x\in K}x\quad\text{for}\quad K\neq\emptyset.$

For a generalized operation Q define

$$\mathop{\tilde{Q}}_{x\in K}x=-\mathop{Q}_{x\in K}(-x).$$

Obviously, if Q is regular then so is $\tilde{Q}$. For every Q,

$$-\mathop{\tilde{Q}}_{x\in K}x=\mathop{Q}_{x\in K}(-x),$$
$$\tilde{Q}_{x\in K}(-x)=-\mathop{Q}_{x\in K}x.$$

The generalized operations Q and $\tilde{Q}$ are said to be dual to each other. Thus, the generalized supremum $\bigvee$ and the generalized infimum $\bigwedge$ form a pair of dual operations.

The result is that we have formed what is called a generalized extended residual lattice

$$\mathrm{L}=(L,\vee,\wedge,\otimes,\to,\{\bar{x}:x\in L\},\{d_j:j\in J_0\},\{Q_j:j\in J_q\}\cup\{\bigvee,\bigwedge\}).$$

In the case we are dealing with, L is either the real line segment $\{0,1\}$ or the $(m+1)$-element chain; J_q is an index set for generalized operations other than $\bigvee,\bigwedge$; and $\{\bar{x}:x\in L\}$ is the set of zero-argument operations, regarded as self-indexing interpretations of logical values.

8.6.3 The Language of the Fuzzy First-Order Predicate Calculus

The alphabet of the fuzzy first-order predicate calculus is composed of:

(j₁) the set of individual variables

$$V=\{x_i:i\in N\};$$

(j₂) the set of individual constants

$$C=\{c_k:k\in K\};$$

(j_3) the set of finite-argument predicates (or relation symbols)

$$\{p_i : i \in I\};$$

in particular, the equality symbol = is considered as a relation symbol;

(j_4) the set of finite-argument function symbols

$$\{f_j{}^{G_j} : j \in J\},$$

indexed by superscripts G_j, $j \in J$;

(j_5) the set of symbols for logical values

$$\{\bar{x} : x \in L\};$$

(j_6) the set of binary connectives $\{\vee, \wedge, \&, \Rightarrow\}$ plus the set of additional connectives

$$\{d_j : j \in J_0\}$$

with $Ar(d_j) = n \geq 1$, $Ex(d_j) = (k_1, k_2, \ldots, k_n)$ (denoting propositional connectives by the same symbols as lattice operations should not cause confusion);

(j_7) the set of symbols denoting generalized quantifiers

$$\{Q_j : j \in J_q\},$$

including the universal quantifier $\forall$ and the existential quantifier $\exists$;

(j_8) the set of auxiliary symbols (parentheses, comma, etc.).

Next, we introduce the concept of a term. We distinguish superindex-free terms (constants and variables) and superindexed ones, arising by substitution of other terms (with or without superindices) into a function symbol. Thus, the set of terms over the alphabet (j_1)–(j_8), denoted by T, is defined as the smallest set T' satisfying the conditions:

(t_1) $V \subseteq T'$, $C \subseteq T'$ (individual variables and constants are superindex-free terms);

(t_2) if $f_j^{G_j}$ is an m-argument function symbol and $t_1, t_2, \ldots, t_m$ are terms, either superindex-free or having the same superindex G_j, then $f_j^{G_j}(t_1, t_2, \ldots, t_m)$ belongs in T' and is a term with superindex G_j.

The set S of well-formed formulas is defined to be the smallest set S' satisfying the conditions:

(f_1) the set $\{\bar{x} : x \in L\}$ is contained in S';

(f$_2$) if $t_1, t_2, \ldots, t_n$ are terms (with superindices or without) and p_i is an n-argument predicate, then $p_i(t_1, t_2, \ldots, t_n)$ belongs to S';

formulas of this type are called atomic or elementary; in particular, if t, u are terms then $t = u$ is an elementary formula;

(f$_3$) if $\alpha, \beta, \alpha_1, \alpha_2, \ldots, \alpha_n$ are in S', then $\alpha \vee \beta, \alpha \wedge \beta, \alpha \& \beta, \alpha \Rightarrow \beta$ are in S', and so is $d_j(\alpha_1, \alpha_2, \ldots, \alpha_n)$, for each $j \in J_0$;

(f$_4$) if $\alpha \in S', x_i \in V$, then $\forall x_i \alpha$, $\exists x_i \alpha$ and $Q_j x_i \alpha$ (for each $j \in J_q$) belong to S'.

For convenience we introduce two more connectives: negation $\sim$ and equivalence $\Leftrightarrow$. For $\alpha, \beta \in S$, the formula $\sim \alpha$ is defined as $\alpha \Rightarrow \bar{0}$, and $\alpha \Leftrightarrow \beta$ is defined as $(\alpha \Rightarrow \beta) \wedge (\beta \Rightarrow \alpha)$.

The notions of free and bound variables are carried over from classical logic. The same concerns substitution, with the only constraint that substitution in a term is allowed provided all terms to substitute have equal superindices or no superindex.

Let us recall that a fuzzy set A in a universe U is defined as a function

$$A : U \rightarrow L,$$

whose value $A(x)$ is viewed as the degree of adherence of the element $x \in U$ to A. It is sometimes convenient to write such a set as

$$\{A(x)/x : x \in U\}.$$

A one-element fuzzy set (fuzzy singleton) is written as $\{A(x)/x\}$.

By an interpretation of the language of the first-order fuzzy predicate calculus we mean the structure

$$\mathrm{D} = (D, \{p_{iD} : i \in I\}, \{g_{jD} : j \in J\}, \{c_k : k \in K\})$$

in which:

- D is some set;

- for $i \in I$, if p_i is an n-argument predicate then $p_{iD} \subseteq D^n$ is an n-argument fuzzy relation in D;

- for $j \in J$, if g_j is an m-argument function symbol then g_{jD} is an m-argument fuzzy function with domain $G_j^m (G_j \lesssim D)$, satisfying the condition:

(g) $G_j(g_{jD}(x_1, x_2, \ldots, x_m)) \geq G_j(x_1) \wedge G_j(x_2) \wedge \ldots \wedge G_j(x_m)$ for $x_1, \ldots, x_m \in D$,

- the c_k (for $k \in K$) are distinguished fuzzy constants, in the sense explained below. Thus, the interpretation of symbols is the following:

(a_1) each individual constant of the language is interpreted as a one-element fuzzy set in D (this concerns, in particular, the distinguished constants c_k);

(a_2) each m-argument function symbol $g_j^{G_j}$ (with superindex G_j) is interpreted as an m-argument fuzzy function g_{jD} with domain G_j^m, $G_j \lessapprox D$;

(a_3) each m-argument fuzzy relation p_{idD} in D.

Every term is interpreted as a fuzzy singleton. And thus, for $x \in L \setminus \{0\}$,

$$\mathrm{D}(\bar{x}) = \{x/d\};$$

if a term t is symbol of an index-free constant, then

$$\mathrm{D}(t) = \{1/d\};$$

if t is the term $g_{jD}(t_{1D}, t_{2D}, \ldots, t_{mD})$, then

$$\mathrm{D}(t) = \{G_j(g_{jD}(\mathrm{D}(t_{1D}), \mathrm{D}(t_{2D}), \ldots, \mathrm{D}(t_{mD})))/g_{jD}(\mathrm{D}(t_{1D}, t_{2D}, \ldots, t_{mD}))\}.$$

We will also write d_x for $\{x/d\}$.

Let D be an interpretation structure, as above. Let $v(\alpha)_{\mathrm{D}}$ denote the value of a formula $\alpha \in S$ in D. The following conditions are imposed on the valuation function:

(a_1) If $\alpha = \bar{x}, x \in L$, then $v(\alpha)_{\mathrm{D}} = x$.

(a_2) If $\alpha = (\bar{x} = \bar{y})$ then

$$v(\alpha)_{\mathrm{D}} = \begin{cases} \bar{1} & \text{if } v(\bar{x})_{\mathrm{D}} = v(\bar{y})_{\mathrm{D}}, \\ \bar{0} & \text{otherwise.} \end{cases}$$

For all $\alpha, \beta \in S$:

(a_3) $v(\alpha \vee \beta)_{\mathrm{D}} = v(\alpha)_{\mathrm{D}} \vee v(\beta)_{\mathrm{D}}$

(a_4) $v(\alpha \wedge \beta)_{\mathrm{D}} = v(\alpha)_{\mathrm{D}} \wedge v(\beta)_{\mathrm{D}}$

(a_5) $v(\alpha \& \beta)_{\mathrm{D}} = v(\alpha)_{\mathrm{D}} \otimes v(\beta)_{\mathrm{D}}$

(a_6) $v(\alpha \Rightarrow \beta)_{\mathrm{D}} = v(\alpha)_{\mathrm{D}} \rightarrow v(\beta)_{\mathrm{D}}$

For $\alpha_1, \alpha_2, \ldots, \alpha_n \in S, j \in J_0$:

(a_7) $v(d_j(\alpha_1, \alpha_2, \ldots, \alpha_n))_{\mathrm{D}} = d_j(v(\alpha_1)_{\mathrm{D}}, v(\alpha_2)_{\mathrm{D}}, \ldots, v(\alpha_n)_{\mathrm{D}})$.

For any $\alpha \in S$:

(a_8) $v(\exists x\alpha)_{\mathrm{D}} = \bigvee_{d \in D} v(\alpha(x/d))_{\mathrm{D}}$

(a_9) $v(\forall x\alpha)_{\mathrm{D}} = \bigwedge_{d \in D} v(\alpha(x/d))_{\mathrm{D}}$

(a_{10}) $v(Q_j x\alpha)_{\mathrm{D}} = \underset{d \in D_j}{\mathrm{Q}} v(\alpha(x/d))_{\mathrm{D}}$

for $j \in J_q$.

For $\alpha, \beta \in S$:

(a_{11}) $v(\alpha \Leftrightarrow \beta)_{\mathrm{D}} = v(\alpha)_{\mathrm{D}} \leftrightarrow v(\beta)_{\mathrm{D}}$

(a_{12}) $v(\sim \alpha)_{\mathrm{D}} = -v(\alpha)_{\mathrm{D}} = v(\alpha)_{\mathrm{D}} \rightarrow \bar{0}$.

Let Q_j be a quantifier. If it corresponds to a regular operation, we can adjoin to our language the symbol $\tilde{Q}_j$ representing the quantifier that corresponds to the dual operation. We call these two quantifiers dual to each other.

A quantifier whose interpretation is a regular operation is called also regular. Obviously, $\forall$ and $\exists$ are regular quantifiers. For the sequel we assume (as Novák does; see [Novák 1987]) that all quantifiers admitted in the language are coupled into pairs of regular dual ones.

Assume D is an interpretation and let $t_G(x)$ be a term with superscript G and a free variable x. Further suppose $\alpha(x)$ is a formula involving a free variable x and u_G is a term without free variables, such that

$$\mathrm{D}(u_G) = \{\delta/d\}$$

is a singleton with label $\bar{d}_\delta$. Then

$$\mathrm{D}(t_G(x/u_G)) = \mathrm{D}(t_G(x/\bar{d}_\delta))$$

and

$$v(\alpha(x/u_G))_{\mathrm{D}} = v(\alpha(x/\bar{d}_\delta))_{\mathrm{D}}.$$

8.6.4 Semantic Consequence Operation

The concept of semantic consequence operation will be now extended to the case of fuzzy logics.

Let $X \subseteq S$ be a fuzzy set of formulas. We define

$$(Cn_{Sem}(X))(\alpha) = \bigwedge\{v(\alpha)_{\mathrm{D}} : X \leq \mathrm{D},\ \mathrm{D} \text{ is a structure for } S\}$$

and call $Cn_{Sem}(X)$ the (fuzzy) set of semantic consequences of X.

If $(Cn_{Sem}(X))(\alpha) \geq \delta$, we write $X \models_\delta \alpha$. A formula α with n free variables $x_1, x_2, \ldots, x_n$ is δ-true in D iff

$$\delta = \bigwedge_{\substack{d_i \in D \\ i=1,\ldots,n}} \{v(\alpha(x_1/\bar{d}_1, \ldots, x_n/\bar{d}_n))\}.$$

We then write

$$\mathrm{D} \models_\delta \alpha.$$

A formula $\alpha \in S$ is δ-tautologous, in symbols $\models_\delta \alpha$, iff

$$\delta = (Cn_{Sem}(\emptyset))(\alpha).$$

If $\models_1 \alpha$, we write simply $\models \alpha$ and call α a tautology.

Suppose we have for a certain $\delta \in L$ the equality

$$\delta = \bigwedge\{\mu : \mathrm{D} \models_\mu \alpha, \mathrm{D} \text{ is a structure for } S\}.$$

We then call δ the degree of truth of α; we also say that the formula α is δ-valid. Obviously, a formula is δ-tautologous if and only if it is δ-valid.

Lemma 8.11

Let $\alpha, \beta \in S$. For any structure D for S:

(a_1) $\models \alpha \Rightarrow \beta$ iff $v(\alpha)_\mathrm{D} \leq v(\beta)_\mathrm{D}$;

(a_2) $\models \alpha \Leftrightarrow \beta$ iff $v(\alpha)_\mathrm{D} = v(\beta)_\mathrm{D}$.

Proof

Obvious. ■

Lemma 8.12

Let Q and $\tilde{Q}$ be a dual pair of regular quantifiers. The following formulas are tautologous:

(a_1) $\models \alpha(x/\delta) \Rightarrow (\exists x)\alpha$

(a_2) $\models (\forall x)\alpha \Rightarrow \alpha(x/\delta)$

(a_3) $\models \sim\sim \alpha \Leftrightarrow \alpha$

(a_4) $\models (Qx)\alpha \Leftrightarrow \sim (\tilde{Q}x) \sim \alpha$

(a_5) $\models (\forall x)(\alpha \Leftrightarrow \beta) \Rightarrow ((Qx)\alpha \Leftrightarrow (Qx)\alpha)$

(a_6) $\models (Qx) \sim \alpha \Leftrightarrow \sim (\tilde{Q}x)\alpha$

(a_7) $\models (\tilde{Q}x) \sim \alpha \Leftrightarrow \sim (Qx)\alpha$

(a_8) $\models \sim (\alpha \Rightarrow \beta) \Leftrightarrow (\alpha \,\&\, \sim \beta)$

(a_9) $\models (\alpha \wedge \beta) \Leftrightarrow \sim(\sim \alpha \vee \sim \beta)$

(a_{10}) $\models (\alpha \Rightarrow \beta) \Leftrightarrow \sim(\alpha \wedge \sim \beta)$

(a_{11}) $\models (\alpha \& \beta) \Rightarrow ((\beta \Rightarrow \gamma) \Rightarrow (\alpha \& \gamma))$

(a_{12}) $\models (\sim \alpha \Rightarrow \beta) \Rightarrow (\sim \beta \Rightarrow \alpha)$

(a_{13}) $\models (\alpha \& \beta) \Rightarrow \alpha$

(a_{14}) $\models (\alpha \& \beta) \Rightarrow \beta$

Proof

Immediate from the definition of a tautology and from the lattice properties. ∎

Lemma 8.13

Let $\alpha' \in S$ be the closure of a formula $\alpha \in S$. Then $\mathrm{D} \models_\delta \alpha$ holds if and only if $\mathrm{D} \models_\delta \alpha'$. In particular, $\models_\delta \alpha$ iff $\models_\delta \alpha'$.

Proof

Obvious. ∎

Now, let $\{f_j : j \in J\}$ and $\{p_i : i \in I\}$ be the sets of function symbols and predicate symbols of the fuzzy predicate calculus. The formulas that follow are taken for axioms of the predicate calculus with identity:

(a_x) $x = x$;

(a_{f_j}) $(x_1 = y_1) \Rightarrow ((x_2 = y_2) \Rightarrow (\ldots \Rightarrow (x_n = y_n) \Rightarrow (f_j(x_1, x_2, \ldots, x_n) = f_j(y_1, y_2, \ldots, y_n))\ldots))$;

(a_{p_i}) $(x_1 = y_1) \Rightarrow ((x_2 = y_2) \Rightarrow (\ldots \Rightarrow (x_n = y_n) \Rightarrow (p_i(x_1, x_2, \ldots, x_n) = p_i(y_1, y_2, \ldots, y_n))\ldots))$

Lemma 8.14

The axioms (a_x), (a_{f_j}), $j \in J$, and (a_{p_i}), $i \in I$, are tautologies of the fuzzy predicate calculus of the first-order.

Proof

We demonstrate the tautologousness of (a_{f_j}). If each of the premises has value 1, then

$$v(f_j(x_1, x_2, \ldots, x_n) = f_j(y_1, y_2, \ldots, y_n))_{\mathrm{D}} = 1$$

and $v(a_{f_j})_{\mathrm{D}} = 1$. If $v(x_i = y_i)_{\mathrm{D}} = 0$ for some i, then also $v(a_{f_j})_{\mathrm{D}} = 1$, since $0 \to \alpha = 1$ for every α. ∎

8.6.5 Syntax of the Fuzzy First-Order Predicate Calculus

Following [Pavelka 1979] and [Novák 1987], we have defined an n-ary rule of inference in a fuzzy calculus as a pair $r = (p, q)$, in which:

p is an n-ary partial operation on S, with values in S;

q is a semi-continuous n-ary operation in L.

We call p and q the syntactic and semantic part of r, respectively.

Semi-continuity of q means that q preserves nonempty unions, in each variable. (This implies that q is isotonic in each variable.)

A fuzzy set $X \subseteq S$ is closed under a rule $r = (p, q)$ if and only if

$$X(p(\alpha_1, \alpha_2, \ldots, \alpha_n)) \geq q(X(\alpha_1), X(\alpha_2), \ldots, X(\alpha_n))$$

holds for all $\alpha_i \in S$ such that p is defined.

A rule $r = (p, q)$ is X-consistent iff

$$v(p(\alpha_1, \alpha_2, \ldots, \alpha_n))_{\mathrm{D}} \geq q(X(\alpha_1), X(\alpha_2), \ldots, X(\alpha_n))$$

holds for every structure D and for any $\alpha_1, \ldots, \alpha_n \in S$ such that $X(\alpha_i) \leq v(\alpha_i)_{\mathrm{D}}, i = 1, 2, \ldots, n$. A rule of inference is consistent iff it is $\emptyset$-consistent.

To write down a rule $r = (p, q)$, we use the patterns

$$r: \frac{\alpha_1, \alpha_2, \ldots, \alpha_n}{p(\alpha_1, \alpha_2, \ldots, \alpha_n)} \quad , \quad \frac{\delta_1, \delta_2, \ldots, \delta_n}{q(\delta_1, \delta_2, \ldots, \delta_n)}$$

where $\delta_i = v(\alpha_i)_{\mathrm{D}}$ for $i = 1, 2, \ldots, n$.

And thus, for instance, the rule

$$p: \frac{\beta}{\alpha \Rightarrow \beta} \quad , \quad q: \frac{\mu}{\delta \rightarrow \mu}$$

where $\delta = (Cn_{Sem}(X))(\alpha)$, is X-consistent. It is called the rule of α, δ-precedence.

The following rules are consistent, provided the quantifiers that occur in them are regular:

(r_1) The rule of detachment (modus ponens):

$$p: \frac{\alpha, \alpha \Rightarrow \beta}{\beta} \quad , \quad q: \frac{\delta, \mu}{\delta \otimes \mu},$$

(r_2) The rule of generalization:

$$p: \frac{\alpha}{(Qx_i)\alpha} \quad , \quad q: \frac{\delta}{\delta}.$$

(r_3) The rule generalization of the consequent:

$$p: \frac{\alpha \Rightarrow \beta}{\alpha \Rightarrow (Qx_i)\beta} \quad , \quad q: \frac{\delta}{\delta},$$

provided that x_i is not a free variable in α.

(r_4) The rule of generalization of the precedent:

$$p : \frac{\alpha \Rightarrow \beta}{(Qx_i)\alpha \Rightarrow \beta}, \quad q : \frac{\delta}{\delta},$$

provided that x_i is not a free variable in α.

(r_5) The rule of distributivity:

$$p : \frac{(Qx_i)(\alpha \& \beta)}{((Qx_i)\alpha) \& \beta}, \quad q : \frac{\delta}{\delta}.$$

8.6.6 Syntactic Consequence Operation

We consider two fuzzy sets in S, denoted by A and X and called the set of logical axioms and the set of specific axioms, respectively. Let R be the set of X-consistent rules of inference. The pair (A, R) will be called the syntax of the fuzzy predicate calculus of the first-order, or simply: the syntax of fuzzy logic.

The symbol $Cn(X)$ or CnX will stand for the set of syntactic consequences of X, defined by

$$(CnX)(\alpha) = \bigwedge\{U(\alpha) : A \subseteq U, X \subseteq U, U \subseteq S, \\ U \text{ closed with respect to every } r \in R\}$$

for $\alpha \in S$.

By a proof of a formula $\alpha \in S$ from X we mean a sequence of formulas $\omega = (\alpha_0, \alpha_1, \ldots, \alpha_n)$ such that:

(a_1) $\alpha_n = \alpha$;

(a_2) for each $i, 0 \leq i \leq n$, either $\alpha_i \in A$ or $\alpha_i \in X$, or else there exists an X-consistent rule $r \in R, r = (p, q)$, such that $\alpha_i = p(\alpha_{i_1}, \alpha_{i_2}, \ldots, \alpha_{i_m})$ for some $i_1, i_2, \ldots, i_m < i$.

The value $w(\omega)$ of a proof $\omega = (\alpha_0, \alpha_1, \ldots, \alpha_n)$ of a formula α is defined by

$$w(\omega) = \begin{cases} A(\alpha_n) & \text{if } \alpha_n \text{ is a logical axiom,} \\ X(\alpha_n) & \text{if } \alpha_n \text{ is a specific axiom,} \\ q(v(\omega_{i_1}), v(\omega_{i_2}), \ldots, v(\omega_{i_m})) & \text{if } \alpha_n \text{ arises} \\ & \text{from formulas } \alpha_{i_1}, \alpha_{i_2}, \ldots, \alpha_{i_m} \\ & \text{by means of a rule } r = (p, q); \end{cases}$$

here $\omega_{i_j} = (\alpha_0, \alpha_1, \ldots, \alpha_{i_j})$.

A proof ω will be often denoted in the following, more explicit manner:

$$\omega = (\alpha_0[\delta_0, r_0], \alpha_1[\delta_1, r_1], \ldots, \alpha_n[\delta_n, r_n]),$$

where $\delta_i = v(\omega_i)$ and r_i is the rule of inference producing formula α_i; if α_i belongs to A or X, we consider r_i to be the identity rule. It is not hard to see that

$$(CnX)(\alpha) = \bigvee\{v(\omega) : \omega \text{ is a proof of } \alpha \text{ from } X\}.$$

A syntax (A, R) is X-consistent iff $A \subseteq Cn\emptyset$ and each rule $r \in R$ is X-consistent. Similarly, (A, R) is consistent when each $r \in R$ is consistent.

8.6.7 An Axiom System for the Fuzzy First-Order Predicate Calculus

The axiom system which we give below has been adopted from [Pavelka 1979] and [Novák 1987].

(ax_1) For every $\delta \in L$, the formula $\bar{\delta}$ with degree of incidence δ is an axiom.

(ax_2) For every $\delta, \mu \in L$, the formula $\bar{\delta} \& \bar{\mu}$ with degree of incidence $\delta \otimes \mu$ is an axiom.

(ax_3) For every $\delta, \mu \in L$, the formula $\bar{\delta} \Rightarrow \bar{\mu}$ with degree of incidence $\delta \rightarrow \mu$ is an axiom.

Moreover, all formulas that follow are axioms, with degree of incidence 1 (characters δ, μ, possibly with subscripts, denote arbitrary elements of L; characters α, β, γ, possibly with subscripts, denote arbitrary formulas in S):

(ax_4) $(\bar{\delta} \wedge \bar{\mu}) \Rightarrow \overline{\delta \wedge \mu}$

(ax_5) $(\bar{\delta} \Rightarrow \bar{\mu}) \Rightarrow \overline{\delta \rightarrow \mu}$

(ax_6) $S_j(\bar{\delta}_1, \bar{\delta}_2, \ldots, \bar{\delta}_n) \Leftrightarrow \overline{O_{S_j}(\delta_1, \delta_2, \ldots, \delta_n)}$

(ax_7) $((\alpha_1 \Leftrightarrow \beta_1)^{k_1} \& (\alpha_2 \Leftrightarrow \beta_2)^{k_2} \& \ldots \& (\alpha_n \Leftrightarrow \beta_n)^{k_n}) \Rightarrow$
$(S_j(\alpha_1, \alpha_2, \ldots, \alpha_n) \Leftrightarrow S_j(\beta_1, \beta_2, \ldots, \beta_n))$,
where $(k_1, k_2, \ldots, k_n) = Ex(S_j)$

(ax_8) $\alpha \Rightarrow \bar{1}$

(ax_9) $\alpha \Rightarrow \alpha$

(ax_{10}) $(\beta \Rightarrow \gamma) \Rightarrow ((\alpha \Rightarrow \beta) \Rightarrow (\alpha \Rightarrow \gamma))$

(ax_{11}) $(\alpha \Rightarrow (\beta \Rightarrow \gamma)) \Rightarrow (\beta \Rightarrow (\alpha \Rightarrow \gamma))$

(ax_{12}) $(\alpha \wedge \beta) \Rightarrow \alpha$

(ax_{13}) $(\alpha \wedge \beta) \Rightarrow \beta$

(ax_{14}) $(\gamma \Rightarrow \alpha) \Rightarrow ((\gamma \Rightarrow \beta) \Rightarrow (\gamma \Rightarrow (\alpha \wedge \beta)))$

(ax_{15}) $\alpha \Rightarrow (\alpha \vee \beta)$

(ax_{16}) $\beta \Rightarrow (\alpha \vee \beta)$

(ax$_{17}$) $(\alpha \Rightarrow \gamma) \Rightarrow ((\beta \Rightarrow \gamma) \Rightarrow ((\alpha \vee \beta) \Rightarrow \gamma))$

(ax$_{18}$) $\alpha \Rightarrow (\beta \Rightarrow (\alpha \& \beta))$

(ax$_{19}$) $(\alpha \Rightarrow (\beta \Rightarrow \gamma)) \Rightarrow ((\alpha \Rightarrow \beta) \Rightarrow \gamma)$

(ax$_{20}$) $(\alpha \& \bar{1}) \Leftrightarrow \alpha$

(ax$_{21}$) $\alpha(\alpha/\gamma) \Rightarrow (\exists x)\alpha$

(ax$_{22}$) $(\forall x)\alpha \Rightarrow \alpha(x/\gamma)$

(ax$_{23}$) $\sim\sim \alpha \Leftrightarrow \alpha$

(ax$_{24}$) $(\bar{Q}x)\alpha \Leftrightarrow \sim (\bar{Q}'x) \sim \alpha$

(ax$_{25}$) $(\forall x)(\alpha \Leftrightarrow \beta) \Rightarrow ((\bar{Q}x)\alpha \Leftrightarrow (\bar{Q}x)\beta)$

(ax$_{26}$) $(\bar{Q}x) \sim \alpha \Leftrightarrow \sim (\bar{Q}'x)\alpha$

(ax$_{27}$) $(\bar{Q}'x) \sim \alpha \Leftrightarrow \sim (\bar{Q}x)\alpha$

(ax$_{28}$) $\sim (\alpha \Rightarrow \beta) \Leftrightarrow (\alpha \;\&\; \sim \beta)$

(ax$_{29}$) $(\alpha \wedge \beta) \Leftrightarrow \sim (\sim \alpha \vee \sim \beta)$

(ax$_{30}$) $(\alpha \Rightarrow \beta) \Leftrightarrow \sim (\alpha \;\&\; \sim \beta)$

(ax$_{31}$) $(\alpha \& \beta) \Rightarrow ((\beta \Rightarrow \gamma) \Rightarrow (\alpha \& \gamma))$

(ax$_{32}$) $(\sim \alpha \Rightarrow \beta) \Leftrightarrow (\sim \beta \Rightarrow \alpha)$

(ax$_{33}$) $\alpha \& \beta \Rightarrow \alpha$

(ax$_{34}$) $\alpha \& \beta \Rightarrow \beta$

(ax$_{35}$) $x_i = x_i$

(ax$_{36}$) $(x_1 = y_1) \Rightarrow ((x_2 = y_2) \Rightarrow ((x_3 = y_3) \Rightarrow \ldots ((x_n = y_n) \Rightarrow$
$(f_j(x_1, x_2, \ldots, x_n) = f_j(y_1, y_2, \ldots, y_n))) \ldots))$

(ax$_{37}$) $(x_1 = y_1) \Rightarrow ((x_2 = y_2) \Rightarrow ((x_3 = y_3) \Rightarrow \ldots ((x_n = y_n) \Rightarrow$
$(p_i(x_1, x_2, \ldots, x_n) \Rightarrow p_i(y_1, y_2, \ldots, y_n))) \ldots))$

(ax$_{38}$) $(\alpha \vee \beta) \Leftrightarrow (\beta \vee \alpha)$

(ax$_{39}$) $(\alpha \wedge \beta) \Leftrightarrow (\beta \wedge \alpha)$

(ax_{40}) $(\alpha \& \beta) \Leftrightarrow (\beta \& \alpha)$

(ax_{41}) Every formula $\alpha \in S$, with degree of incidence equal to 0, is an axiom.

The set

$$R = \{r_{\alpha,\delta} : \delta \in L\} \cup \{r_i : i = 1, 2, 3, 4, 5\}$$

is the complete set of rules of inference ($r_{\alpha,\delta}$ denoting the rule of α, δ-precedence).

8.6.8 Fuzzy First-Order Theories

Let (A, R) be a fuzzy predicate calculus of the first order. By a fuzzy theory over (A, R) we mean a triple

$$\mathrm{T} = (A, A_{\mathrm{T}}, R)$$

in which A_{T} is any fuzzy set in S, called the set of specific axioms of the theory.

Suppose D is a structure for the language S. We call D a model for $\mathrm{T} = (A, A_{\mathrm{T}}, R)$ if

$$A_{\mathrm{T}}(\alpha) \leq v(\alpha)_{\mathrm{D}}$$

holds for every $\alpha \in S$; in symbols, $\mathrm{D} \models \mathrm{T}$. Since $A(\alpha) \leq v(\alpha)_{\mathrm{D}}$, we have

$$(CnA_{\mathrm{T}})(\alpha) = \bigwedge\{v(\alpha)_{\mathrm{D}} : \mathrm{D} \models \mathrm{T}\}.$$

If

$$(CnA_{\mathrm{T}})(\alpha) = \delta,$$

we call α a δ-theorem of T and write $\mathrm{T} \vdash_{\overline{\delta}} \alpha$; we also say that α is derivable in T in degree δ. When $\mathrm{T} \vdash_{\overline{1}} \alpha$, we write simply $\mathrm{T} \vdash \alpha$ and call α a theorem of T.

The following two facts follow immediately from the definitions:

(a_1) $CnA_{\mathrm{T}} \subseteq Cn_{Sem}A_{\mathrm{T}}$;

(a_2) If $\mathrm{T} \vdash_{\mu} \alpha$ and $\delta \leq \mu$ then $\mathrm{T} \vdash_{\delta} \alpha$.

Lemma 8.15

If $\mathrm{T} \vdash_{\overline{\delta}} \alpha$ then $\mathrm{D}(\alpha) \geq \delta$ holds for every model D of theory T.

Proof

Since $\mathrm{T} \vdash_{\overline{\delta}} \alpha$ and $\mathrm{D} \models_{\mu} \alpha$ imply $\delta \leq \mu$, the claim is proved. ∎

Let T be a first-order fuzzy theory. The element

$$\mu = \bigvee\{\varphi : \varphi = \delta_1 \otimes \delta_2, \mathrm{T} \vdash_{\overline{\delta_1}} \alpha, \mathrm{T} \vdash_{\overline{\delta_2}} \sim \alpha \quad \text{for} \quad \alpha \in S\}$$

will be called the degree of inconsistency of T (the theory T is then said to be μ-inconsistent); the element $\delta = -\mu$ is called its degree of consistency.

If the theory T is consistent and if $T\vdash_{\delta} \sim \alpha$ then $T\vdash_{-\delta} \alpha$. It can however happen that $T\vdash_{\delta_1} \alpha$ and $T\vdash_{\delta_2} \sim \alpha$ occur simultaneously, with positive degrees $\delta_1 \neq 0$ and $\delta_2 \neq 0$ (situations of this type are nothing unusual in the disputes of our everyday life).

Lemma 8.16

Suppose T is μ-inconsistent. Then

$$\mu \leq \bigvee\{\varphi : T\vdash_{\varphi} \alpha \,\&\sim \alpha, \alpha \in S\}.$$

And conversely, if there exists $\delta > 0$ such that $T\vdash_{\delta} \alpha \,\&\sim \alpha$ for some $\alpha \in S$, then T is δ-inconsistent.

Proof

Since T is μ-inconsistent, there exists α such that $T\vdash_{\delta_1} \alpha$ and $T\vdash_{\delta_2} \sim \alpha$, and $\delta_1 \otimes \delta_2 \leq \mu$. Let p_α and $p_{\sim\alpha}$ be proofs of α and $\sim \alpha$, and let $Val(p_\alpha) = \delta_1', Val(p_{\sim\alpha}) = \delta_2'$. The sequence

$(*)\quad p = (p_\alpha, p_{\sim\alpha}, \alpha \Rightarrow (\sim \alpha \Rightarrow \alpha \,\&\sim \alpha), \sim \alpha \Rightarrow (\alpha \,\&\sim \alpha), \alpha \,\&\sim \alpha)$

is a proof, of value $\delta_1' \otimes \delta_2'$. Writing

$$\mu' = \bigvee\{Val(p) : p \text{ is a proof of form } (*)\}$$

we have

$$T\vdash_{\varphi} \alpha \,\&\sim \alpha \quad \text{wherever} \quad \mu' \leq \varphi$$

and

$$\mu = \bigvee\{\mu' : \alpha \in S\}.$$

This proves the first part of the lemma.

For the second part, assume $T\vdash_{\delta} \alpha \,\&\sim \alpha$ for some $\alpha \in S$ and $\delta > 0$, and assume that T is δ-consistent. Then $D \models_\mu \alpha \,\&\sim \alpha$ holds for every model D and every μ with $\delta \leq \mu$. Yet,

$$D(\alpha \,\&\sim \alpha) = D(\alpha) \otimes D(\sim \alpha) = D(\alpha) \otimes (-D(\alpha)) = 0,$$

a contradiction. ∎

Theorem 8.21

If a fuzzy theory T has a model then T is consistent.

Proof

If $D \models_\delta \alpha$ then $D \models_{-\delta} \sim \alpha$. Hence $T\vdash_{\mu} \alpha$ and $T\vdash_{\varphi} \sim \alpha$ wherever $\mu \leq \delta$ and $-\varphi \leq -\delta$. So $\mu \otimes \varphi \leq 0$, showing that T is consistent. ∎

The axiom system of Section 8.6.7 implies that in a fuzzy theory T every formula is derivable, at least in degree 0.

A fuzzy theory T is called complete iff it is consistent and, for each $\alpha \in S$, if $T\vdash_{-\delta} \sim \alpha$ then $T\vdash_{\delta} \alpha$. The notions of extension, conservative extension and simple extension are defined in fuzzy theories analogously to the corresponding notions in classical logic. The analogues of theorems on extensions and on canonical models and the consistency and completeness theorems are also valid. For details we again refer to [Novák 1987].

9 Approximation Logics

9.1 Introduction

In computer science, one is often faced with a problem whose solution requires the use of approximate methods. It suffices to mention the recognition of images, automatic classification, or inductive reasoning. In response to this demand, the theory of rough sets has arisen (its principles formulated in the works by Z. Pawlak), giving rise in turn to rough logics.

The objective of rough logic, also called approximation logic, is to provide a method of inference in situations where the object of inference is not known precisely and can only be approached in a more or less adequate manner (is thus only "roughly" known). The degree of accuracy in defining (approximating) objects or sets is measured in rough logics in terms of so-called indiscernibility relations. These are certain equivalence relations that carry the information as to which objects are (for a given observer) not distinguishable, i.e., belong to a common equivalence class, and can thus be considered as providing an approximation to objects defined more precisely. A logic can be equipped with one or several relations of this kind; when there are several, they can be viewed as degrees of precision attributed to various observers or research experts in their investigation of the space of samples in question.

Rough logics, also called approximation logics, are the object of this chapter. Section 9.2 introduces the basic concepts. In Section 9.3 we deal with the case in which the indiscernibility relations constitute a chain, i.e., are linearly ordered with respect to their "precision." The results presented are taken from the papers by H. Rasiowa and A. Skowron [Rasiowa and Skowron 1984, 1985].

Section 9.4 is devoted to logics whose indiscernibility relations form a partially ordered structure. The approach to the question is rooted in the many-valued logics of Post. Approximation logics of the type dealt with here were created by H. Rasiowa [Rasiowa 1986].

For both these types of rough logics, we present a deduction system; that is to say, we state the axioms and rules of inference, and we formulate the completeness theorems for the respective systems.

9.2 Rough Sets

The underlying ground for approximation logics is the theory of rough sets, due to Z. Pawlak [Pawlak 1981, 1982, 1984]. Therefore we commence with a brief survey of that theory.

Let A be any nonempty set and let I be an equivalence relation in A. A subset $T \subseteq A$ is said to be I-open if and only if, for any $x, y \in A$, the conditions $x \in T$ and xIy imply $y \in T$.

The following facts are immediate corollaries of this definition:

(p₁) For each $x \in A$ the coset $[x]_I = \{y : xIy\}$ is I-open.

(p₂) If a set T is I-open, then so is $A \setminus T$.

(p₃) For a family of I-open sets $\{T_j : j \in J\}$, the union $\bigcup_{j \in J} T_j$ and the intersection $\bigcap_{j \in J} T_j$ are I-open.

Denoting by T_I the family of all I-open subsets of A we see that (A, T_I) is a topological space.

Given a pair $U = (A, I)$, where I is an equivalence relation in A, consider for every subset $X \subseteq A$ the sets

(a₁) $\underline{U}_I(X) = \{x \in A : [x]_I \subseteq X\}$,

(a₂) $\overline{U}_I(X) = \{x \in A : [x]_I \cap X \neq \emptyset\}$,

(a₃) $Fr_I(X) = \overline{U}_I(X) \setminus \underline{U}_I(X)$.

They are called, respectively, the lower approximation, the upper approximation and the boundary (or fringe) of X in (A, I).

It is easy to see that $\underline{U}_I(X)$, $\overline{U}_I(X)$ and $Fr_I(X)$ are I-open sets. Obviously, $\underline{U}_I(X)$ and $\overline{U}_I(X)$ can be regarded as the interior of X and the closure of X in the topology T_I.

An equivalence relation $I \subseteq A^2$ induces an equivalence relation S_I in $P(A)$, the power set of A (i.e., the set of all subsets of A) as follows:

(g₁) XS_IY iff $\underline{U}_I(X) = \underline{U}_I(Y)$ and $\overline{U}_I(X) = \overline{U}_I(Y)$.

Each equivalence class

$$[X]_{S_I} = \{Y \subseteq A : XS_IY\}$$

is called a rough set modulo relation I, or short: an I-rough set.

Let M denote the set of all pairs (B, C) of subsets of A such that

(b₁) $B \cap C = \emptyset$;

(b₂) B, C are I-open sets;

(b₃) for every $x \in C$ there exists $y \in B$ such that $x \neq y$ and xIy.

It is not hard to see that the mapping $i: P(A)/S_I \to M$ given by

$$i([X]_{S_I}) = (\underline{U}_I(X), Fr_I(X))$$

is a bijection.

Consequently, a rough set can be equally well defined as a quadruple $P = (A, I, B, C)$ where I is an equivalence in A and (B, C) is a pair of subsets of A satisfying conditions (b_1), (b_2), (b_3). Then I is referred to as a relation of indiscernibility, and the pair (A, I) is called an approximation space.

Suppose $U = (A, I)$ is an approximation space, in the sense just defined, and let n be a positive integer. Consider the pair $U^n = (A^n, I^n)$, with $I^n \subseteq A^n \times A^n$ defined as follows:

$$((u_1, u_2, \ldots, u_n), (v_1, v_2, \ldots, v_n)) \in I^n$$

if and only if $(u_j, v_j) \in I$ for each j, $1 \leq j \leq n$. Then U^n is an approximation space.

Equivalence classes modulo I^n,

$$[(u_1, u_2, \ldots, u_n)]_{I^n},$$

will be called elementary n-ary relations in U; the empty set is also admitted as a rough relation. Unions of n-ary rough relations will be called definable relations in the approximation space U. The family of all definable n-ary relations in U will be denoted by $def(U^n)$.

Let $R \subseteq A^n$. By $\overline{U}(R)$ and $\underline{U}(R)$, we denote the least definable relation containing R and the greatest definable relation contained in R, respectively. These relations will be called the upper and lower approximations of R in U.

In view of these definitions:

(c$_1$) $(u_1, u_2, \ldots, u_n) \in \overline{U}(R)$ if and only if there exists $v_1, v_2, \ldots, v_n \in A$ such that $(u_j, v_j) \in I$ for $j = 1, 2, \ldots, n$ and $(v_1, v_2, \ldots, v_n) \in R$;

(c$_2$) $(u_1, u_2, \ldots, u_n) \in \underline{U}(R)$ if and only if, for any $v_1, v_2, \ldots, v_n \in A$ the conditions $(u_j, v_j) \in I$ for $j = 1, 2, \ldots, n$ force $(v_1, v_2, \ldots, v_n) \in R$.

Evidently, a relation $R \subseteq A^n$ is definable in U if and only if $\underline{U}(R) = \overline{U}(R)$.

Finally, we introduce the following terminology (concerning relations $R \subseteq A^n$):

(d$_1$) R is approximately definable in U iff $\underline{U}(R) \neq \emptyset$ and $\overline{U}(R) \neq U^n$;

(d$_2$) R is outer-undefinable in U iff $\underline{U}(R) \neq \emptyset$ and $\overline{U}(R) = U^n$;

(d$_3$) R is inner-undefinable in U iff $\underline{U}(R) = \emptyset$ and $\overline{U}(R) \neq U^n$;

(d$_4$) R is totally-undefinable in U iff $\underline{U}(R) = \emptyset$ and $\overline{U}(R) = U^n$.

9.3 Rough Logics with a Chain of Indistinguishability Relations

9.3.1 Basic Concepts

The notions of the preceding section are generalized to the case of many indistinguishability relations, arranged into a chain.

Let A be a nonempty set, let ξ be an ordinal number, and suppose we are given a family of equivalence relations $I_j \subset A \times A$, indexed by ordinals $j < \xi$, such that $I_{j+1} \subset I_j$ for $0 \le j < j+1 < \xi$. Let $V_j = (A, I_j)$ for $j < \xi$. The system

$$U = (A, (I_j)_{0 \le j < \xi})$$

will be again called an approximation space, and the I_j will be called indiscernibility relations. Thus, we are dealing with the case in which the relations of indistinguishability form a chain.

When $\xi = \omega + 1$, we set $I_\omega = \bigcap_{0 \le j < \omega} I_j$.

For a positive integer n consider the system

$$U^n = (A^n, (I_j^n)_{0 \le j < \xi})$$

where each I_j^n is the relation in $A^n \times A^n$ defined as follows:

$$((u_1, \ldots, u_n), (u'_1, \ldots, u'_n)) \in I_j^n \quad \text{iff} \quad (u_i, u'_i) \in I_j \quad \text{for} \quad i = 1, \ldots, n.$$

Every such system U^n will be also called an approximation space.

For each j with $0 \le j < \xi$, equivalence classes modulo I_j^n are called I_j-elementary n-argument relations in U; the empty set is also admitted. Unions of I_j-elementary n-argument relations are called I_j-definable relations in U. The family of all those relations is denoted by $\mathrm{Def}(U_j^n)$.

Let $R \subset A^n$ be any subset.

The upper approximation of R in U, denoted by $\overline{U}_j R$, is defined as the least n-argument I_j-definable relation containing R.

The lower approximation of R in U, denoted by $\underline{U}_j R$, is defined as the greatest n-argument I_j-definable relation contained in R.

Thus,

(1) $(u_1, \ldots, u_n) \in \overline{U}_j R$ if and only if there exists $u'_1, \ldots, u'_n \in A$ such that $(u_i, u'_i) \in I_j$ for $i = 1, \ldots, n$ and $(u'_1, \ldots, u'_n) \in R$;

(2) $(u_1, \ldots, u_n) \in \underline{U}_j R$ if and only if, for any $u'_1, \ldots, u'_n \in A$, the conditions $(u_i, u'_i) \in I_j$ for $i = 1, \ldots, n$ imply $(u'_1, \ldots, u'_n) \in R$;

(3) $\overline{U}_{j+1}R \subseteq \overline{U}_jR$,
$\underline{U}_jR \subseteq \underline{U}_{j+1}R$
for $0 \leq j < j+1 < \xi$.

Given a relation $R \subset A^n$, we say that:

(d$_1$) R is I_j-definable iff $\underline{U}_jR = \overline{U}_jR$;

(d$_2$) R is approximately I_j-definable iff $\underline{U}_jR \neq \emptyset$ and $\overline{U}_jR \neq A^n$;

(d$_3$) R is outer-I_j-undefinable iff $\underline{U}_jR \neq \emptyset$ and $\overline{U}_jR = A^n$;

(d$_4$) R is inner-I_j-undefinable iff $\underline{U}_jR = \emptyset$ and $\overline{U}_jR \neq A^n$;

(d$_5$) R is totally I_j-undefinable iff $\underline{U}_jR = \emptyset$ and $\overline{U}_jR = A^n$.

Example [Rasiowa 1989]

Let $U = (R^+, I_0)$ where R^+ is the set of positive real numbers and I_0 is the relation defined by

$$(x,y) \in I_0 \quad \text{iff} \quad \text{entier}(x) = \text{entier}(y).$$

Thus

$$U^2 = (R^+ \times R^+, I_0^2),\ ((x,y),(z,t)) \in I_0^2 \quad \text{iff} \quad (x,z) \in I_0,\ (y,t) \in I_0.$$

The coset modulo I_0^2 containing a pair (m,n), $m,n \in N$, is the square

$$[(m,n)] = \{(x,y) : m \leq x < m+1,\ n \leq y < n+1\}.$$

Let Id be the identity relation in R^+ and $\leq$ be the usual ordering of the real line. Then:

$$\underline{U}_0 \leq \neq \emptyset \quad \text{and} \quad \overline{U}^0 \leq \neq\ R^+ \times R^+;$$

hence $\leq$ is approximately definable;

$$\underline{U}_0 Id = \emptyset \quad \text{and} \quad \overline{U}^0 Id \neq R^+ \times R^+;$$

hence Id is inner-I_0-undefinable.

The set

$$Z = \{(x,y) : x \in N \cup \{0\}, y \in N \cup \{0\}\}$$

is totally I_0-undefinable, whereas

$$S = \{(x,y) : 0 \leq x < 1,\ 0 \leq y < 1\} \cup Z$$

is outer-I_0-undefinable.

The approximation space U can be enriched by introducing an infinite family of relations, for example as follows:

$$U = (R^+, (I_j)_{0 \leq j < \omega}),$$

where $(x,y) \in I_j$ (i.e., x and y are considered to be related by I_j) when

$$n + \frac{m}{2^j} \leq x < n + \frac{m+1}{2^j} \quad \text{iff} \quad n + \frac{m}{2^j} \leq y < n + \frac{m+1}{2^j}.$$

9.3.2 Approximate Logical Systems

We begin by defining the language we are going to use. This will be the language of the first-order predicate calculus L_ξ, in which:

V is the set of individual variables, denoted by $x, y, z, x_1, x_2, \ldots$;
P_j are binary predicates denoting indiscernibility relations, $0 \leq j < \xi$;
$\sigma_i^{n_i}$ are n_i-argument predicates, $n_i \in N, i = 1, \ldots, m$;
$\varphi_i^{n_i'}$ are n_i'-argument function symbols, called also functors,

$$n_i' \in N \cup \{0\},\ i = 1, \ldots, k;$$

$\underline{U}_j$ are one-argument propositional connectives, $0 \leq j < \xi$;
$\vee, \wedge, \neg, \Rightarrow$ are the classical connectives;
$\forall, \exists$ are the quantifiers.

The set of terms is denoted by T.

The set of atomic formulas contains all formulas of the form:

(1) $P_j(t_1, t_2)$ with $t_1, t_2 \in T$, $0 \leq j < \xi$,

or

(2) $\sigma_i^{n_i}(t_1, \ldots, t_{n_i})$ with $t_1, \ldots, t_{n_i} \in T$, $i = 1, \ldots, m$.

The set of all formulas is defined in the usual way and is denoted by S_ξ. Define

$$\overline{U}_j \alpha = \neg \underline{U}_j \neg \alpha \quad \text{for} \quad \alpha \in S_\xi,\ 0 \leq j < \xi.$$

The semantics of L_ξ is defined by a model and a valuation (which will enable us to describe satisfiability).

By a model we mean a relational system

$$M = (A, (I_j)_{0 \leq j < \xi}, R_1^{n_1}, \ldots, R_m^{n_m}, f_1^{n_1'}, \ldots, f_k^{n_k'}),$$

in which:

$U = (A, (I_j)_{0 \leq j < \xi})$ is an approximation space;
R^{n_i} are n_i-argument relations in A, $n_i \in N$, $i = 1, \ldots, m$;
$f_i^{n_i'}$ are n_i'-argument functions in A, $n_i' \in N$, $i = 1, \ldots, k$.

By a valuation we mean an arbitrary function $v : V \to A$.

The symbol Val will stand for the set of all valuations; thus $Val = A^V$.

A valuation $v \in Val$ extends according to the following rules (for $t \in T$, $v(t)$ denotes the value given to t by v):

(1) $v(\varphi_i^k(t_1, \ldots, t_k)) = f_i(v(t_1), \ldots, v(t_k))$;

(2) $v(P_j(t_1, t_2)) = 1$ iff $(v(t_1), v(t_2)) \in I_j$, for $0 \leq j < \xi$;

(3) $v(\sigma_i^{n_i}(t_1, \ldots, t_{n_i})) = 1$ iff $(v(t_1), \ldots, v(t_{n_i})) \in R_i^{n_i}$, for $i = 1, \ldots, m$;

(4) $v(\neg\alpha) = 1$ iff $v(\alpha) = 1$ fails to hold;

(5) $v(\alpha \vee \beta) = 1$ iff $v(\alpha) = 1$ or $v(\beta) = 1$;

(6) $v(\alpha \wedge \beta) = 1$ iff $v(\alpha) = 1$ and $v(\beta) = 1$;

(7) $v(\alpha \Rightarrow \beta) = 1$ iff either $v(\alpha) = 1$ fails or $v(\beta) = 1$ holds;

(8) $v((\forall x)\alpha) = 1$ iff $v_u(\alpha)$ holds for each $u \in A$; here v_u is defined by: $v_u(x) = u, v_u(y) = v(y)$ for all $y \in V, y \neq x$;

(9) $v((\exists x)\alpha) = 1$ iff $v((\forall x)\neg\alpha) = 1$ fails to hold;

(10) $v(\underline{U}_j\alpha(x_1, \ldots, x_n)) = 1$ iff, for any $u_1, \ldots, u_n \in A$, the conditions $(v(x_i), u_i) \in I_j$ for $i = 1, \ldots, n$ imply $v_{u_1,\ldots,u_n}(\alpha(x_1, \ldots, x_n)) = 1$; here $v_{u_1,\ldots,u_n}$ are defined by:

$$v_{u_1,\ldots,u_n}(x_i) = u_i \quad \text{for } i = 1, \ldots, n,$$
$$v_{u_1,\ldots,u_n}(y) = v(y) \quad \text{for } y \neq x_1, \ldots, x_n,\ y \in V.$$

A valuation v is said to satisfy a formula α if $v(\alpha) = 1$. Given α, we will denote the set of all valuations which satisfy α by α_M:

$$\alpha_M = \{v \in Val : v(\alpha) = 1 \text{ in model } M\}.$$

We say that:

(1) α is satisfiable in model M iff $\alpha_M \neq \emptyset$;

(2) α is true in model M iff $\alpha_M = Val$;

(3) α is true iff it is true in every model.

Now, let $\alpha(x_1, \ldots, x_n)$ be a formula containing free variables $x_1, \ldots, x_n$. Define

$$\alpha_{M^{xx}} = \{(u_1, \ldots, u_n) \in A^n: \text{ there exists a valuation } v \in \alpha_m \text{ such that } u_1 = v(x_1), \ldots, u_n = v(x_n)\}.$$

Lemma 9.1

For any formula $\alpha(x_1, x_2, \ldots, x_n)$,

(a$_1$) $\alpha(x_1, x_2, \ldots, x_n)$ is satisfiable in M if and only if $\alpha(x_1, x_2, \ldots, x_n)_{M^{xx}} \neq \emptyset$;

(a$_2$) $\alpha(x_1, x_2, \ldots, x_n)$ is true in M if and only if $\alpha(x_1, x_2, \ldots, x_n)_{M^{xx}} = A^n$. ■

We introduce some further terminology:

(d$'_1$) $\alpha(x_1, \ldots, x_n)_{M^{xx}}$ is I_j-definable iff $\underline{U}_j\alpha(x_1, \ldots, x_n)_{M^{xx}} = \overline{U}_j\alpha(x_1, \ldots, x_n)_M$.

(d$'_2$) $\alpha(x_1, \ldots, x_n)_{M^{xx}}$ is approximately I_j-definable in U iff $\underline{U}_j\alpha(x_1, \ldots, x_n)$ is satisfiable in M and $\neg\overline{U}_j\alpha(x_1, \ldots, x_n)$ is satisfiable in M.

(d$'_3$) $\alpha(x_1, \ldots, x_n)_{M^{xx}}$ is outer-I_j-undefinable in U iff $\underline{U}_j\alpha(x_1, \ldots, x_n)$ is satisfiable in M and $\overline{U}_j\alpha(x_1, \ldots, x_n)$ is a tautology in M.

(d$'_4$) $\alpha(x_1, \ldots, x_n)_{M^{xx}}$ is inner-I_j-undefinable in U iff $\neg\underline{U}_j\alpha(x_1, \ldots, x_n)$ is a tautology in M and $\neg\overline{U}_j\alpha(x_1, \ldots, x_n)$ is satisfiable in M.

(d$'_5$) $\alpha(x_1, \ldots, x_n)_{M^{xx}}$ is totally undefinable in U_j iff $\neg\underline{U}_j\alpha(x_1, \ldots, x_n)$ and $\overline{U}_j\alpha(x_1, \ldots, x_n)$ are true in M.

Instead of I_j-definable, we also say definable in U_j.

Lemma 9.2

Consider $\alpha(x_1, \ldots, x_n)_{M^{xx}} \subseteq A^n$ and let $U_j = (A_j, I_j)$.

(11) $\alpha(x_1, \ldots, x_n)_{M^{xx}}$ is definable in U_j iff $v(\underline{U}_j\alpha(x_1, \ldots, x_n) \Leftrightarrow \overline{U}_j\alpha(x_1, \ldots, x_n)) = 1$ for every v.

(12) $\alpha(x_1, \ldots, x_n)_{M^{xx}}$ is approximately definable in U_j iff $v((\exists x_1, \ldots, \exists x_n \, \underline{U}_j\alpha(x_1, \ldots, x_n)) \wedge (\exists x_1, \ldots, \exists x_n \, \overline{U}_j\alpha(x_1, \ldots, x_n))) = 1$ for every v.

(13) $\alpha(x_1, \ldots, x_n)_{M^{xx}}$ is outer-undefinable in U_j iff $v((\exists x_1, \ldots, \exists x_n \, \underline{U}_j\alpha(x_1, \ldots, x_n)) \wedge (\forall x_1, \ldots, \forall x_n \, \overline{U}_j\alpha(x_1, \ldots, x_n))) = 1$ for every v.

(14) $\alpha(x_1, \ldots, x_n)_{M^{xx}}$ is inner-undefinable in U_j iff $v((\forall x_1, \ldots, \forall x_n \, \neg\underline{U}_j\alpha(x_1, \ldots, x_n)) \wedge (\exists x_1, \ldots, \exists x_n \, \overline{U}_j\neg\alpha(x_1, \ldots, x_n))) = 1$ for every v.

(15) $\alpha(x_1, \ldots, x_n)_{M^{xx}}$ is totally undefinable in U_j iff
$v((\forall x_1, \ldots, \forall x_n \neg \underline{U}_j \alpha(x_1, \ldots, x_n)) \wedge (\forall x_1, \ldots, \forall x_n \overline{U}_j \alpha(x_1, \ldots, x_n))) = 1$
for every v.

Proof

Immediate from the definitions. ■

The next stage is to construct the deduction system for the rough logic L_ξ. The system is composed of axioms and rules of inference.

The axioms are:

(1°) the axiom schemes of the classical propositional calculus;

(2°) $\underline{U}_j(\alpha \Rightarrow \beta) \Rightarrow (\underline{U}_j\alpha \Rightarrow \underline{U}_j\beta)$

(3°) $\underline{U}_j\alpha \Rightarrow \alpha$

(4°) $P_j(x, x)$

(5°) $P_j(x, y) \Rightarrow (P_j(z, y) \Rightarrow P_j(x, z))$

(6°) $P_j(x, y) \Rightarrow (\underline{U}_j\, \alpha(x) \Rightarrow \underline{U}_j\, \alpha(y))$

(7°) $\forall y_1 \ldots \forall y_n(((P_j(x_1, y_1) \wedge \ldots \wedge P_j(x_n, y_n)) \Rightarrow \alpha(y_1, \ldots, y_n)) \Rightarrow \underline{U}_j\, \alpha(x_1, \ldots, x_n))$,
for every formula α with n free variables;

(8°) $\alpha \Rightarrow \underline{U}_j\alpha$ for every formula α without free variables;

(9°) $P_{j'}(x, y) \Rightarrow P_j(x, y)$ wherever $0 \leq j \leq j' < \xi$.

The rules of inference are:

(1) $\dfrac{\alpha, \alpha \Rightarrow \beta}{\beta}$,

(2) $\dfrac{\alpha}{\underline{U}_j\alpha}$ for each j, $0 \leq j < \xi$,

(3) $\dfrac{\{\alpha \Rightarrow P_j(t_1, t_2)\}, 0 \leq j < \omega}{\alpha \Rightarrow P_\omega(t_1, t_2)}$ when $\xi = \omega + 1$,

(4) the rule of substitution,

(5) the rules of insertion and reduction of quantifiers.

Let $X \subseteq S$. As before, we write $\alpha \in CnX$ to indicate that α is derivable from X.

Lemma 9.3

(i) If $(\alpha \Rightarrow \beta) \in CnX$, then $(\underline{U}_j\alpha \Rightarrow \underline{U}_j\beta) \in CnX$

(ii) $P_j(x_1, y_1) \Rightarrow \ldots \Rightarrow (P_j(x_n, y_n) \Rightarrow (\underline{U}_j\alpha(x_1, \ldots, x_n) \Leftrightarrow \underline{U}_j\alpha(y_1, \ldots, y_n)) \ldots) \in Cn\emptyset$

(iii) $\underline{U}_j\alpha(x_1, \ldots, x_n) \Leftrightarrow \forall_{y_1}, \ldots, \forall_{y_n}(P_j(x_1, y_1) \wedge \ldots \wedge P_j(x_n, y_n) \Rightarrow a(y_1, \ldots, y_n)) \in Cn\emptyset$

(iv) $\underline{U}_j\alpha \Rightarrow \underline{U}_j\underline{U}_j\alpha \in Cn\emptyset$

(v) $\alpha \Rightarrow \underline{U}_j\neg\underline{U}_j\neg\alpha \in Cn\emptyset$

(vi) $\underline{U}_{j'}\alpha \Rightarrow \underline{U}_j\alpha \in Cn\emptyset$ for $0 \leq j \leq j' < \xi$

(vii) $\underline{U}_j\underline{U}_{j'}\alpha \Leftrightarrow \underline{U}_j\alpha \in Cn\emptyset$ for $0 \leq j \leq j' < \xi$

(viii) $\underline{U}_{j'}\underline{U}_j\alpha \Leftrightarrow \underline{U}_j\alpha \in Cn\emptyset$ for $0 \leq j \leq j' < \xi$

(ix) $(\forall x)\underline{U}_j\alpha(x) \Leftrightarrow \underline{U}_j(\forall x)\alpha(x) \in Cn\emptyset$

(x) $(\exists x)\underline{U}_j\alpha(x) \Rightarrow \underline{U}_j(\exists x)\alpha(x) \in Cn\emptyset$

(xi) $(\underline{U}_j\alpha \vee \underline{U}_j\beta) \Rightarrow \underline{U}_j(\alpha \vee \beta) \in Cn\emptyset$

(xii) $\underline{U}_j(\alpha \wedge \beta) \Leftrightarrow (\underline{U}_j\alpha \wedge \underline{U}_j\beta) \in Cn\emptyset$. ■

Theorem 9.1 (the soundness theorem)
If $\alpha \in CnX$ then, in each model M, if $v(\beta) = 1$ for $\beta \in X$, then $v(\alpha) = 1$. ■

Theorem 9.2 (the existence of a model)
If $\alpha \notin CnX$ then there exists a model M such that every $\beta \in X$ is true in M, whereas

$$v(\alpha) = 1 \quad \text{fails to hold}$$

for a certain valuation v. ■

The proofs of these two theorems are given in [Rasiowa 1989].

The logic L_ξ can be enriched by adjoining to its syntax the following formation rule:

(*) If $\alpha(\overline{x})$ is a formula, $\overline{x} = (x_1, \ldots, x_n)$ denoting the free variables that occur in α, then $\underline{U}_{\underline{i}}\alpha(\overline{x})$ and $\overline{U}_{\underline{i}}\alpha(\overline{x})$ are formulas; here $\underline{i} = (i_1, \ldots, i_k)$, $0 \leq i_j < \xi$ for $j = 1, \ldots, k$.

The semantics is extended accordingly by the rules:

(1) $v(\underline{U}_{\underline{i}}\alpha(\overline{x})) = 1$ iff $v'(\alpha(\overline{x})) = 1$ holds
for every v' such that $v'(y) = v(y)$
for $y \notin \{x_1, \ldots, x_k\}$ and
$(v(x_j), v'(x_j)) \in I_j$ for $j = 1, \ldots, k$;

(2) $v(\overline{U}_{\underline{i}}\alpha(\overline{x})) = 1$ iff $v'(\alpha(\overline{x})) = 1$ holds
for a certain v' such that $v'(y) = v(y)$
for $y \notin \{x_1, \ldots, x_k\}$ and
$(v(x_j), v'(x_j)) \in I_j$ for $j = 1, \ldots, k$.

The operators $\underline{U}_j, \overline{U}_j$ are a particular case of $\underline{U}_{\underline{i}}$, $\overline{U}_{\underline{i}}$; namely,

$$\underline{U}_j\alpha(x_1, \ldots, x_k) = \underline{U}_{\underline{i}}\alpha(x_1, \ldots, x_k) \quad \text{for} \quad \underline{i} = (j, \ldots, j) \quad (k \text{ times}).$$

The axiom schemes of L_ξ can be modified so as to provide a complete axiomatization of the extended logic; see [Rasiowa 1991].

9.3.3 Approximation Theories

Consider the theory in L_1 with the following axiom system:

(0°) the axioms for the equality predicate (=)

(1°) $\neg(E(x) \wedge C(x))$

(2°) $E(x) \vee C(x)$,

(3°) $In(x, y) \Rightarrow (E(x) \wedge C(y))$

(4°) $E(x) \Rightarrow (\exists y!)\, In(x, y)$

(5°) $P_0(x, x') \Leftrightarrow (In(x) = In(x'))$, where $In(x) = y \Leftrightarrow In(x, y)$

(6°) $\rho_i^n(x_1, \ldots, x_n) \Rightarrow (E(x_1) \wedge \ldots \wedge E(x_n))$

(6'°) $x_0 = \varphi_i^n(x_1, \ldots, x_n) \Rightarrow (E(x_0) \wedge \ldots \wedge E(x_n))$

(7°) certain additional specific axioms imposed on P_0.

Intuitively,

$E(x)$ says that x is an element (an object),

$C(x)$ says that x is the information corresponding to an equivalence class of P_0,

$In(x, y)$ says that y is the information corresponding to the equivalence class $[x]$ of P_0.

The set of all formulas in L_1 formed without the use of In, C, E will be denoted by S_E.

A formula $\alpha(x_1, \ldots, x_k)$ with free variables $x_1, \ldots, x_k$ will be called a C-formula if each variable x_i $(i = 1, \ldots, k)$ occurs within subformulas of the form $x_i = In(x_i')$ alone.

The set of all C-formulas will be denoted by S_C.

Here are some examples of C-formulas:

$$(1)\ (\exists x_1'), \ldots, (\exists x_n')(x_1 = In(x_1') \wedge \ldots \wedge x_n = In(x_n') \wedge\ \rho_i^n(x_1', \ldots, x_n'))$$

$$(2)\ (\forall x_1'), \ldots, (\forall x_n')(x_1 = In(x_1') \wedge \ldots \wedge x_n = In(x_n') \Rightarrow \rho_i^n(x_1', \ldots, x_n'))$$

$$(3)\ (\exists x_0'), \ldots, (\exists x_n')(x_0 = In(x_0') \wedge \ldots \wedge x_n = In(x_n') \wedge\ x_0' = \varphi_i^n(x_1', \ldots, x_n'))$$

$$(4)\ (\forall x_0'), \ldots, (\forall x_n')(x_0 = In(x_0') \wedge \ldots \wedge x_n = In(x_n') \Rightarrow x_0' = \varphi_i^n(x_1', \ldots, x_n'))$$

Let S_C^0 denote the set of all formulas in S_C composed of formulas of types (1)–(4) with the only aid of connectives $\wedge, \vee, \neg, \Rightarrow$ and quantifiers.

Type (1)–(4) formulas represent an approximation to concepts such as predicates and functors. Approximation properties are expressed by formulas of the set S_C^0.

In approximate inference reasonings, the key point is the formation of suitable approximations to given properties. We now introduce the concept which very well serves this purpose.

By an approximating translation for a theory T in model M we mean a partial function $t : S_E - \circ \rightarrow S_C^0$ such that, for $\alpha \in S_E, \beta \in S_C^0$,

$$t(\alpha) = \beta \quad \text{holds iff } In(\alpha_{M^{**}}) = \beta_{M^{**}} \text{ and}$$
$$In(\alpha_{M^{**}}) = \{(In(a_1), \ldots, In(a_k)) : (a_1, \ldots, a_k) \in \alpha_{M^{**}}\}.$$

A partial function $t : S_E - \circ \rightarrow S_C^0$ is said to be an approximating translation for T if it is so in every model M.

In [Rasiowa 1989] the problem of approximations to proposition sets is considered.

9.4 Approximation Logics with Partially Ordered Sets of Indiscernibility Relations

In the preceding section we presented approximation (rough) logics whose indistinguishability relations constitute a chain. Now we discuss the case in which the set of those relations is partially ordered.

We begin by defining the notions to serve as a basis for that more general logic.

9.4.1 Plain Semi-Post Algebras

Let $T = (T, \leq)$ be a partially ordered set.

By a T-ideal we mean a nonempty subset S of T such that the conditions $s \in S$ and $t \leq s$ force $t \in S$.

Let LT be the set whose elements are the empty set and all T-ideals. Obviously, LT is closed under unions and intersections, hence it is a complete lattice.

An important example of such a situation is this.

Consider a certain set of pieces of information T'. Let t, s be elements of T'; we write $t \leq s$ when "s implies t," or when, speaking imprecisely, information t is weaker than s. We then say that t approximates s.

In this situation, we may take for T the set of equivalence classes modulo: $(t \leq s$ and $s \leq t)$.

Since $(LT, \cup, \cap)$ is a complete lattice, we can define the implication $(\Rightarrow)$ and negation $(\neg)$ by

1) $t \Rightarrow s = \bigcup\{z \in LT : t \cap z \leq s\}$,

2) $\neg t = t \Rightarrow \emptyset$.

The resulting system $(LT, \cup, \cap, \Rightarrow, \neg)$ is a Heyting algebra, with $\bigwedge = \emptyset$ and $\bigvee = T$.

We now introduce the approximation operators d_t, indexed by elements $t \in T$:

$$d_t s = \begin{cases} \bigvee & \text{if } t \in s, \\ \bigwedge & \text{otherwise.} \end{cases}$$

The system

$$\mathrm{LT} = (LT, \cup, \cap, \Rightarrow, \neg, (d_t)_{t \in T}, (s)_{s \in LT})$$

is referred to as the basic plain semi-Post algebra (psP-algebra) of type T.

Let $ELT = \{e_s\}_{s \in LT}$. Let as agree that

$$\begin{aligned} e_s \leq e_t \quad &\text{iff} \quad s \subseteq t, \\ e_s = e_t \quad &\text{iff} \quad s = t. \end{aligned}$$

Then $(ELT, \leq)$ becomes a complete lattice.

Let $e_\vee = e_T$ and $e_\wedge = e_\emptyset$. We define implication and negation by

$$e_t \Rightarrow e_s = \bigcup\{e_u : e_t \cap e_u \leq e_s\},$$
$$\neg e_t = e_t \Rightarrow e_\wedge \qquad (t, s, u \in LT).$$

The operators d_t are defined just as they were in LT:

$$d_t e_s = \begin{cases} e_\vee & \text{if } t \in s, \\ e_\wedge & \text{otherwise.} \end{cases}$$

The algebra ELT $= (ELT, \cup, \cap, \Rightarrow, \neg, \ldots)$ is isomorphic to LT, hence is also a plain semi-Post algebra of type T.

In general, we define a plain semi-Post algebra over T $= (T, \leq)$ as a system

$$\mathrm{P} = (P, \cup, \cap, \Rightarrow, \neg, (d_t)_{t\in T}, (e_s)_{s\in LT})$$

satisfying the following conditions (for $a, b \in P,\;\; t, u \in T,\; s \in LT$):

(0) $(P, \cup, \cap, \Rightarrow, \neg)$ is a Heyting algebra, with unit $e_\vee$ and zero $e_\wedge$ $(= \neg e_\vee)$;

(1) $d_t(a \cup b) = d_t a \cup d_t b$;

(2) $d_t(a \cap b) = d_t a \cap d_t b$;

(3) $d_t d_u a = d_u a$;

(4) $d_t e_s = \begin{cases} e_\vee & \text{if } t \leq s, \\ e_\wedge & \text{otherwise;} \end{cases}$

(5) $d_t a \cup \neg d_t a = e_\vee$;

(6) $a = \bigcup_{t\in T} (d_t a \cap e_t)$.

Evidently, every basic psP-algebra is an instance of a psP-algebra.

Fundamental properties of psP-algebras

(i) If $t \leq s$ then $d_s a \leq d_t a$ for every $a \in P$.

(ii) For $t \in T$, $a \in P$, the elements $d_t a$ have complements and constitute a Boolean algebra $(B_P, \cup, \cap, \Rightarrow)$, where
$B_P = \{d_t a : t \in T, a \in P\} \cup \{\neg d_t a : t \in T, a \in P\}$.

It follows from (i) and (ii) that $(d_t)_{t\in T}$ is a descending sequence of Boolean elements in B_P.

(iii) If $s \le u$ in LT then $e_s \le e_u$;

if $s \ne u$ then $e_s \ne e_u$;

(iv) $a \le b$ iff $d_t a \le d_t b$ for all $t \in T$;

(v) $a = b$ iff $d_t a = d_t b$ for all $t \in T$;

(vi) $d_t(a \Rightarrow b) = \bigcap_{u \le t} (d_u a \Rightarrow d_u b)$;

(vii) $d_t(\neg a) = \bigcap_{u \le t} \neg d_u a$.

Let $B(U)$ be the set of all subsets of a given set U and let $DS_T B(U)$ be the family of all descending T-sequences of sets in $B(U)$; i.e.,

$$DS_T B(U) = \{Y_t : t \in T\}$$

where $Y_t \subseteq U$ for $t \in T$ and $Y_s \subseteq Y_t$ wherever $t \le s$.

Write

$$E_\vee = (X_t)_{t \in T} \text{ where } X_t = U \text{ for all } t \in T,$$
$$E_\wedge = (Z_t)_{t \in T} \text{ where } Z_t = \emptyset \text{ for all } t \in T,$$
$$E_s = (Y_t)_{t \in T} \text{ where } Y_t = \begin{cases} U & \text{if } t \le s, \\ \emptyset & \text{otherwise,} \end{cases}$$
$$Y \cup Z = (Y_t \cup Z_t)_{t \in T},$$
$$Y \cap Z = (Y_t \cap Z_t)_{t \in T},$$
$$Y \Rightarrow Z = (X_t)_{t \in T} \text{ where } X_t = \bigcap_{u \le t} (\neg Y_u \cup Z_t),$$
$$\neg Y = (X_t)_{t \in T} \text{ where } X_t = \bigcap_{u \le t} \neg X_u,$$
$$D_s Y = (X_t)_{t \in T} \text{ where } X_t = Y_s \text{ for all } t \in T.$$

Let $Y = (Y_t)_{t \in T}$, $Z = (Z_t)_{t \in T}$. Define:

$$Y \le Z \quad \text{iff} \quad Y_t \subseteq Z_t \quad \text{for all} \quad t \in T.$$

With this semi-ordering,

$$DS_T B(U) = (DS_T B(U), \cup, \cap, \Rightarrow, \neg, (D_t)_{t \in T}, (E_s)_{s \in LT})$$

becomes a psP-algebra of type T. It is called a complete psP-algebra to indicate that it is composed of all decreasing T-sequences over $B(U)$.

The following theorems are valid for psP-algebras.

Theorem 9.3

Let P be a psP-algebra of type T.

The algebra

$$\mathrm{ELT} = (ELT, \cup, \cap, \Rightarrow, \neg, (d_t)_{t\in T}, (e_s)_{s\in LT})$$

is a complete sublattice of P and we have for each $S \subseteq LT$:

$$(\mathrm{P})\bigcup_{s\in S} e_s = (\mathrm{ELT})\bigcup_{s\in S} e_s,$$
$$(\mathrm{P})\bigcap_{s\in S} e_s = (\mathrm{ELT})\bigcap_{s\in S} e_s. \quad \blacksquare$$

Theorem 9.4

For any set $\{a_w\}_{w\in W}$ of elements of P and for every $t \in T$,

$$(\mathrm{P}) \bigcup_{w\in W} d_t a_w = (B_P) \bigcup_{w\in W} d_t a_w,$$
$$(\mathrm{P}) \bigcap_{w\in W} d_t a_w = (B_P) \bigcap_{w\in W} d_t a_w. \quad \blacksquare$$

Theorem 9.5

For every $a \in \mathrm{P}$ and any set $\{a_w\}_{w\in W}$ of elements of P,

$$a = (\mathrm{P}) \bigcup_{w\in W} a_w \quad \text{iff} \quad d_t a = (B_P) \bigcup_{w\in W} d_t a_w \text{ for all } t \in T,$$
$$a = (\mathrm{P}) \bigcap_{w\in W} a_w \quad \text{iff} \quad d_t a = (B_P) \bigcap_{w\in W} d_t a_w \text{ for all } t \in T. \quad \blacksquare$$

Theorem 9.6

For every psP-algebra P of type T there exists a monomorphism $h : \mathrm{P} \to DS_T B(U)$ with the following property: if (Q) is any countable set of infinite joins and meets in P:

$$Q = \{a_n : n \in N\} \cup \{b_n : n \in N\},$$
$$a_n = (\mathrm{P}) \bigcup_{j\in J_n} a_{n_j}, \; b_n = (\mathrm{P}) \bigcap_{k\in K_n} b_{n_k},$$

then h preserves the following joins and meets:

$$d_t a_n = (B_p) \bigcup_{j\in J_n} d_t a_{n_j},$$
$$d_t b_n = (B_p) \bigcap_{k\in K_n} d_t b_{n_k},$$

for each t and all all infinite joins and meets in (Q). $\blacksquare$

Now we investigate the set ELT^U, i.e., the set of all functions $f : U \to ELT$; we will call them psP-functions of type T. To every such f we assign the Boolean functions

$$d_t f : U \to \{e_\vee, e_\wedge\}, \quad \text{for } t \in T,$$

defined by

$$d_t f(u) = \begin{cases} e_\vee & \text{if} \quad f(u) = e_s \quad \text{and} \quad t \leq s, \\ e_\wedge & \text{if} \quad f(u) = e_s \quad \text{and} \quad \text{it is not the case that} \quad t \leq s. \end{cases}$$

Hence,

$$f(u) = \bigcup_{t \in T} (d_t f(u) \cap e_t).$$

Each $d_t f$ is thus a characteristic function of a subset of U. We will denote this subset by $Sd_t f$:

$$Sd_t f = \{u \in U : d_t f(u) = e_\vee\}.$$

Hence, given f, we have

$$Sd_s f \subseteq Sd_t f \quad \text{wherever} \quad t \leq s,$$

which means that $(Sd_t f)_{t \in T}$ is a descending T-system of sets in the field $B(U)$.

Write $S(f) = \{u \in U : f(u) = e_\vee\}$. Then

$$S(f) = \bigcap_{t \in T} Sd_t f;$$

so $(Sd_t f)_{t \in T}$ is a sequence approximating $S(f)$.

We are now in position to define rough logics of type T, with T a partially ordered set.

9.4.2 Approximation Logic of Type T

Let L_T be the language of predicates of first-order. The syntax of L_T consists of:

- the set of individual variables, V;
- propositional constants: $e_t, t \in T$;
- predicate symbols: $p_i(x_1, \ldots, x_n), i = 1, \ldots, m$;
- connectives: $\vee, \wedge, \Rightarrow, \neg, (d_t)_{t \in T}$;
- quantifiers: $\forall, \exists$.

Propositional constants and predicate symbols are jointly referred to as atomic formulas. The symbol $\alpha \Leftrightarrow \beta$ stands for: $(\alpha \Rightarrow \beta) \wedge (\beta \Rightarrow \alpha)$.

By a model for L_T in a universe $U \neq \emptyset$ we mean a structure

$$M = (U, (S_{1t})_{t \in T}, \ldots, (S_{mt})_{t \in T}); \text{ where}$$

with every n_i-argument predicate p_i there is associated a decreasing T-sequence $(S_{it})_{t \in T}$ of n_i-argument relations in U.

Each S_{it} should be understood as the n_i-argument relation corresponding in model M to the two-valued predicate $d_t p_i$ approximating p_i.

If $t \leq s$ then $S_{is} \subseteq S_{it}$ for $i = 1, \ldots, m$.

Let Val be the set of all valuation functions. A valuation v is said to satisfy a formula α in model M if $v(\alpha) = e_\vee$; in symbols, $M, v \models \alpha$.

Valuation of predicate symbols is realized by means of functions

$$f_i : U^{n_i} \to \{e_s\}_{s \in LT}$$

such that

$$d_t f_i(u_1, \ldots, u_{n_i}) = \begin{cases} e_\vee & \text{if } (u_1, \ldots, u_{n_i}) \in S_{it}, \\ e_\wedge & \text{otherwise.} \end{cases}$$

For atomic formulas,

$$v(\underline{e}_s) = e_s,$$
$$v(p_i(x_1, \ldots, x_{n_i})) = f_i(v(x_1), \ldots, v(x_{n_i})).$$

For composite formulas, v obeys the following rules:

$$v(d_t(p_i(x_1, \ldots, x_{n_i}))) = e_\vee \quad \text{iff} \quad (v(x_1), \ldots, v(x_{n_i})) \in S_{it};$$
$$v(d_t \underline{e}_s) = e_\vee \quad \text{iff} \quad t \leq s;$$
$$v(d_t d_s \alpha) = e_\vee \quad \text{iff} \quad v(d_s \alpha) = e_\vee;$$
$$v(d_t \alpha) = d_t(v(\alpha));$$
$$v(d_t(\alpha \vee \beta)) = e_\vee \quad \text{iff} \quad v(d_t \alpha) = e_\vee \text{ or } v(d_t \beta) = e_\vee;$$
$$v(d_t(\alpha \wedge \beta)) = e_\vee \quad \text{iff} \quad v(d_t \alpha) = e_\vee \text{ and } v(d_t \beta) = e_\vee;$$

$$v(d_t(\alpha \Rightarrow \beta)) = e_\vee \quad \text{iff, for every } s \leq t, \text{ either } v(d_s \alpha) = e_\vee \text{ fails to hold or } v(d_s \beta) = e_\vee \text{ holds;}$$
$$v(d_t \neg \alpha) = e_\vee \quad \text{iff, for every } s \leq t, \ v(d_s \alpha) = e_\vee \text{ fails to hold;}$$
$$v(d_t (\forall x) \alpha(x)) = e_\vee \quad \text{iff} \quad v_u(d_t \alpha(x)) = e_\vee \text{ for every } u \in U, \text{ where } v_u \text{ is defined by } v_u(x) = u, \ v_u(y) = y \quad \text{for } y \neq x, \ y \in V;$$
$$v(d_t (\exists x) \alpha(x)) = e_\vee \quad \text{iff} \quad \text{there exists } u \in U \text{ such that } v_u(d_t \alpha(x)) = e_\vee;$$
$$v(\alpha) = e_\vee \text{ in a model } M \quad \text{iff, for every } t \in T, \ v(d_t \alpha) = e_\vee \text{ in } M.$$

Example

Regard $T = (T, \leq)$ as a partially ordered set of agents. Intuitively, $t \leq s$ says that the set recognition ability of t is less than that of s. Predicates $p_1, \ldots, p_m \in L_T$ are viewed as propositional functions representing sets (relations) which are to be recognized.

Each agent $t \in T$ associates with $p_1, \ldots, p_m$ his predicates $d_t p_1, \ldots, d_t p_m$, which approximate to $p_1, \ldots, p_m$, from the viewpoint of t.

If $t \leq s$ then $d_s p_i \Rightarrow d_t p_i$ is satisfied by every valuation in the model M.

Axiomatization of logic LT:

0° The substitutions of axioms of the intuitionistic propositional calculus;

1° $d_t(\alpha \vee \beta) \Leftrightarrow (d_t\alpha \vee d_t\beta)$, for $t \in T$;

2° $d_t(\alpha \wedge \beta) \Leftrightarrow (d_t\alpha \wedge d_t\beta)$, for $t \in T$;

3° $d_t d_s \alpha \Leftrightarrow d_s\alpha$, for $s, t \in T$;

4° $d_t \underline{e}_s$ for $t \leq s$,

$\neg d_t \underline{e}_s$ if $t \leq s$ does not hold, $s \in LT$;

5° $d_t\alpha \vee \neg d_t\alpha$, for $t \in T$;

6° $(d_t\alpha \wedge \underline{e}_t) \Rightarrow \alpha$, for $t \in T$.

Rules of inference

(1) The rules of inference of the intuitionistic predicate calculus;

(2) $\dfrac{\alpha \Rightarrow \beta}{d_t\alpha \Rightarrow d_t\beta}$, for $t \in T$;

(3) $\dfrac{\{d_t\alpha \Rightarrow d_t\beta : t \in T\}}{\alpha \Rightarrow \beta}$.

If T is an infinite set, rule (3) is infinitistic.

As usual, we write $\alpha \in CnX$ to indicate that a formula α is derivable from a set of formulas X.

Here are some properties of this symbol:

1. If $\alpha \in CnX$, then $d_t\alpha \in CnX$ for every $t \in T$;
2. If $d_t\alpha \in CnX$ for every $t \in T$, then $\alpha \in CnX$;
3. $\underline{e}_\vee \in Cn\emptyset$;
4. If $t \leq s$, for $t, s \in T$, then $d_s\alpha \Rightarrow d_t\alpha \in Cn\emptyset$;
5. If $s \leq t$, then $d_t(\alpha \Rightarrow \beta) \Rightarrow (d_s\alpha \Rightarrow d_s\beta) \in Cn\emptyset$;
6. If $s \leq t$, then $d_t\neg\alpha \Rightarrow \neg d_s\alpha \in Cn\emptyset$;
7. If $d_w\gamma \Rightarrow (d_s\alpha \Rightarrow d_s\beta) \in CnX$, for $s \leq t$, then
$$d_w\gamma \Rightarrow d_t(\alpha \Rightarrow \beta) \in CnX;$$
8. If $(d_t\alpha \wedge \underline{e}_t) \Rightarrow \beta \in CnX$ for all $t \in T$, then $\alpha \Rightarrow \beta \in CnX$.

Theorem 9.7 (the completeness theorem)

For every set of formulas X and every formula α in the language L_T,

$$\alpha \in CnX \quad \text{iff} \quad \text{each model for } X \text{ is also a model for } \alpha. \blacksquare$$

Theorem 9.8 (Craig's Lemma, an interpolation theorem)

Suppose α, β are formulas of the language L_T; suppose α is a closed formula and $\alpha \Rightarrow \beta \in Cn\emptyset$. Then there exists in L_T a closed formula γ which contains only the predicates that appear both in α and β, and such that $\alpha \Rightarrow \gamma \in Cn\emptyset$ and $\gamma \Rightarrow \beta \in Cn\emptyset$.

Hence, in particular, if α and β have no common predicates then γ is one of the constants e_t, $t \in T$. ■

9.4.3 Approximation Logics of Type T with Many Indiscernibility Relations

Finally, we extend the language L_T by adjoining to it two equivalence predicates eq, eq^* and two families of connectives, $(C_t)_{t \in T}$ (feasibility connectives) and $(I_t)_{t \in T}$ (necessity connectives), plus C_T and I_T. The extended language will be denoted by ML_T.

Let $U \neq \emptyset$ be a universe space. The following structure M is a model for ML_T over U:

$$M = (U, (\equiv_t)_{t \in T}, \equiv_T, (S_{1t})_{t \in T}, \ldots, (S_{mt})_{t \in T}),$$

with $(\equiv_t)_{t \in T}$ a descending T-sequence of equivalence relations in U, represented by predicate eq; each particular $\equiv_t$ is called the indiscernibility relation associated with t. Relation $\equiv_T$ is defined by

$$\equiv_T = \bigcap_{t \in T} \equiv_t$$

(simultaneous indiscernibility by all agents).

The introduction of new predicates and connectives requires an extension of the valuation functions v. Thus:

For atomic formulas,

$$(1)\ v(d_t eq(u_1, u_2)) = \begin{cases} e_\vee & \text{if } u_1 \equiv_t u_2, \\ e_\wedge & \text{otherwise,} \end{cases}$$

$$(2)\ v(eq^*(u_1, u_2)) = v(d_t eq^*(u_1, u_2)) = \begin{cases} e_\vee & \text{if } u_1 \equiv_T u_2, \\ e_\wedge & \text{otherwise,} \end{cases}$$

$$(3)\ v(C_w d_t f(u)) = \begin{cases} e_\vee & \text{if there exists } u' \in U \\ & \text{such that } u' \equiv_w u \text{ and } d_t f(u') = e_\vee, \\ e_\wedge & \text{otherwise,} \end{cases}$$

(4) $v(C_w f(u)) = \bigcup_{t \in T}(C_w d_t f(u) \cap e_t) = \bigcup_{t \in T}(d_t C_w f(u) \cap e_t)$,

(5) $v(I_w d_t f(u)) = \begin{cases} e_\vee & \text{if, for any } u' \in U, \text{ the condition} \\ & u' \equiv_w u \text{ implies } d_t f(u') = e_\vee, \\ e_\wedge & \text{otherwise,} \end{cases}$

(6) $v(I_w f(u)) = \bigcup_{t \in T}(I_w d_t f(u) \cap e_t) = \bigcup_{t \in T}(d_t I_w f(u) \cap e_t)$;

For composite formulas,

(1)′ $v(d_t C_w\ \alpha(x_1, \ldots, x_n)) = e_\vee$ iff there exist $u_1, \ldots, u_n \in U$ such that $u_i \equiv_w v(x_i)$ for $i = 1, \ldots, n$ and $v'(d_t \alpha(x_1, \ldots, x_n)) = e_\vee$, where $v'(x_i) = u_i$, $v'(x) = v(x)$ for other x,

where $x_1, \ldots, x_n$ are all free variables of α;

(2)′ $v(d_t I_w\ \alpha(x_1, \ldots, x_n)) = e_\vee$ iff, for any $u_1, \ldots, u_n \in U$, the conditions $u_i \equiv_w v(x_i)$ for $i = 1, \ldots, n$ imply $v'(d_t \alpha(x_1, \ldots, x_n)) = e_\vee$, with v' defined as above,

(3)′ $v(d_t C_w\ \alpha) = e_\vee$ iff $v(d_t \alpha) = e_\vee$ wherever α has no free variables;

(4)′ $v(d_t I_w\ \alpha) = e_\vee$ iff $v(d_t \alpha) = e_\vee$ wherever α has no free variables,

where in (1)′–(4)′, $t, w \in T \cup \{T\}$.

In the extended logic ML_T we assume the additional axioms:

(1°) $eq(x, x)$

(2°) $eq(x, y) \Rightarrow eq(y, x)$

(3°) $eq(x, y) \wedge eq(y, z) \Rightarrow eq(x, z)$

(4°) $eq^*(x, y) \Rightarrow d_t eq(x, y)$, for $t \in T$

(5°) $eq^*(x, y) \Leftarrow d_t eq^*(x, y)$, for $t \in T$

(6°) $I_w \alpha \Leftrightarrow \alpha$ wherever α has no free variables;

(7°) $\forall y_1, \ldots, \forall y_n((eq^*(x_1, y_1) \wedge \ldots \wedge eq^*(x_n, y_n)) \Rightarrow d_t\alpha(y_1, \ldots, y_n))$
$\Leftrightarrow I_T d_t\alpha(x_1, \ldots, x_n)$,

for every α with n free variables $x_1, \ldots, x_n$;

(8°) $\forall y_1, \ldots, \forall y_n((deq(x_1, y_1) \wedge \ldots \wedge deq(x_n, y_n)) \Rightarrow d_t\alpha(y_1, \ldots, y_n))$
$\Leftrightarrow Isd_t\alpha(x_1, \ldots, x_n)$,

for every α with n free variables $x_1, \ldots, x_n$;

(9°) $d_t I_w \alpha \Leftrightarrow I_w d_t \alpha$;

(10°) $d_t C_w \alpha \Leftrightarrow C_w d_t \alpha$;

(11°) $C_w d_t \alpha \Leftrightarrow \neg I_w \neg d_t \alpha$;

where in (7°)–(11°), $w \in T \cup \{T\}$ and $t \in T$.

Finally, we accept the additional rule of inference

$$\frac{\{\alpha \Rightarrow d_t eq(x, y) : t \in T\}}{\alpha \Rightarrow eq^*(x, y)}$$

The logic ML_T thus constructed is a complete system. That is to say, the completeness theorem holds:

Theorem 9.9

For every set of formulas X and every formula α in the language ML_T,

$\alpha \in CnX$ iff each model for X is also a model for α. ∎

Readers interested in a more recent treatement of the subject of approximation logics should consult the survey article [Rasiowa 1991].

10 Probability Logics

10.1 Introduction

Probability logics have been an object of interest for many scientists, in their striving to construct a language in which probability could be assigned to logical formulas, along with random events. Probability of that kind is often referred to as logical probability; it has been studied and developed by Łukasiewicz, Keynes, Jeffreys, Reichenbach, Carnap, Popper, Ajdukiewicz, Czezowski, and Suszko, among others.

An interesting viewpoint on probability in propositional calculus is that of Łukasiewicz. In Section 10.2 we outline his ideas and method. It seems very natural and can well serve as an introduction to further work on probability logics. Section 10.3 shows how standard concepts of the usual first-order logic are carried over to probability logic, considered as an algebraic system, in which the notions of interpretation, model, and ultraproduct are introduced. Logical values assigned to formulas (and defined as certain probabilities) are arbitrary numbers in the interval [0,1]; this justifies our view of this theory as a many-valued logic. The results presented in this section come from a paper by D. Scott and P. Krauss [Scott and Krauss 1966].

A different construction of probability logic has been proposed by E. W. Adams [Adams 1966]. His logic is rooted in Kolmogorov's axiom system; we present it in Section 10.4. Instead of transferring the notions of classical logic to a more general setting, the originator of that theory has focused attention on "reasonable" assignment of logical values to composite formulas; in particular, to formulas involving conditional statements.

The two approaches, of Scott-Krauss and of Adams, share a common feature; namely, the concept of probabilistic consequence. It is of course defined in a different fashion in each of the two theories; the first definition is model-theoretic, while the second refers to the calculus of probabilities. It is this second approach which seems to supply simpler tools, hence is easier to handle in practical applications. We present examples (due to Adams) that motivate this viewpoint.

The concept of probability logic has been worked out also in articles by J. Y. Halpern [Halpern 1990] and by R. Fagin, J. Y. Halpern, and N. Megiddo [Fagin, Halpern and Megiddo 1990]. Their approach is oriented toward the methods of probabilistic inference. It will be presented in the second volume of our book, devoted to automated reasoning in various many-valued logics.

10.2 Łukasiewicz' Idea of Logical Probability

Logical values of propositional functions were for Łukasiewicz a starting point to further considerations. Propositional functions (called by Łukasiewicz indefinite propositions) involve a precisely established domain of a finite number of individuals. The logical value of a propositional function is given by the ratio of the number of elements satisfying it to the cardinality of the whole domain. A propositional function is said to be false if it is satisfied by no element and true if it is satisfied by all elements of the domain. Accordingly, the function assumes value 0 or 1. (It ought to be mentioned that the term "propositional function" is used here in a slightly different meaning from that usually accepted.)

It is thus evident that there exist propositional functions of values δ other than 0 and 1, necessarily with $0 < \delta < 1$. Then δ is just the probability, or likelihood degree of $\alpha(x)$ (the propositional function under consideration). We call δ the logical value of α.

Let α, β be any propositional functions and let $v(\alpha), v(\beta)$ denote their logical values. The symbols $\sim \alpha$ and $\alpha \wedge \beta$, $\alpha \vee \beta$, $\alpha \Rightarrow \beta$, $\alpha \Leftrightarrow \beta$ denote, respectively, the negation of α and the conjunction, disjunction, implication, equivalence of α and β.

Lukasiewicz assumes the following axioms for his probability calculus (α, β denoting propositional functions):

(a_1) if α is false then $v(\alpha) = 0$;

(a_2) if α is true then $v(\alpha) = 1$;

(a_3) if $v(\alpha \Rightarrow \beta) = 1$ then $v(\alpha) + v(\sim \alpha \wedge \beta) = v(\beta)$.

Further properties, holding for any propositional functions α, β, are an easy consequence of the axioms:

(w_1) if $v(\alpha \Leftrightarrow \beta) = 1$ then $v(\alpha) = v(\beta)$;

(w_2) $v(\alpha) + v(\sim \alpha) = 1$;

(w_3) $v(\alpha \wedge \beta) + v(\sim \alpha \wedge \beta) = v(\beta)$;

(w_4) $v(\alpha) + v(\sim \alpha \wedge \beta) = v(\alpha \vee \beta)$;

(w_5) $v(\alpha \vee \beta) = v(\alpha) + v(\beta) - v(\alpha \wedge \beta)$;

(w_6) if $v(\alpha \wedge \beta) = 0$ then $v(\alpha \vee \beta) = v(\alpha) + v(\beta)$.

The analogy with the usual probability calculus is apparent. It is also obvious that probability (thus understood) can be only attributed to propositional functions and not to propositions.

10.3 An Algebraic Description of Probability Logic

10.3.1 Syntax

In our presentation of probability systems we will be working with two first-order languages, denoted L and $L^{(\omega)}$. It is assumed that $L^{(\omega)}$ contains a countable set of individual variables p_n, indexed by $n < \omega$, whereas L has uncountably many individual variables p_ξ, indexed by $\xi < \omega_1$, the least uncountable ordinal. Atomic formulas in both languages are of the form $Rp_\xi p_\eta$ (where R is a relation symbol) or $p_\xi = p_\eta$ (where $=$ denotes identity). Other formulas are composed from atomic ones using the connectives $\wedge, \vee, \neg$ (conjunction, disjunction, negation) and quantifiers $\forall, \exists$ (universal and existential). Additionally, L admits the use of countable conjunctions and disjunctions, denoted by the symbols $\bigwedge$ and $\bigvee$.

Non-logical vocabulary can be adjoined in the form of sets T of individual constants; the resulting systems will be denoted by $L^{(\omega)}(T)$ and $L(T)$.

The symbols S and $S(T)$ will stand for the sets of formulas of L and $L(T)$; by $s(T)$ we will denote the set of quantifier-free formulas of $L(T)$. The analogous notation $S^\omega, S^\omega(T), s^\omega(T)$ is employed for the language $L^{(\omega)}$ or $L^{(\omega)}(T)$.

Let α be a formula, let X be a set of formulas, and suppose that α is satisfied in every model in which all formulas from X are satisfied. Then α is said to be a consequence of X; in symbols, $X \models \alpha$ or $\alpha \in Cn_{Sem}X$.

A formula is a tautology (is said to be true) if it is a consequence of the empty set; in symbols, $\models \alpha$.

The deduction system is created in the standard way. We write $\alpha \in CnX$, or $X \vdash \alpha$, when α is derivable (provable) from the set X. A formula derivable from the empty set is called a theorem; this is written as $\alpha \in Cn\emptyset$, or $\vdash \alpha$.

Theorem 10.1 (completeness theorem)

(a) For any set $X \subseteq S^\omega$ and any formula $\alpha \in S^\omega$, $\alpha \in CnX$ iff $\alpha \in Cn_{Sem}X$.

(b) For all $\alpha \in S$, $\alpha \in Cn\emptyset$ iff $\alpha \in Cn_{Sem}\emptyset$. ∎

If $\{\alpha\} \vdash \beta$ and $\{\beta\} \vdash \alpha$, we write $\alpha \leftrightarrow \beta$.

Formulas α and β are equivalent when $\vdash (\alpha \leftrightarrow \beta)$. This relation partitions the set of all formulas into classes, which constitute a Boolean algebra called the Lindenbaum–Tarski formula algebra. This algebra is σ-complete, in the infinite case. It is isomorphic to the set field of models (a σ-field, in the infinite case).

Let $S/\vdash$, $S(T)/\vdash$, $s(T)/\vdash$ denote the Tarski algebras of the respective sets of formulas. Then $S/\vdash$ and $s(T)/\vdash$ are σ-subalgebras of $S(T)/\vdash$.

Let $\alpha/\vdash$ be the coset of α.

Formulas α and β are equivalent relative to a set X if $\alpha \leftrightarrow \beta \in CnX$; this relation induces the Lindenbaum–Tarski quotient algebras $S/X \vdash$, $S(T)/X \vdash$ and $s(T)/X \vdash$.

The algebra $S^\omega/X \vdash$ is isomorphic to the field of models (which fails to hold in the case of an infinite language).

By $\alpha/X \vdash$ we denote the coset of α modulo a set X. The mapping which sends $\alpha/\vdash$ to $\alpha/X \vdash$ is a σ-homomorphism from $S/\vdash$ to $S/X \vdash$.

10.3.2 Semantics

Definition 10.1

A probability system is defined as an ordered quintuple:

$$\mathfrak{A} = (A, R, id, \mathrm{A}, m),$$

in which

(1) A is a nonempty set;

(2) (A, m) is a measure algebra; that is, A is a Boolean σ-algebra and M is a strictly positive probability measure on A;

(3) R is a function from $A \times A$ into A;

(4) id is a function from $A \times A$ into A, with the following substitution property: for all $a, a', b, b' \in A$,

(a) $id(a, a) = 1$,

(b) $\sim id(a, a') \cup \sim id(b, b') \cup \sim id(a, b) \cup id(a', b') = 1$,

(c) $\sim id(a, a') \cup \sim id(b, b') \cup \sim R(a, b) \cup R(a', b') = 1$. ∎

If $\mathfrak{A}$ is a probability system, then A is a complete algebra and $\mathfrak{A}$ is an algebraic model, in the traditional meaning of the word.

If $id(a, b) = 0$ wherever $a \neq b$, $a, b \in A$, then $\mathfrak{A}$ is called a strict identity system and we write $\mathfrak{A} = (A, R, \mathrm{A}, m)$. If, additionally, m is a two-valued measure and A is the two-element Boolean algebra, then (A, m) may be identified with a standard model.

When $\mathfrak{A}$ is a strict identity system, the cardinality of A is also called the cardinality of $\mathfrak{A}$, denoted $| \mathfrak{A} |$. In the general case, the cardinality of $\mathfrak{A}$ modulo a subset $A' \subseteq A$ is defined as the cardinality of the set of cosets modulo relation $\approx$ ($a \approx b$ iff $id(a, b) = 1$) which intersect A'.

Now assume that $T_{\mathfrak{A}} = \{t_a : a \in A\}$ is a set of new individual constants ($t_a \neq t_b$ when $a \neq b$).

The valuation $v: S(T_{\mathfrak{A}}) \to \mathrm{A}$ is defined as follows:

(1) $v(t_a = t_b) = id(a, b)$,

(2) $v(Rt_a t_b) = R(a, b)$,

(3) $v(\neg\alpha) = 1 - v(\alpha)$,

(4) $v(\bigvee_{i<\xi} \alpha_i) = \bigvee_{i<\xi} v(\alpha_i)$,

(5) $v(\bigwedge_{i<\xi} \alpha_i) = \bigwedge_{i<\xi} v(\alpha_i)$,

(6) $v(\exists p\alpha) = \bigvee_{x \in A} v(\alpha(t_x))$,

(7) $v(\forall p\,\alpha) = \bigwedge_{x \in A} v(\alpha(t_x))$.

Lemma 10.1

(i) For $\alpha, \beta \in S(T_{\mathfrak{A}})$, if $\alpha \leftrightarrow \beta \in Cn\emptyset$ then $v(\alpha) = v(\beta)$.

(ii) v induces a homomorphism of $S(T_{\mathfrak{A}})/\vdash$ into A. ∎

Setting

$$\mu_{\mathfrak{A}}(\alpha/\vdash) = m(v(\alpha)) \quad \text{for} \quad \alpha \in S(T_{\mathfrak{A}})$$

we define a probability measure $\mu_{\mathfrak{A}}$ on $S(T_{\mathfrak{A}})/\vdash$. We will write simply $\mu_{\mathfrak{A}}(\alpha)$ rather than $\mu_{\mathfrak{A}}(\alpha/\vdash)$. The equality $\mu_{\mathfrak{A}}(\alpha) = \delta$ reads: "in the probability system $\mathfrak{A}$, formula α is satisfied (holds) with probability δ".

Thus the definition of $\mu_{\mathfrak{A}}$ is the canonical extension of the usual definition of validity.

Lemma 10.2

Let $\alpha \in S(T_{\mathfrak{A}})$. If $\alpha(p)$ holds for some p, then

$$(*) \qquad \mu_{\mathfrak{A}}(\exists x\,\alpha) = \sup_{F \in A^{(\omega)}} \mu_{\mathfrak{A}}\left(\bigvee_{x \in F} \alpha(t_x)\right),$$

where $A^{(\omega)}$ denotes the family of all finite subsets of A. ∎

The proofs of both lemmas can be found in [Scott and Krauss 1966].

Now, it turns out that probability $\mu_{\mathfrak{A}}$ can be restricted to $s(T_{\mathfrak{A}})/\vdash$ without affecting the system $\mathfrak{A}$ in any substantial way. In view of the definition of v, we may assume that A is the σ-algebra generated by the union of the images of $A \times A$ under id and R.

Clearly, if we restrict the definition of v to conditions (1)–(5), Lemma 10.1 retains its validity, with $S(T_{\mathfrak{A}})$ replaced by $s(T_{\mathfrak{A}})$.

Since m is strictly positive, we have the set equality

$$\{\alpha \in s(T_{\mathfrak{A}}) : \mu_{\mathfrak{A}} = 0\} = \{\alpha \in s(T_{\mathfrak{A}}) : v(\alpha) = 0\}.$$

Thus the probability system $\mathfrak{A}$ is determined by the pair $(T_{\mathfrak{A}}, \mu_{\mathfrak{A}})$, with $\mu_{\mathfrak{A}}$ restricted to $s(T_{\mathfrak{A}})/\vdash$. Generally, in fact, a probability system can be regarded as the pair (T, m), where T is the set of new individual constants and m is a probability on $s(T)/\vdash$. For a justification of this rough statement we again refer to [Scott and Krauss 1966].

A probability system (T, m) with strict identity (that is, such that $m(t = t') = 0$ if $t \neq t'$; $t, t' \in T$) is called a probability model.

Theorem 10.2

Let (T, m) be a probability system. There exists a unique probability m^* on $S(T)/\vdash$ which extends m and satisfies the Gaifman condition:

If $\alpha(p)$ holds for some p, then

$$m^*(\exists p\, \alpha) = \sup_{F \in T^\omega} m^* \left(\bigvee_{t \in F} \alpha(t) \right).$$

Proof

In [Scott and Krauss 1966]. ∎

Suppose (T_1, m_1) and (T_2, m_2) are probability systems. Let m_i^* be as in Theorem 10.2 and let $\overline{m_i}$ be the restriction of m_i^* to $S(T_i)/\vdash$, for $i = 1, 2$. We introduce:

Definition 10.2

(i) (T_1, m_1) is a subsystem of (T_2, m_2), in symbols $(T_1, m_1) \subseteq (T_2, m_2)$, iff $T_1 \subseteq T_2$ and m_1 is the restriction of m_2 to $s(T_1)/\vdash$.

(ii) (T_1, m_1) is an L-subsystem of (T_2, m_2), in symbols $(T_1, m_1) \leq (T_2, m_2)$, iff $T_1 \subseteq T_2$ and m_1^* is the restriction of m_2^* to $S(T_1)/\vdash$.

(iii) Systems (T_1, m_1) and (T_2, m_2) are L-equivalent, in symbols

$$(T_1, m_1) \equiv (T_2, m_2) \quad \text{iff} \quad \overline{m_1} = \overline{m_2};$$

here $\overline{m_1}$ denotes the measure m_1 restricted to $S/\vdash$ and $\overline{m_2}$ is defined analogously. ∎

10.3.3 Constructions

Certain general constructions involving probability systems prove useful and are easily applied in practical situations. This concerns, in particular, the notions of independent unions, ultraproducts and symmetric probability systems. We now briefly outline these concepts.

Independent unions. Let I be an index set. For each $i \in I$ let L_i be an infinite language whose only non-logical constant is a binary predicate R_i. Let S_i be the set of all formulas of L_i. By T we denote as usual the set of new individual constants. Suppose that for each $i \in I$, (T, m_i) is a probability system with a probability measure m_i on $S_i(T)/\vdash$. The (T, m_i) are assumed to be strict identity systems. Consider the infinite language L containing all predicates R_i, $i \in I$, as non-logical constants and let S be the set of formulas of L. Assume that all $s_i(T)/\vdash$ are isomorphic with respect to σ-fields of sets of models, $s(T)/\vdash$ is isomorphic to the product $\prod\limits_{i\in I} s_i(T)/\vdash$, and define $m = \prod\limits_{i\in I} m_i$ as the product measure on $s(T)/\vdash$, induced by the family $\{m_i : i \in I\}$.

The resulting probability system (T, m) is called the independent union of systems (T, m_i) and is denoted by $\sum\limits_{i\in I}(T, m_i)$.

Corollary 10.1

For each $i \in I$, if $\alpha \in S_i(T)$ then $m^*(\alpha) = m_i^*(\alpha)$.

Proof

By transfinite induction. ∎

Now, let X and Y be some sets, let A and B be fields of subsets of X and Y, respectively, let $\overline{\mathrm{A}}$ and $\overline{\mathrm{B}}$ denote the σ-fields generated by A and B, and let μ and ν be probability measures on $\overline{\mathrm{A}}$ and $\overline{\mathrm{B}}$.

Lemma 10.3

Suppose λ is a probability on the product σ-field $\overline{\mathrm{A}} \times \overline{\mathrm{B}}$, such that

$$\lambda(A \times B) = \mu(A)\nu(B) \quad \text{for all} \quad A \in \mathrm{A}, B \in \mathrm{B}.$$

Then

$$\lambda(A \times B) = \mu(A)\nu(B) \quad \text{for all} \quad A \in \overline{\mathrm{A}}, B \in \overline{\mathrm{B}}. \quad ∎$$

Corollary 10.2

For each $n < \omega$, let $i_n \in I$ and let $\alpha_n \in S_{i_n}(T)$. Then

$$m^*(\bigwedge_{n<\omega} \alpha_n) = \prod_{n<\omega} m^*(\alpha_n).$$

Proof

Based on the continuity of m^* and on the proof of Theorem 10.2. ∎

Utilizing independent unions we can facilitate an analysis of a probability system with a complicated relation structure. For instance, let $\mathfrak{A}_1 = (A, R, \mathrm{A}, m)$ be a probability system and suppose there is given an ordering $<$ in A. We can introduce the system $\mathfrak{A}_2 = (A, <)$ and consider the independent union of $\mathfrak{A}_1$ and $\mathfrak{A}_2$; the values of probabilities are then computed with the aid of Corollaries 10.1 and 10.2.

Ultraproducts are constructed along the same lines as they are in logical systems.

Let L be a language with a predicate R, let I be an index set and let (T_i, m_i) be a probability system, for each $i \in I$. Consider the Cartesian product of the set family $\{T_i : i \in I\}$,

$$T = \prod_{i \in I} T_i.$$

For $\alpha \in S(T)$ and $i \in I$ denote by $\alpha \mid i$ the projection of α to the i-th coordinate; each $t \in T$ that occurs in α is replaced in $\alpha \mid i$ by the suitable $t_i \in T_i$ $(t = (t_i)_{i \in I})$. If $\vdash \alpha$, $\alpha \in S(T)$, then $\vdash \alpha \mid i$ for each $i \in I$.

Suppose there is given a probability λ on the power set of I. For every $\alpha \in s(T)$ define

$$m(\alpha) = \int_I m_i(\alpha \mid i) d\lambda(i).$$

Lemma 10.4

(i) For $\alpha, \beta \in s(T)$, if $\alpha \leftrightarrow \beta \in Cn\emptyset$ then $m(\alpha) = m(\beta)$.

(ii) For $\alpha \in s(T)$, if $\alpha \in Cn\emptyset$ then $m(\alpha) = 1$.

(iii) m is a probability measure on $s(T)/\vdash$. ∎

Corollary 10.3

For every $\alpha \in S(T)$,

$$m^*(\alpha) = \int_I m_i^*(\alpha \mid i) d\lambda(i). \quad ∎$$

This shows that the ultraproduct probability is a kind of weighted mean of the respective probabilities in the component systems.

Symmetric probability systems. Let (T, m) be a probability system and let $\Pi \in T^T$ be a permutation of T (one-to-one map "onto"). A permutation Π is finite if $\Pi(t) = t$ except for finitely many $t \in T$.

We call (T, m) a symmetric system if $m(\alpha) = m(\alpha^\Pi)$ holds for every $\alpha \in s^\omega(T)$ and every permutation Π; here α^Π denotes the formula obtained from α by replacing each $t \in T$ occurring in α by $\Pi(t)$.

Lemma 10.5

In a symmetric probability system (T, m),

$$m^*(\alpha) = m^*(\alpha^\Pi)$$

holds for every $\alpha \in s(T)$ and every permutation Π of T. ∎

Lemma 10.6

In a symmetric probability system (T, m),

$$m^*(\alpha) = m^*(\alpha^\Pi)$$

holds for every $\alpha \in S(T)$ and every permutation Π of T. ∎

Symmetric systems are in a sense a contrary of relation systems. In relation systems, probability is as concentrated as possible, while in symmetric ones it is uniformly spread over all of $S(T)$.

10.3.4 Probabilistic Consequence

We now discuss the concept of probabilistic consequence in terms of the theory outlined in the preceding subsections. However, it should be emphasized that this approach is not the only one possible. In the next section we present another view on the problem of deduction in a probabilistic setting, based on a different construction of probability systems. To begin with, we have to settle what is meant by (probabilistic) assertion and what laws it is subject to. For this purpose we introduce a language, which we denote by M. This is a first-order language of the algebra of real numbers. M has countably many individual variables λ_n, $n < \omega$. Non-logical constants in M are: a binary predicate $\leq$, binary function symbols $+$ and $\cdot$, and individual constants 0, $+1$, -1. The language contains logical functors $\wedge$ (conjunction), $\vee$ (disjunction), $\neg$ (negation), $\forall$ (universal quantifier), $\exists$ (existential quantifier). Formulas and propositions are defined in the standard way.

We say that M is interpreted in the standard relation system

$$\mathfrak{R} = (R, \leq, +, \cdot, 0, +1, -1),$$

where R denotes the set of real numbers. Every proposition of M true in $\mathfrak{R}$ will be called a theorem.

A quantifier-free formula of M will be called an algebraic formula.

Every algebraic formula is equivalent, in the sense of real-number algebra, to an inequality of the form $p \geq 0$ or $p > 0$, where p is a polynomial with integer coefficients.

An algebraic formula is said to be closed (or open) if it is equivalent to a conjunctive (or disjunctive) expression whose literals have the form of inequalities of type $p \geq 0$ (or $p > 0$, respectively).

A probabilistic assertion is defined as an ordered $(n+1)$-tuple $\langle \alpha, \alpha_0, \alpha_1, \ldots \ldots, \alpha_{n-1} \rangle$ in which α is an algebraic formula involving n free variables, $\alpha_0, \ldots \ldots, \alpha_{n-1} \in S$. When α is a closed (or open) formula, the assertion is called closed (or open, respectively).

A probability system (T, m) is a probabilistic model for $\langle \alpha, \alpha_0, \ldots, \alpha_{n-1} \rangle$ if the n-tuple of real numbers $(m(\alpha_0), \ldots, m(\alpha_{n-1}))$ satisfies formula α in $\mathfrak{R}$.

Let $\sum$ be a set of probabilistic assertions and let β be a probabilistic assertion. We say that β is a probabilistic consequence of $\sum$ iff every system which is a probabilistic model for all assertions in $\sum$ is a probabilistic model for β as well. An assertion which is a probabilistic consequence of the empty set is called a probabilistic law.

We wish to investigate polynomial inequalities in variables $\mu(\alpha_0), \ldots, \mu(\alpha_{n-1})$, where $\alpha_0, \ldots, \alpha_{n-1} \in S$ and μ is a probability on $S/\vdash$. However, our definition of assertion is inadequate for dealing with polynomials of arbitrary real coefficients. This would require extension of M to a language M′ with uncountably many individual constants, representing all real numbers. It is not hard to show that closed assertions of the language M′ are equivalent to countable sets of assertions of M′ with rational coefficients. This is no longer true for open assertions. Yet we stick to our definitions because of their simplicity, and also because of the lack of other proposals for solving this problem.

A system (T, m) is called a probabilistic model for a set of formulas X if $\bar{m}(\alpha) = 1$ for all $\alpha \in X$.

In general, if μ is a probability on $S/\vdash$, then μ induces a set X of probabilistic assertions in the following fashion:

For every $\alpha \in S$ we select sequences of rational numbers p_n/q_n and p'_n/q'_n so that

$$\frac{p_n}{q_n} \leq \frac{p_{n+1}}{q_{n+1}} \leq \mu(\alpha) \leq \frac{p'_{n+1}}{q'_{n+1}} \leq \frac{p'_n}{q'_n}$$

for $n < \omega$, and

$$\lim_{n\to\infty} \frac{p_n}{q_n} = \mu(\alpha) = \lim_{n\to\infty} \frac{p'_n}{q'_n}.$$

Let

$$X_\alpha = \{\langle q_n\lambda_0 - p_n \geq 0, \alpha\rangle : n < \omega\} \cup \{\langle -q'_n\lambda_0 + p'_n \geq 0, \alpha\rangle : n < \omega\}$$

and define

$$X = \bigcup_{\alpha \in S} X_\alpha.$$

We say that (T, m) is a probabilistic model for μ iff $\bar{m}(\alpha) = \mu(\alpha)$ for all $\alpha \in S$. Hence, (T, m) is a probabilistic model for μ if and only if it is a probabilistic model for the set X (defined above).

A set $X \subseteq S$ is consistent if there is no formula $\alpha \in S$ such that $(\alpha \wedge \neg\alpha) \in CnX$.

Not every consistent set has a model. Consider for example a doubly-indexed set of propositional constants $c_{\xi n}$, $\xi < \omega_1, n < \omega$, and define

$$X = \left\{ \bigvee_{n<\omega} c_{\xi n} : \xi < \omega_1 \right\} \cup \{\neg(c_{\xi n} \wedge c_{\xi' n}) : \xi < \xi' < \omega_1,\ n < \omega\}.$$

Clearly, X is consistent. Now assume there exists a σ-additive probability measure μ on $S/\vdash$ such that $\mu(\alpha) = 1$ for $\alpha \in X$. Then for every $\xi < \omega_1$ there exists $n < \omega$ such that $\mu(c_{\xi n}) > 0$. So there exists $n < \omega$ such that $\mu(c_{\xi n}) > 0$ holds for infinitely many ξ. And this yields a contradiction, in view of $\mu([c_{\xi n} \wedge c_{\xi' n}]) = 0$ for $\xi \neq \xi'$. Consequently X admits no probability model.

The problem of the existence of models for complete consistent sets of formulas has been investigated by C. Ryll-Nardzewski (among others); we refer to his works for various interesting examples in this area.

The question of whether or not a given set $X \subseteq S$ has a probabilistic model admits a further algebraic interpretation.

A Boolean algebra A is said to have the Kelley property if A $\setminus \{0\}$ is a countable union of sets with a positive number of intersections [Sikorski 1964].

Proposition 10.1

A set $X \subseteq S$ has a probabilistic model if and only if X has a complete algebraic model, weakly distributive, with the Kelley property.

Proof

Follows immediately from Kelley's theorem [Kelley 1959]. ∎

It is a well known fact for usual logical calculi that every countable consistent set $X \subseteq S$ has a countable model [Adams 1966]. A direct analogue of this is valid in probability logic.

Theorem 10.3

(i) Let μ be a probability on $S/\vdash$ and let $X \subseteq S$ be a countable set. Then there exists a countable probabilistic model (T, m) such that $\bar{m}(\alpha) = \mu(\alpha)$ for $\alpha \in X$.

(ii) For each $n < \omega$ consider the formula

$$\Theta_n = \exists p_0, \ldots, \exists p_{n-1} \forall p_n \bigvee_{i<n} p_i = p_n.$$

If $\mu(\Theta_n) \in \{0,1\}$ for all $n < \omega$, the model can be required to possess strict identity. ∎

Example

We give an example of a probability μ on $S/\vdash$ which has a probabilistic model, but not a symmetric one.

Consider $(\omega_1, <)$, the system of countable ordinal numbers with the usual ordering. For $\alpha \in S$ define:

$$\mu(\alpha) = 1 \quad \text{iff} \quad \alpha \text{ is satisfied in } (\omega_1, <).$$

Each $\xi < \omega_1$ is definable in language L by means of a formula α_ξ with a single free variable p, such that, for any $\eta < \omega_1$,

$$\eta \text{ satisfies } \alpha_\xi \text{ in } (\omega_1, <) \text{ iff } \eta = \xi.$$

Hence $\mu(\exists p \alpha_\xi) = 1$ and

$$\text{if } \eta < \xi < \omega_1 \text{ then } \mu(\exists p\, \alpha_\eta \wedge \alpha_\xi) = 0.$$

Now assume that μ has a symmetric probability model (T, m). Then $\bar{m}(\exists p\, \alpha_\xi) = 1$ for every $\xi < \omega_1$. Hence, for every $\xi < \omega_1$ there exists $t \in T$ such that $m^*(\alpha_\xi(t)) > 0$. Since (T, m) is symmetric, we have by Lemma 10.6

$$m^*(\alpha_\xi(t')) = m^*(\alpha_\xi(t))$$

for every $t' \in T$. Therefore

$$m^*(\alpha_\xi(t)) > 0$$

holds for each $\xi < \omega_1$ and every $t \in T$. Fix $t \in T$. There is $\varepsilon > 0$ such that

$$m^*(\alpha_\xi(t)) \geq \varepsilon$$

for infinitely many ξ. And since

$$\bar{m}(\exists p\, \alpha_\eta \wedge \alpha_\xi) = 0$$

for $\eta < \xi < \omega_1$, we get

$$m^*(\alpha_\eta(t) \wedge \alpha_\xi(t)) = 0,$$

which is impossible.

However, for countable languages the following theorem is valid:

Theorem 10.4

Let X be a set of probabilistic assertions of a countable language $\mathrm{L}^{(\omega)}$. Then X has a countable probability model if and only if X has a countable symmetric probability model. ∎

The next problems that arise are whether one can find a method of deductive generation of probabilistic consequences of a given set of probabilistic assertions, or a method of generation of probabilistic laws.

Though there is little hope for a positive answer to the first question, we can try to make some relevant remarks in this matter.

Theorem 10.5

Let $\langle \alpha, \alpha_0, \ldots, \alpha_{n-1} \rangle$ be a probabilistic assertion of uncountable language L. Let $\lambda_0, \ldots, \lambda_{n-1}$ be the free variables of formula α. Assume

$$\vdash \neg(\alpha_i \wedge \alpha_j) \quad \text{for} \quad i \neq j$$

and

$$\vdash \bigvee_{i<n} \alpha_i.$$

Let $I = \{i < n : \vdash \neg \alpha_i\}$. Then assertion $\langle \alpha, \alpha_0, \ldots, \alpha_{n-1} \rangle$ is a probabilistic law in L if and only if the formula

$$\forall \lambda_0, \ldots, \forall \lambda_{n-1} \left(\left(\bigwedge_{i \in I} \lambda_i = 0 \wedge \bigwedge_{i<n} \lambda_i \geq 0 \wedge \lambda_0 + \ldots + \lambda_{n-1} = 1 \right) \rightarrow \alpha \right)$$

is a theorem of real number algebra. ∎

For every probabilistic assertion $\langle \beta, \beta_0, \ldots, \beta_{m-1} \rangle$ we can find an equivalent assertion $\langle \alpha, \alpha_0, \ldots, \alpha_{n-1} \rangle$ satisfying the conditions of Theorem 10.5. The two assertions are to be equivalent in the following sense: every probability model for $\langle \beta, \beta_0, \ldots, \beta_{m-1} \rangle$ is a probability model for $\langle \alpha, \alpha_0, \ldots, \alpha_{n-1} \rangle$, and vice versa.

It is also possible to generate all probabilistic assertions of language L which fulfill the conditions of Theorem 10.5. And since the theorems of real algebra are decidable (Tarski), we are able to decide, with respect to every probabilistic assertion of L, whether or not the respective formula is a theorem in $\mathfrak{R}$ [Scott and Krauss 1966].

Thus the theorem gives a positive answer to the second question.

At present, we do not know an answer to the question about the existence of a method for generating probabilistic consequences of a given set of assertions; we also do not know conditions ensuring that a set of probabilistic assertions possesses a model, in the case of the uncountable language L.

10.4 Axiomatic Approach to Probability Logic

10.4.1 Syntax

In this section L is assumed to be a finite language; i.e. a language generated by finite sets of atomic formulas. The letters T and F stand for logical truth and falsity; these are of course atomic formulas. Other atomic formulas will be denoted by Greek characters $\delta, \mu, \varphi, \psi$. Furthermore, we distinguish formulas of two types namely, truth-functional and conditional.

Truth-functional (for short, tf-) formulas are those composed from atomic formulas with use of the connectives $\wedge$ (conjunction), $\vee$ (disjunction) and $-$ (negation); single atomic formulas are also regarded as truth-functional. Conditional formulas have the form $\alpha \rightarrow \beta$ where α, β are tf-formulas.

We adopt the following notation: small Greek characters $\alpha, \beta, \gamma, \ldots$ to denote tf-formulas; capitals A, B, C, D to denote conditional formulas; capitals X, Y, Z to denote set of formulas.

The symbol $\alpha - \beta$ will stand for $\alpha \wedge -\beta$.

A set of formulas X is said to imply tautologically a formula A if this agrees with the rules of logic (possibly counterintuitively; e.g., "φ" tautologically implies "$-\varphi \rightarrow \psi$").

We will moreover use the following metalinguistic notation:

1. If $A = \beta$ then $\sim A = -\beta$, and if $A = \alpha \rightarrow \beta$ then $-A = \alpha \rightarrow -\beta$.

2. If $A = \beta$ then $Ant(A) = T, Cons(A) = \beta$ and $Cond(A) = T \rightarrow \beta$; and if $A = \alpha \rightarrow \beta$ then $Ant(A) = \alpha, Cons(A) = \beta$ and $Cond(A) = \alpha \rightarrow \beta$.

Now, let XD be a set of truth-functional formulas of L, pairwise contradictory and such that every formula, not tautologically equivalent to F, is tautologically equivalent to a disjunction of formulas from XD.

Assuming that X is a finite set of formulas, the sets XD are formed as follows:

Definition 10.3

Let L be a finite language; let XD be an XD-set for L and X be a finite set of formulas.

Let $\alpha \in XD$ and let A be such that $Ant(A) = \alpha$, $Cons(A) = \beta$.

(i) γ belongs to α iff γ tautologically implies α;

(ii) $XD(A)$ is the set of all γ ($\gamma \in XD$) belonging to α;

$\mathrm{XD}(A)$ is the set of all γ ($\gamma \in XD$) belonging to $\alpha - \beta$;

(iii) $XD(S)$ is the set of all $XD(B)$ where $B \in X$;

$\mathrm{XD}(S)$ is the set of all $\mathrm{XD}(B)$ where $B \in X$;

(iv) X is a null-set iff $XD(X) = \mathrm{XD}(X)$. ■

10.4.2 Probability and Probabilistic Consequence

We define probability by Kolmogorov's axioms, with the additional requirement that every formula A with $Ant(A)$ of probability null shall be assigned probability one ($P(A) = 1$ if $P(Ant(A)) = 0$).

And thus:

Definition 10.4

Let L be the language.

By a probability on L we mean a function $P : \mathrm{L} \rightarrow R$ satisfying the following conditions (for any truth-functional formulas α, β):

(i) $0 \leq P(\alpha) \leq 1, \quad P(T) = 1$;

(ii) if β is a tautological consequence of α, then $P(\alpha) \leq P(\beta)$;

(iii) if $-\beta$ is a tautological consequence of α, then $P(\alpha \vee \beta) = P(\alpha) + P(\beta)$;

(iv) if $P(\alpha) \neq 0$ then $P(\alpha \rightarrow \beta) = P(\alpha \wedge \beta)/P(\alpha)$;
if $P(\alpha) = 0$ then $P(\alpha \rightarrow \beta) = 1$. ■

A sequence of probabilities $P_1, P_2, \ldots$ is called uniform if the limit $\lim_{n\rightarrow\infty} P_n(A)$ exists for every A.

Suppose $P_1, P_2, \ldots$ is a uniform sequence of probabilities and suppose $P_n(\alpha) = 0$ for infinitely many n. Since $\lim_{n\rightarrow\infty} P_n(\alpha \rightarrow 0)$ has to exist, and since each $P_n(\alpha \rightarrow 0)$ must be equal to either 1 or 0 (according as $P_n(\alpha)$ is zero or not), it follows that $P_n(\alpha) = 0$ except for finitely many n.

Thus, as long as we are concerned with a finite language, uniform sequences of probabilities $P_1, P_2, \ldots$ have the property that, for every α, the equality $P_n(\alpha) = 0$ holds either for finitely many n or for all but finitely many n.

Definition 10.5

Let L be the language, X be a set of formulas, and let A be a formula of L.

(i) A is a reasonable consequence of X, in symbols $X \Vdash A$, iff the following condition is satisfied: to every $\varepsilon > 0$ we can find $\sigma > 0$ so that, for any probability P, if the inequality $P(B) > 1 - \sigma$ holds, for all $B \in X$, then $P(A) > 1 - \varepsilon$.

(ii) A is a strict consequence of X, in symbols $X \vdash A$, iff, for any probability P, the equality $P(B) = 1$ holding for all $B \in X$ implies $P(A) = 1$. ■

Hence, if $P_1, P_2, \ldots$ is a uniform sequence of probabilities and X is a finite set of formulas, and if $\lim_{n\to\infty} P_n(B) = 1$ for each $B \in X$ while $\lim_{n\to\infty} P_n(A) < 1$, then A is not a reasonable consequence of X.

Intuitively, A is a reasonable consequence of X if, wherever all formulas in X hold with high probability, so does A. The notion of strict consequence can be identified with tautological consequence.

Theorem 10.6

Let L be the language, X a set of formulas, and A a formula. Then:

(1) $X \vdash A$ if and only if A is a tautological consequence of X.

(2) If $X \Vdash A$ then $X \vdash A$.

(3) Relation $\Vdash$, restricted to finite sets, is a deduction relation; that is, if X and X' are finite sets of formulas, then:

   (a) if $A \in X$ then $X \Vdash A$;

   (b) if $X' \Vdash B$ for all $B \in X$ and if $X \Vdash A$, then $X' \Vdash A$;

   (c) suppose X' and A' arise from X and A by replacing a truth-functional formula α by an atomic formula σ (other that T or F); then $X \Vdash A$ entails $X' \Vdash A'$. ∎

This theorem shows that the operation of reasonable consequence has the minimum properties required from a deduction relation. Yet it lacks the compactness property. This means that there can exist infinite sets of formulas X such that $X \Vdash A$ holds for a certain formula A, without $X' \Vdash A$ holding for any finite subset $X' \subset X$.

Example

Let $X = \{B_i = \alpha_i \vee \alpha_{i+1} \to \alpha_{i+1} \wedge -\alpha_i : i = 1, 2, \ldots\}$ where α_i are distinct atomic formulas. Consider $A = \alpha_1 \to F$. If $P(B_i) > \frac{2}{3}$ for $i = 1, 2, \ldots$, then $P(\alpha_1) \leq \frac{1}{2} P(\alpha_{i+1})$ for all i. Hence $P(\alpha_1) = 0$ because $P(\alpha_1 \to 0) = 1$.

A high probability of all formulas in X ensures that A has high probability too ($X \Vdash A$, according to Definition 10.5). Yet, if for some finite subset X' of X we assign $P(\alpha_1) > 0$, and hence $P(\alpha_1 \to F) = 0$, imposing large probabilities on all formulas in X', then $X' \Vdash A$ shall not hold.

Definition 10.6

Let X be a set of formulas. We define the set of probabilistic consequences of X as the smallest set X' containing X and such that, for any truth-functional formulas α, β, γ:

(1) if α is tautologically equivalent to β and $(\alpha \to \gamma) \in X'$, then $(\beta \to \gamma) \in X'$;

(2) $\alpha \in X'$ iff $(1 \to \alpha) \in X'$;

(3) if α implies β tautologically, then $(\alpha \to \beta) \in X'$;

(4) if $(\alpha \to \gamma) \in X'$ and $(\beta \to \gamma) \in X'$, then $(\alpha \vee \beta \to \gamma) \in X'$;

(5) if $(\alpha \vee \beta \to \gamma) \in X'$ and $(\beta \to -\gamma) \in X'$, then $(\alpha \to \gamma) \in X'$;

(6) if $(\alpha \to \beta \wedge \gamma) \in X'$, then $(\alpha \to \beta) \in X'$;

(7) if $(\alpha \to \beta) \in X'$ and $(\alpha \to \gamma) \in X'$, then $(\alpha \to \beta \wedge \gamma) \in X'$;

(8) if $(\alpha \to \beta) \in X'$ and $(\alpha \wedge \beta \to \gamma) \in X'$, then $(\alpha \to \gamma) \in X'$. ■

Theorem 10.7

Let X be a set of formulas and A be a formula. If A is a probabilistic consequence of X then A is a reasonable consequence of X.

Proof

This is a step-by-step verification that the rules (1)–(8) obey the conditions defining reasonable consequence. ■

Theorem 10.8

Let $\gamma, \alpha_1, \ldots, \alpha_n, \beta_1, \ldots, \beta_n$ be truth-functional formulas. Then:

(1) If γ is a tautological consequence of β_1, then $(\alpha_1 \to \gamma)$ is a probabilistic consequence of $(\alpha_1 \to \beta_1)$.

(2) Each of the formulas $(\alpha_1 \to \beta_1)$ and $(\alpha_1 \to (\alpha_1 \wedge \beta_1))$ is a probabilistic consequence of the other one.

(3) $(\alpha_1 \vee \alpha_2) \to -(\alpha_1 - \beta_1)$ is a probabilistic consequence of $(\alpha_1 \to \beta_1)$.

(4) $(\alpha_1 \vee \ldots \vee \alpha_n) \to (-(\alpha_1 - \beta_1) \wedge \ldots \wedge -(\alpha_n - \beta_n))$ is a probabilistic consequence of $((\alpha_1 \to \beta_1) \wedge \ldots \wedge (\alpha_n \to \beta_n))$.

(5) If $(\alpha_2 \wedge \beta_2)$ is a tautological consequence of $(\alpha_1 \wedge \beta_1)$ and $(\alpha_1 \to \beta_1)$ is a tautological consequence of $(\alpha_2 - \beta_2)$, then $(\alpha_1 \to \beta_2)$ is a probabilistic consequence of $(\alpha_1 - \beta_1)$. ■

10.4.3 Completeness of Probability Logic

According to Theorem 10.7, the condition that a formula A be a reasonable consequence of a formula set X is sufficient in order that A be a probabilistic consequence of X. We are now going to show that this condition is also necessary, thus obtaining the completeness theorem for probability logic. To achieve this, we introduce some further notions and constructions.

The first new concept to enter the stage will be that of *P*-ordering of the set of *tf*-formulas (*tf* is short for truth-functional). This is a weak order relation, denoted by $\leq$ and defined as follows.

Definition 10.7

Let L be the language.

(1) By a *P*-ordering of L we mean a binary relation $\leq$ in the set of *tf*-formulas, satisfying the following conditions (for any *tf*-formulas α, β, γ):

(i) Either $\alpha \leq \beta$ or $\beta \leq \alpha$; if $\alpha \leq \beta$ and $\beta \leq \gamma$, then $\alpha \leq \gamma$;

(ii) If α implies β tautologically, then $\alpha \leq \beta$;

(iii) $\alpha \vee \beta \leq \gamma$ iff $\alpha \leq \gamma$ and $\beta \leq \gamma$;

(iv) $\alpha \leq \beta \vee \gamma$ iff $\alpha \leq \beta$ or $\alpha \leq \gamma$.

We write $\alpha < \beta$ iff $\alpha \leq \beta$ but not $\beta \leq \alpha$.

(2) Let $\leq$ be a P-ordering. A *tf*-formula α is said to hold in $\leq$ iff $-\alpha < \alpha$. ∎

Theorem 10.9

Suppose the language L is finite and let $\leq$ be a binary relation on *tf*-formulas of L. Then $\leq$ is a P-ordering if and only if there exists a weak ordering $\leq_0$ in the set $XD_+ = XD \cup \{0\}$ such that:

(1) $F \leq_0 \gamma$ for all $\gamma \in XD_+$ and $F <_0 \gamma$ for some $\gamma \in XD_+$;

(2) if α, β are any *tf*-formulas and $\gamma, \sigma \in XD_+$ are maximal elements occurring in α, β (respectively), then $\alpha \leq \beta$ iff $\gamma \leq_0 \sigma$. ∎

Definition 10.8

Let L be the language, let $\leq$ be a P-ordering in L and let $P_1, P_2, \ldots$ be a uniform sequence of probabilities in L. The sequence $P_1, P_2, \ldots$ is said to be associated with $\leq$ if, for any *tf*-formulas α and β,

$$\alpha \leq \beta \quad \text{iff} \quad \lim_{n \to \infty} P_n(\alpha \vee \beta \to \beta) > 0.$$

The limit is positive if and only if the limit of $P_n(\beta)/P_n(\alpha)$ is positive; this to some extent explains the notation $\leq$. ∎

Theorem 10.10

Suppose the language L is finite. Let A be a formula and X be a finite set of formulas. Under these assumptions:

(1) If $P_1, P_2, \ldots$ is a uniform sequence of probabilities, then there exists a unique P-ordering $\leq$ such that $P_1, P_2, \ldots$ is associated with $\leq$.

(2) If $\leq$ is a P-ordering, then there exists a uniform sequence of probabilities associated with $\leq$.

(3) If $\leq$ is a P-ordering and $P_1, P_2, \ldots$ is a uniform sequence associated with $\leq$, then A holds in $\leq$ if and only if $\lim_{n\to\infty} P_n(A) = 1$.

(4) If $X \Vdash A$, then A holds in all P-orderings in which all formulas of X hold.

Definition 10.9

Let L be finite and let X be a finite set of formulas of L.

(1) The immediate reduction of X is the set $Red(X) \subseteq X$ consisting of all formulas $A \in X$ such that $XD(A) \subseteq \mathrm{XD}(X)$.

(2) The reducing sequence of X is the sequence $X_1, \ldots, X_p$ of subsets of X defined by the conditions $X_1 = X$, $X_{i+1} = Red(X_i)$ and $Red(X_p) = X_p$.

(3) The ordinal partition of the set XD, generated by X, is the sequence $XD_1, \ldots, XD_{p+1}$ of subsets of XD defined by the conditions,

$XD_1 = XD \setminus \mathrm{XD}(X_1)$,

$XD_{i+1} = \mathrm{XD}(X_i) \setminus \mathrm{XD}(X_{i+1}) \quad \text{for} \quad i = 1, \ldots, p-1$,

$XD_{p+1} = \mathrm{XD}(X_p)$,

where $X_1, \ldots, X_p$ is the reducing sequence of X.

(4) Suppose $\mathrm{XD}(X) \neq XD$ and let $XD_1, \ldots, XD_{p+1}$ be the ordinal partition of XD, generated by X. The standard P-ordering of L associated with X is defined by the conditions: for $\gamma\sigma \in XD$, if $\gamma \in XD_i$, $\sigma \in XD_j$, then:

$$\gamma \leq F \ \text{ iff } \ i = p+1,$$
$$\gamma \leq \sigma \ \text{ iff } \ j \leq i.$$

Obviously, the ordinal partition is indeed a partition of XD. ■

Theorem 10.11

Let L be finite; let X be a finite set of formulas and let $XD_1, \ldots, XD_{p+1}$ be the ordinal partition of XD generated by X. Then:

(1) $XD_1, \ldots, XD_{p+1}$ is a partition of XD.

(2) If $\mathrm{XD}(X)$ is not equal to XD, then all formulas of X hold in the standard P-ordering of L associated with X. ■

This apparatus provides a tool for the proof of the equivalence between the notions of probabilistic consequence and reasonable consequence. A formula turns out to be a consequence of a set X, in any sense, if and only if it holds in all P-orderings in which all formulas of X hold.

For a detailed proof we refer to [Adams 1966]. There also appear other equivalent statements, serving as links in the cycle of proofs of those major equivalences. Here is the full completeness theorem, as given in [Adams 1966]:

Theorem 10.12

Let L be a finite language; let A be a formula and X be a finite set of formulas. Write $X' = X \cup \{-A\}$; let $X'_1, \ldots, X'_p$ be the reducing sequence for X', let $XD'_1, \ldots, XD'_{p+1}$ be the ordinal partition of XD generated by X', and write $X'_0 = X'_p \setminus \{-A\}$.

Under these conditions, each of the following statements implies the other ones:

(1) A is a probabilistic consequence of X;

(2) A is a reasonable consequence of X;

(3) A holds in every P-ordering in which all formulas of X hold;

(4) $XD(A) \subseteq XD(X'_p)$;

(5) $\mathrm{XD}(A) \subseteq \mathrm{XD}(X_0)$ and $XD(X_0) \setminus \mathrm{XD}(X_0) \subseteq XD(A) \setminus \mathrm{XD}(A)$;

(6) for a certain subset $X'' \subseteq X$,
$\mathrm{XD}(A) \subseteq \mathrm{XD}(X'')$ and $XD(X'') \setminus \mathrm{XD}(X'') \subseteq XD(A) \setminus \mathrm{XD}(A)$. ∎

The last theorem in this section displays the relevance of reasonable consequence to strong implication.

Theorem 10.13

Let L be a finite language, let A be a formula and X be a set of formulas. Then A is a reasonable consequence of X if and only if there exists a subset X' of X such that X' strongly implies A. ∎

Example

Let L be the language generated by two atomic formulas p, q (beside F and T). Assume that XD contains the four formulas

$$\begin{aligned}
\alpha &= -p \wedge -q, \\
\beta &= p \wedge -q, \\
\gamma &= p \wedge q, \\
\delta &= -p \wedge q,
\end{aligned}$$

whose mutual interrelations are illustrated in the table given below.

Let $X = \{p \to q,\ -q \to -p,\ q \to -p,\ p \vee q \to p\}$.

First, to construct the immediate reduction of X, we seek formulas $A \in X$ such that $XD(A) \subseteq \mathrm{XD}(X)$. All formulas $A \in X$ together with the corresponding sets $XD(A)$, $\mathrm{XD}(A)$ are explicitly written down in the following table:

	formula A	$XD(A)$	$\mathrm{XD}(A)$
1	$p \to q$	$\{\beta, \gamma\}$	$\{\beta\}$
2	$-q \to -p$	$\{\alpha, \beta\}$	$\{\beta\}$
3	$q \to -p$	$\{\gamma, \delta\}$	$\{\gamma\}$
4	$p \vee q \to p$	$\{\beta, \gamma, \delta\}$	$\{\delta\}$

Thus $Red(X)$ is constituted by the formulas 1, 3 and 4.

In order to determine the reducing sequence, we iterate this procedure. Thus,

$$X_1 = X,$$
$$X_2 = Red(X_1) = \{p \to q,\ q \to -p,\ p \vee q \to p\}.$$

It is seen that $\mathrm{XD}(X_2) = \{\beta, \gamma, \delta\}$ and so $XD(A) \subseteq \mathrm{XD}(X_2)$ for all $A \in X_2$, which means that $Red(X_2) = X_2$. The process stops.

Now we have to construct the ordinal partition of XD generated by X. By definition, XD_1 denotes the set of those XD sets which are outside $\mathrm{XD}(X_1)$. Since $\mathrm{XD}(X_1) = \mathrm{XD}(X) = \{\beta, \gamma, \delta\}$, we see that $XD_1 = \{\alpha_1\}$. Next, $XD_2 = \mathrm{XD}(X_1) \setminus \mathrm{XD}(X_2)$. And since $\mathrm{XD}(X_1) = \mathrm{XD}(X) = \{\beta, \gamma, \delta\}$, we get $XD_2 = \emptyset$ and $XD_3 = \mathrm{XD}(X_2) = \{\beta, \gamma, \delta\}$. Hence, the ordinal partition of XD generated by X is the sequence of sets

$$XD_1 = \{\alpha\},\ XD_2 = \emptyset,\ XD_3 = \{\beta, \gamma, \delta\}.$$

Now we are able to determine the standard P-ordering of L by reversing the order of sets as they appear in the ordinal partition:

$$F < \beta < \gamma < \delta < \alpha < T.$$

Thus, to fix the order relation holding between p and q, we observe that the XD sets belonging to p and q (simultaneously) are equal, and hence p and q are equivalent. On the other hand, the maximal XD set belonging to p is strictly smaller than the maximal XD set belonging to $-p$ (i.e., $\{\alpha\}$), and therefore $p < -p$.

Likewise, it can be shown that

$$p \leq F,\ -q \wedge -(-p) < -q \wedge -p,\ q \leq F,\ p \vee q \leq F.$$

It follows that all formulas of X hold in the standard ordering.

10.4.4 Applications

Let us now have a look at the consequences of the completeness theorem (Theorem 10.12) and the way they can be applied in practice.

There are two drawbacks in the method of verifying, on the basis of the above definition, whether or not the conclusion is a reasonable consequence of its premises. Firstly, it can happen that the conclusion is a reasonable consequence of its premises without being strongly implied by them (roughly speaking, reasonable consequences are more easily achieved than strongly implied conclusions). And secondly, this method provides less information on the interrelation between the premises on the one hand and the conclusion on the other than the method resorting to the completeness theorem; this can be of importance for formal proofs of the properties of formulas.

Using Theorem 10.13 as a basis one can state several metatheorems concerning reasonable inference:

Theorem 10.14
Let A, B be formulas and let X be a finite set of formulas.

(1) $X \Vdash A$ iff $X, \sim A \Vdash Ant(A) \rightarrow 0$.

(2) If $X = \emptyset$ or A is a tf-formula, then $x \Vdash A$ iff X tautologically implies A.

(3) If $X, B \Vdash A$ and $X, \sim B \Vdash A$, then $X \Vdash A$.

(4) If X consists entirely of tf-formulas, then $X \Vdash A$ iff either A is a tautology or X tautologically implies $Ant(A) \wedge Cons(A)$.

(5) If $X = \{\alpha_1 \rightarrow \beta_1, \ldots, \alpha_n \rightarrow \beta_n\}$ and if $X \Vdash A$, while $X' \Vdash A$ does not hold for any proper subset $X' \subset X$, then A is a reasonable consequence of the set
$$C(S) = (\alpha_1 \vee \ldots \vee \alpha_n) \rightarrow -(\alpha_1 - \beta_1) \wedge \ldots \wedge -(\alpha_n - \beta_n).$$

(6) If $A = p \rightarrow \beta$ (where p is an atomic formula not occurring in X) then $X \Vdash A$ iff either $p \rightarrow \beta$ is a tautology or $X \Vdash -\beta \rightarrow F$. ∎

Theorem 10.15
Let L be a finite language, A a formula, X a finite set of formulas and α a tf-formula. Under these assumptions:

(1) If $X \Vdash A$ does not hold, then for every $\varepsilon > 0$ there exists probability P such that $P(B) > 1 - \varepsilon$ for all $B \in X$, and yet $P(A) < \varepsilon$.

(2) If $X \Vdash A$ and X has n elements then, for every $\varepsilon > 0$ and every probability P, if $P(B) > 1 - \varepsilon$ for all $B \in X$ then $P(A) > 1 - n\varepsilon$.

(3) If $X \Vdash -\alpha \to 0$ and X has n elements then, for every probability P, if $P(B) > 1 - \frac{1}{n}$ for all $B \in X$ then $P(\alpha) = 1$. ∎

This theorem provides some information which can be helpful in estimating the probability of the occurrence of the conclusion given that the premises satisfy certain conditions.

10.4.5 Unreasonable Inference

The analysis in the previous subsection shows that a tautologically valid conclusion need not be a reasonable consequence of the premises. The following is perhaps the most interesting question in this context: what interrelation between the premises and the conclusion, in a tautologically valid inference, entails unreasonableness? That is, under what circumstances does it happen that, despite tautological validity, the premises have high probability while the conclusion has low probability?

Suppose X is the set of premises and A is the conclusion. The first question to pose is this: in order that all formulas in X be satisfied and A be unsatisfied, what properties should the ordering relation $\leq$ have? We illustrate this by an example of unreasonable inference, in which $-p \vee q$ is the premise and $p \to q$ is the conclusion. Let p and q be as in the example in Section 10.4.3. Then $-p \vee q$ is satisfied in $\leq$ iff $\beta < \alpha \vee \gamma \vee \delta$, whereas $p \to q$ is satisfied in $\leq$ iff $\beta < \gamma$ or $\gamma \leq 0$. Hence $p \to q$ does not hold in $\leq$ iff $\alpha \leq \gamma$ and $0 < \beta \vee \gamma$. On account of the definition of P-orderings we can say that $-p \vee q$ is satisfied in $\leq$ and $p \to q$ is not if and only if the inequalities $0 < \beta$, $\gamma \leq \beta$ and $\beta < \alpha \vee \delta$ hold simultaneously.

Thus, as $P(\alpha)$ has to be small as compared to $P(\beta)$, we can infer that:

(1) $P(p) > 0$;

(2) $P(\beta)$ is not small relative to $P(\gamma)$;

(3) $P(\beta)$ is small relative to $P(\alpha \vee \beta) = P(-p)$.

We wish to point out by a counterexample that this inference is unreasonable. Now, it is enough to set (as E. W. Adams [Adams 1966] does):

p = "Mr. Jones will have an accident on his way to work"

and

q = "Mr. Jones will arrive on time for work".

Then

$$-p \vee q = \text{"Either Mr. Jones will not have an accident on his way to work or he will arrive on time for work"},$$
$$p \to q = \text{"If Mr. Jones has an accident on his way to work, then he will arrive on time for work"}.$$

It is readily seen that $-p \vee q$ is highly probable while $p \to q$ is highly improbable. Thus, in this case, it is not reasonable to derive $p \to q$ from $-p \vee q$.

In most situations, however, this scheme of inference is quite legitimate. For suppose that the given set X implies $\alpha \to \beta$ tautologically but $X \Vdash \alpha \to \beta$ does not hold; assume also that $P(B) > 1 - \varepsilon$ for all $B \in X$ but $P(\alpha \to \beta) \leq \frac{1}{2}$; then we have $P(\alpha) < 2n\varepsilon$. Usually, when we are convinced that $-p$ holds, we assert $-p \vee q$. So the probability $P(p \to q)$ should be not less than $\frac{1}{2}$; the reasoning can be regarded as well justified.

References

Ackermann, R. (1967): Introduction to Many-Valued Logics. London: Routledge and Kegan Paul

Adams, E. W. (1965): The logic of conditionals, Inquiry, 8, 166–197

Adams, E. W. (1966): Probability and the logic of conditionals. In: J.Hintikka, P.Suppes (eds.): Aspects of inductive logic. Amsterdam: North-Holland, 265–316

Anderson, J. M., Johnstone, H. W., Jr. (1962): Natural deduction. The Logical Basis of Axiom Systems. Belmont, CA: Wadsworth

Anderson, A. R., Belnap, N. D. (1975): Entailment. Vol. I, Princeton University Press

Angell, R. B. (1962): A propositional logic with subjunctive conditionals. Journal of Symbolic Logic, 27, 327–343

APF (1963): Proceedings of a colloquium on modal and many-valued logics, Helsinki, August 1962. Acta Philosophica Fennica, 16, 1–290

Åqvist, L. (1962): Reflections on the Logic of Nonsense. Theoria, 28:2. 138–158

Asser, G. (1953) Die endlichwertigen Łukasiewiczschen Aussagenkalküle. Bericht über die Mathematiker-Tagung in Berlin, Januar 1953. Berlin, 15–18

Baldwin, J. F. (1979): Fuzzy logic and fuzzy reasoning. Int. J. Man-Machine Studies, 11, 465–480

Baldwin, J. F. (1987): Evidential support logic programming. Fuzzy Sets and Systems, 24:1, 1–26

Bandler, W., Kohout, L. (1985): Probabilistic versus fuzzy production rules in expert systems. Int. J. Man-Machine Studies, 22, 347–353

Barcan, R. C. (1946): A functional calculus of first order based on strict implication. Journal of Symbolic Logic, 11, 1–16

Bell E. T. (1943): Polynomials on a finite discrete range. Duke Mathematical Journal, 10, 33–47

Bellman R. E., Zadeh, L. A. (1977): Local and fuzzy logics. In: Dunn, Epstein (eds.): Modern uses of multiple-valued logic. Reidel, 105–165

Bellman, R. E., Giertz, M. (1973): On the analytic formalism of the theory of fuzzy sets. Information Sciences, 5, 149–156

Belluce, L. P., Chang, C. C. (1963): A weak completeness theorem for infinite-valued first-order logic. Journal of Symbolic Logic, 28, 43–50

Belluce, L. P. (1964): Further results on infinite-valued predicate logic. Journal of Symbolic Logic, 29, 69–78

Belnap, N. D. (1977): A useful four-valued logic. In: G. Epstein, J. M. Dunn (eds.): Modern Uses of Multiple-Valued Logic. Reidel, 30–56

Bergmann, G. (1949): The finite representations of S5. Methodos, 1, 217–219

Bernstein, B. A. (1928): Modular representations of finite algebras. Proceedings of the International Mathematical Congress, Toronto, August 1924, 1, 207–216

Beyth-Marom R. (1982): How probable is probable? A numerical taxonomy translation of verbal probability expressions. Journal of Forecasting, 1, 257–269

Birkhoff G., von Neumann, J. (1936): The logic of quantum mechanics. Annals of Mathematics, 37, 823–843

Birkhoff G. (1948): Lattice Theory, American Mathematical Society Colloquium Publications. New York

Bloom, S. L., Brown, D. J. (1973): Classical abstract logics. Dissertationes Mathematicae, 102, 43–52

Bocheński, I. M. (1959): A Précis of Mathematical Logic. Dordrecht: Reidel

Bocheński, I. M., Menne, A. (1962): Grundriss der Logistik. Paderborn: Schöningh

Bochvar, D. A. (1939): On a 3-valued logical calculus and its application to the analysis of contradictions. Matématičéskij sbornik, 4, 287–308 (In Russian)

Bochvar, D. A. (1943): On the consistency of a 3-valued calculus. Matématičéskij sbornik, 12, 353–369 (In Russian)

Bochvar, D. A. (1944): On the question of paradoxes of the mathematical logic and theory of sets. Matématičéskij sbornik, 15, 369–384 (In Russian)

Bonissone, P. (1985): Selecting uncertainty calculi and granularity: An experiment in trading-off precision and complexity. KBS Working Paper. Schenectady, New York: General Electric Research and Development Center

Bonissone, P. (1986): Plausible reasoning: coping with uncertainty in expert systems. Research Report No.86CRD053. Schenectady, New York: General Electric Corporate Research and Development

Bonissone, P. (1987): Summarizing and propagating uncertain information with triangular norms. Intern. J. Approximate Reasoning, 1, 71–101

Borkowski, L., Słupecki, J. (1958): The logical works of J.Lukasiewicz. Studia Logica, 8, 7–56

Brouwer, L. E. J. (1925): Intuitionistische Zerlegung mathematischer Grundbegriffe. Jahresbericht der deutschen Mathematiker-Vereinigung, 33, 251–256. English translation in J. van Heijenoort (ed.): From Frege to Gödel: A Source Book in Mathematical Logic, 1879–1931. Cambridge, MA: 1967, 334–341

Brouwer, L. E. J. (1925): Über die Bedeutung des Satzes vom ausgeschlossenen Dritten in der Mathematik, insbesondere in der Funktionen-theorie. Journal für die reine und angewandte Mathematik, 154, 1–7

Bull, R. A. (1964): An axiomatization of Prior's modal calculus Q. Notre Dame Journal of Formal Logic, 5, 211–214

Butler, R. J. (1955): Aristotle's sea fight and three-valued logic. Philosophical Review, 64, 264–274

Carnap, R. (1943): The Formalization of Logic. Harvard: University of Harvard Press

Carnap, R. (1950): Logical Foundations of Probability. Chicago: University of Chicago Press, 2d ed.,

Carnielli, W. A. (1987): Systematization of finite many-valued logics throught the method of tableaux. Journal of Symbolic Logic, 52:2, 473-493

Cat Hao, Ng. (1973): Generalized Post algebras and their applications to some infinitary many-valued logics. Dissertationes Mathematicae 107, 1–76

Cat Hao, Ng., Rasiowa, H. (1987): Semi-Post algebras. Studia Logica 46, 2, 147–158

Cat Hao, Ng., Rasiowa, H. (1989): Plain semi-Post algebras as a posets-based generalization of Post algebras and their representability, Studia Logica, 48(4)

Caton, C. E. (1963): A stipulation of a modal propositional calculus in terms of modalized truth-values. Notre Dame Journal of Formal Logic, 4, 224–226

Chandrasekharan, K. (1944): Partially ordered sets and symbolic logic. The Mathematics Student, 12, 14–24

Chang, C. C. (1958): Proof of an axiom of Łukasiewicz. Transactions of the American Mathematical Society, 87, 55–56

Chang, C. C. (1958): Algebraic analysis of many-valued logics. Transactions of the American Mathematical Society, 88, 467–490

Chang, C. C. (1959): A new proof of the completeness of the Łukasiewicz axioms. Transactions of the American Mathematical Society, 93, 74–80

Chang, C. C. (1961): Theory of models of infinite-valued logics, I–IV. Abstracts in the Notices of American Mathematical Society, 8, 68–141.

Chang, C. C. (1963): Logic with positive and negative truth values. Acta Philosophica Fennica, 16, 19–39

Chang, C. C. (1963): The axiom of comprehension in infinite-valued logic. Mathematica Scandinavica, 13, 9–30

Chang, C. C., Horn, A. (1961): Prime ideal characterization of generalized Post algebras. Proceedings of the Symposium of Pure Mathematics, 2, 43–48

Chang, C. C., Keisler, H. J. (1962): Model theories with truth values in a uniform space. Bulletin of the American Mathematical Society, 68, 107–109

Chang, C. C. (1968): Fuzzy topological spaces. J.Math. Anal. Appl. 24:1, 182–190

Chang, C. C. (1973): Modal model, theory. In: Proc. Cambridge Summer School in Mathematical Logic. Lecture Notes in Mathematics, 337. Berlin: Springer 1973, 599–617

Chang, C. C., Keisler, H. J. (1966): Continuous model theory. New York: Princeton University Press

Chang, C. C., Keisler, H. J. (1973): Model Theory. Amsterdam: North-Holland

Chang, C. L., Lee, R. C. T. (1973): Symbolic logic and mechanical theorem proving.

Chang, S. K. (1971): Fuzzy programs, theory and applications. Proc. Brooklyn Polytechnical Institute Symp. on Computers and Automata, 21

Chen, K. H., Raś, Z., Skowron, A. (1984): Attributes and rough properties in information systems, Proc. of the 1984 Conf. on Information Sciences and Systems, March 1984, Princeton, 362–366

Christensen, N. E. (1957): Further comments on two-valued logic. Philosophical Studies, 8, 9–15

Church, A. (1953): Non-normal truth-tables for the propositional calculus. Buletin de la Sociedad Matematica Mexicana, 10, 41–52

Church, A. (1956): Introduction to Mathematical Logic. Princeton University Press

Church, A. (1956): Laws of thought. Encyclopaedia Britannica, 22, 157

Cignoli, R. (1982): Proper n-valued Łukasiewicz algebras as S-algebras of Łukasiewicz n-valued propositional calculi, Studia Logica XLI, 1, Wrocław: Ossolineum, North-Holland

Clay, R. E. (1962): Note on Słupecki T-functions. Journal of Symbolic Logic, 27, 53–54

Clay, R. E. (1962): A simple proof of functional completeness in many-valued logics based on Łukasiewicz's C and N. Notre Dame Journal of Formal Logic, 3, 114–117

Clay, R. E. (1963): A standard form for Łukasiewicz many-valued logics. Notre Dame Journal of Formal Logic, 4, 59–66

Cleave, J. P. (1974): The notion of logical consequence in the logic of inexact predicates. Zeitschrift fur Mathematische Logik und Grundlagen der Matematik, 20, 307–324

Cohen, J. (1951): Three-valued ethics. Philosophy, 26, 208–227

Comey, D. D. (1965): An extensive review of V.A.Smirnov's paper: The logical views of N.A.Vasiliev. Journal of Symbolic Logic, 30, 368–370

Costa de Beauregard, O. (1945): Extension d'une théorie de M.J. de Neumann au cas des projecteurs non commutables. Comptes rendus hebdomadaires des seances de l'Academie des Sciences, 21, 230–231

Crawley, P., Dilworth, R. P. (1973): Algebraic theory of lattices. Englewood Cliffs, NJ: Prentice Hall

Cresswell, M. J., Hughes, G. E. (1968): An introduction to modal logic. London: Methuen

Czelakowski, J. (1980): Model-theoretic methods in methodology of propositional calculi. Warszawa: Polish Academy of Sciences, Institute of Philosophy and Sociology

Czogola E., Drewniak, J. (1984): Associative monotonic operations in fuzzy set theory. Fuzzy Sets and Systems, 12, 249–269

van Dalen, D. (1986): Intuitionistic logic. In: D.Gabbay, F.Guenthner (eds.): Handbook of Philosophical Logic, Part 3, Dordrecht: Reidel, 225–339

Dempster, A. P. (1967): Upper and lower probabilities induced by a multivalued mapping. Annals of Mathematical Statistics, 38, No.2, 325–339

Destouches-Février, P. (1949): Logique et théories physiques. Synthése, 7, 400–410

Destouches-Février, P. (1951): La structure des théories physiques. Paris: Presses Universitaires de France

Destouches-Février, P. (1952): Applications des logiques modales en physique quantique. Theoria, 1, 167–169

Dienes, P. (1949): On Ternary Logic. Journal of Symbolic Logic, 14, 85–94

Dienes, Z. P. (1949): On an implication function in many-valued systems of logic. Journal of Symbolic Logic, 14, 95–97

D'Ottaviano, I. M. L. (1982): On trivalued models theory. Ph.D. thesis, Brazil: University of Campinas (In Portuguese)

Doob, J. L. (1947): Probability in function space. Bulletin of the American Mathematical Society, 53, 15–30

Dreben, B. (1960): Relation of m-valued quantificational logic to 2-valued quantificational logic. Summaries of talks presented at the Summer Institute for Symbolic Logic, Cornell University, Princeton University Press, 303–304

Driankov, D. (1988): Towards a Many-Valued Logic of Quantified Belief. Department of Information Science, Linköping University, Linköping

Dubois, D., Prade, H. (1980): New results about properties and semantics of fuzzy set-theoretic operators. In: P.P.Wang, S.K.Chang (eds.): Fuzzy Sets: Theory and Applications to Policy Analysis and Information Systems. New York: Plenum Press, 59–75

Duda, R. O., Hart, P. E., Nilsson, J. (1976): Subjective Bayesian methods for rule-based inference systems. AFIPS Conference Proceedings, New York, 1075–1082

Dumitriu, A. (1943): Logica Polivalenta (Many-valued Logic). Bucharest: Viata Literara

Dummett, M. A. E. (1959): A propositional calculus with denumerable matrix. Journal of Symbolic Logic, 24, 97–106

Dunn, J. M. (1976): Intuitive semantics for first degree entailments and coupled trees. Journal of Philosophical Studies, 29, 149–168

Dunn, J. M., Epstein, G. (eds.) (1977): Modern Uses of Multiple-Valued Logic. Dordrecht: Reidel

Dunn, J. M. (1986): Relevance logic and entailment. In: D.Gabbay, F.Guenther (eds.): Handbook of Philosophical Logic, Part 3, Dordrecht: Reidel, 117–224

Dwinger, P. (1966): Notes on Post Algebras I,II. Indag. Math., 28, 462–478

Dwinger, P. (1977): A survey of the theory of Post algebras and their generalizations. In: Dunn, Epstein (eds.): Modern Uses of Multiple-Valued Logic, 53–75

Ehrenfeucht, A., Mostowski, A. (1961): A compact space of models of axiomatic theories. Bull. Acad. Polon. Sc. Ser. Sci. Math. Astron. Phys., 9, 369–373

Epstein, G., Horn, A. (1974): P-algebras, An abstraction from Post algebras. Algebra Universalis, 4, 195–206

Epstein, G. (1960): The lattice theory of Post algebras. Trans. Amer. Math. Soc., 95, 300–317

Epstein, G., Frieder, G., Rine, D. C. (1974): The development of multiple-valued logic as related to computer science. Computer, 7(9), 20–32

Evans, T., Hardy, L. (1957): Sheffer stroke functions in many-valued logics. Portugaliae Mathematica, 16, 83–93

Evans, T., Schwartz, P. B. (1958): On Słupecki T-functions. Journal of Symbolic Logic, 23, 267–270

Fagin, R., Halpern, J. Y., Megiddo, N. (1990): A logic for reasoning about probabilities. Information and Computation, 87, 78–128

Fenstad, J. E. (1964): On the consistency of the axiom of comprehension in the Łukasiewicz infinite-valued logic. Mathematica Scandinavica, 14, 64–74

Fenstad, J. E. (1965): A limit theorem in polyadic probabilities. Proc. Logic Colloquium, Leicester

Février, P. (1959): Logical structure of physical theories. In L.Henkin (ed.): The Axiomatic Method. Amsterdam, 376–389

Feyerabend, P. (1958): Reichenbach's interpretation of quantum mechanics. Philosophical Studies, 9, 49–59

Finn, V. K. (1972): An Axiomatization of some propositional calculi and their algebras. Vsiechsoiuznyi Institut Naučeskoi Informacii Academii Nauk SSSR, Moskva (In Russian)

Fitting, M. C. (1969): Intuitionistic Logic, Model Theory and Forcing. Amsterdam: North-Holland

Fitting, M. C. (1973): Model-existence theorems for modal and intuitionistic logics. Journal of Symbolic Logic, 38, 613–627

Fletcher, T. J. (1963): Models of many-valued logics. American Mathematical Monthly, 70, 381–391

Foxley, E. (1962): The determination of all sheffer functions in 3-valued logic, using a logical computer. Notre Dame Journal of Formal Logic, 3, 41–50

Frege, G. (1879): Begriffsschrift, eine der arithmetischen nachgebildete Formelsprache des reinen Denkens. Halle

Frege, G. (1892): Über Sinn und Bedeutung. Zeitschrift für Philosophische Kritik C, 25–50

Gaifman, H. (1964): Concerning measures on first-order calculi. Israel J.Math., 2, 1–18

Gainess, B. R. (1975): Fuzzy reasoning and the logics of uncertainty. EES–MMS–UNC 75, Dept. of Electrical Engineering Science, University of Essex

Gainess, B.R. (1976): Foundations of fuzzy reasoning. International Journal of Man-Machine Studies, 8, 623–668

Gainess, B.R. (1976): General Fuzzy Logics. Proceedings of the 3rd European Meeting on Cybernetics and Systems Research, Vienna

Gainess, B.R. (1978): Fuzzy and probability uncertainty logics. Information and Control, 38, 158–169

Garson, J.W. (1984): Quantification in modal logic. D.Gabbay, F.Guenthner (ed.): Handbook of Philosophical Logic, Part 2. Dordrecht: Reidel, 249–307

Gavrilov, G. P. (1959): Certain conditions for completeness in countable-valued logic. Doklady Akademii Nauk SSSR, 128, 21–24 (In Russian)

Gazalé, M.J. (1957): Multi-valued switching functions. Summaries of Talks Presented at the Summer Institute for Symbolic Logic, Cornell University, 147

Geach, P. T. (1949): If's and and's. Analysis, 9, 58–62

Giles, R. (1974): A non-classical logic for physics. Studia Logica, 33, 397–416

Giles, E. (1979): A formal system for fuzzy reasoning. Fuzzy Sets and Systems 2, 233–257

Giles, R. (1984): A resolution logic for fuzzy reasoning. Research Report No.1984–18, Queen's University, Dept. of Mathematics and Statistics, Kingston, Ontario, Canada

Ginsberg, M. L. (1984): Analyzing Incomplete Information. Technical Report 84–17, KSL, Stanford University

Ginsberg, M. L. (1984): Non-monotonic reasoning using Dempster's rule. In Proceedings of the Fourth National Conference on Artificial Intelligence, 126–129

Ginsberg, M. L. (1985): Does probability have a place in non-monotonic reasoning? In Proceedings of the Ninth International Joint Conference on Artificial Intelligence, 107–110

Ginsberg, M. L. (1985): Implementing probabilistic reasoning. In Proceedings 1985 Workshop on Uncertainty in Artificial Intelligence, Los Angeles, CA, 84–90

Ginsberg, M. L. (1986): Possible worlds planning. In Proceedings of the 1986 Workshop on Planning and Reasoning about Action, Timberline, Oregon

Ginsberg, M. L., Smith, D. E. (1986): Reasoning About Action I: A Possible Worlds Approach. Technical Report 86-65, KSL, Stanford University

Gniedenko, V. M. (1962): Determination of orders of pre-complete classes in three-valued logic. Problemy Kibernetiki, 8, 341-346

Gniedenko, B. V. (1962): The Theory of Probability. Translated from Russian by B.R.Seckler. New York: Chelsea

Goddard, L. (1960): The Exclusive 'Or'. Analysis, 20, 97-106

Goguen, J. A. (1967): L-fuzzy sets. J. Math. Anal. Appl., 18, 145-174

Goguen, J. A. (1969): The logic of inexact concepts. Synthese, 19, 325-373

Goguen, J. A. (1974): Concept representation in natural and artificial languages: axioms, extensions and applications for fuzzy sets. International Journal of Man-Machine Studies, 6, 513-561

Gottwald, S. (1981): Fuzzy-Mengen und ihre Anwendungen. Ein Überblick, Elektronische Informationsverarbeitung und Kybernetik, 17, 207-235

Gödel, K. (1930): Die Vollständigkeit der Axiome des logischen Funktionenkalküls. Monatshefte für Mathematik und Physik, 37, 349-360

Gödel, K. (1932): Zum intuitionistischen Aussagenkalkül. Anzeiger der Akademie der Wissenschaften Wien, mathematisch, naturwissenschaftliche Klasse, 69, 65-66

Gödel, K. (1933): Eine Interpretation des intuitionistischen Aussagenkalküls. Ergebnisse eines mathematischen Kolloquiums, 4, 33-40

Gödel, K. (1933): Zum intuitionistischen Aussagenkalküls. Ergebnisse eines mathematischen Kolloquiums, 4, 40

Gödel, K. (1935): Eine Eigenschaft der Realisierungen des Aussagenkalküls. Ergebnisse eines mathematischen Kolloquiums, 5, 20-21

Götlind, E. (1951): A Leśniewski-Mihailescu-theorem for m-valued propositional calculi. Portugaliae mathematica, 10, 91-102

Götlind, E. (1952): Some Sheffer functions in n-valued logic. Portugaliae Mathematica, 11, 141-149

Grätzer, G. (1968): Universal Algebra. Princeton: van Nostrand

Grätzer, G. (1978): General Lattice Theory. Basel: Birkhäuser

Green, C. (1969): Application of theorem proving to problem solving. Proc. of the First International Joint Artificial Intelligence Conf., Washington, D.C.

Greniewski, M. (1956): Utilisation des logiques trivalentes dans la théorie des mécanismes automatiques. I. Réalisation des fonctions fondamentales par des circuits, Comunicarile Academiei Republicii Populare Romine, Bucharest, 6, 225-229

Greniewski, H. (1957): 2^{n+1} logical values. Studia filozoficzne, 2, 82-116, and 3, 3-28 (In Polish)

Grigolia, R. (1977): Algebraic analysis of Łukasiewicz-Tarski's n-valued logical systems. Selected Papers on Łukasiewicz Sentential Calculi, R.Wójcicki, G.Malinowski (ed.), Ossolineum, Wrocław, 81–92

Grosof, B. N. (1984): Default reasoning as circumscription. In Proceedings 1984 Non-monotonic Reasoning Workshop, American Association for Artificial Intelligence, New Paltz, NY, 115-124

Günther, G. (1958): Die aristotelische Logik des Seins und die nicht-aristotelische Logik der Reflexion. Zeitschrift für philosophische Forschung, 12, 360–407

Habermann, E. (1936): The concept of polarity and many-valued logic. Przegląd Filozoficzny, 39, 438–441 (In Polish)

Hailperin, T. (1937): Foundations of probability in mathematical logic. Philosophy of Science, 4, supplement, 125–150

Halldén, S. (1949): On the decision-problem of Lewis' calculus S5, Norsk matematisk tidsskrift, 31, 89–94

Halldén, S. (1949): The logic of nonsense. Uppsala Universitets arsskrift, 9, 132

Halpern, J. Y. (1990): An analysis of first-order logics of probability. Artificial Intelligence, 46, 311–350

Hanks, S., McDermott, D. (1980): Default reasoning, nonmonotonic logics and the frame problem. In Proceedings of the Fifth National Conference on Artificial Intelligence, 328–333

Hanson, W. H. (1963): Ternary threshold logic. IEEE Transactions on Electronic Computers, EC–12, 191–197

Hartshorne, C. (1964): Deliberation and excluded middle. Journal of Philosophy, 61, 476–477

Hay, L. S. (1963): Axiomatization of the infinite-valued predicate calculus. Journal of Symbolic Logic, 28, 77–86

Helmer O., Oppenheim, P. (1945): A syntactical definition of probability and degree of confirmation. Journal of Symbolic Logic, 10, 25–60

Henkin, L. (1949): The completeness of the first-order functional calculus, Journal of Symbolic Logic, 14, 159–166

Henle, P. (1951): n-valued Boolean algebra, Structure, Method, and Meaning. Essays in Honor of Henry M.Sheffer. New York, Liberal Arts Press, 68–73

Herbrand, J. (1930): Recherches sur la theoria de la demonstration. Traveaux de la Societe des Sciences de Varsoria, 33

Hewitt, E., Savage, L. (1955): Symmetric measures on Cartesian products. Trans. Am. Math. Soc., 80, 470–501

Heyting, A. (1930): Die Formalen Regeln der intuitionistischen Logik. Sitzungsberichte der Preussischen Akademie der Wissenschaften zu Berlin, 42–46

Heyting, A. (1956): Intuitionism: An Introduction. Amsterdam: North-Holland

Hilbert, D., Bernays, P. (1934): Grundlagen der Mathematik. I, Berlin

Hilbert, D., Bernays, P. (1939): Grundlagen der Mathematik. II, Berlin

Hoo, T. (1949): m-valued sub-system of $(m+n)$-valued propositional calculus. Journal of Symbolic Logic, 14, 177–181

Ito, M. (1955): On the "lattice of n-valued functions" [n-valued logic]. Technology Reports of the Kyushu University, 28, 96–101 (In Japanese)

Ito, M. (1956): On the general solution of the n-valued function-lattice [logical] equation in one variable. Technology Reports of the Kyushu University, 28, 239–243 (In Japanese)

Ito, m. (1956): On the general solution of the n-valued function-lattice [logical] equation in several variables. Technology Reports of the Kyushu University, 28, 243–246 (In Japanese)

Jaśkowski, S. (1936): Recherches sur le systeme de la logique intuitioniste. Actes du Congres International de Philosophie Scientifique. Part 6, Philosophie des mathematiques, Paris, 59–61. Translated in S.McCall (ed.) Polish Logic: 1920–1939 Oxford: Oxford University Press 1967, 259–263

Jeffreys, H. (1961): Theory of Probability. London: Oxford University Press

Jobe, W. H. (1962): Functional completeness and canonical forms in many-valued logics. Journal of Symbolic Logic, 27, 409–422

Johansson, I. (1936): Der Minimalkalkül, ein reduzierter intuitionischer Formalismus. Compositio Mathematicae, 4, 119–136

Joja, A. (1960): About tertium non datur. Acta Logica, Bucharest, 1, (1958), 11

Kalicki, J. (1950): Note on truth-tables. Journal of Symbolic Logic, 15, 174–181

Kalicki, J. (1950): A test for the existence of tautologies according to many-valued truth-tables. Journal of Symbolic Logic, 15, 182–184

Kalicki, J. (1954): On equivalent truth-tables of many-valued logics. Proceedings of the Edinburgh Mathematical Society, 10, 56–61

Kalicki, J. (1954): An undecidable problem in the algebra of truth-tables. Journal of Symbolic Logic, 19, 172–176

Kalmár, L. (1934): Über die Axiomatisierbarkeit des Aussagenkalküls. Acta Scientiorum Mathematicurum, 7, 222–243

Kanger, S. (1957): Probability in logic. Stockholm Studies in Philosophy, Stockholm, 1–47

Karp, C. R. (1964): Languages with expressions of infinite length. Amsterdam: North-Holland

Kattsoff, L. O. (1937): Modality and probability. Philosophical Review, 46, 78–85

Katz, M. (1984): Controlled-error theories of proximity and dominance. In: H.J.Skala, S.Termini, E.Trillas (eds.): Aspects of Vagueness. Dordecht: Reidel

Kauf, D. K. (1954): A comment on Hochberg's reply to storer. Philosophical Studies, 5, 57–58

Keisler, J. H. (1970): Logic with the added quantifier there exists uncountable many. Annals of Mathematical Logic, 1, 1–94

Kelley, J. H. (1959): Measures in Boolean algebras. Pacific J.Math., 9, 1165–1177

Kirin, V. G. (1963): On the polynomial representation of operators in the n-valued propositional calculus. Glasnik Matematičko-Fizički i Astronomski. Društvo Matematičara i Fizičara Hrvatske: Serija II, 18, 3–12

Kirkerud, B. (1982): Undefinedness in Assertion Languages. Preprint 74, University of Oslo

Kleene, S. C. (1952): Introduction to Metamathematics. New York: Van Nostrand

Kolmogorov, A. N. (1932): Zur Deutung der intuitionistischen Logik. Mathematische Zeitschrift, 35, 58–65

Kolmogorov, A. N. (1960): The foundations of probability. Translated by M.Morrison, New York: Chelsea

Kotarbiński, T. (1957): Outlines of the history of Logic. Łódź, Zakład Narodowy im Ossolińskich, Wrocław

Kóczy, L. T., Hajnal, M. (1977): A new attempt to axiomatize fuzzy algebra with an application example. Probl. Control and Inf. Theory, 6, 47–66

Körner, S. (1966): Experience and Theory. London: Routledge and Kegan Paul

Kreisel, G. (1962): On weak completeness of intuitionistic predicate logic. Journal of Symbolic Logic, 27, 139–158

Kripke, S. A. (1963): Semantical analysis of modal logic. I. Normal Modal Propositional Calculi, Zeitschrift für mathematische Logik und Grundlagen der Mathematik, 9, 67–96

Kripke, S. A. (1965): Semantical analysis of modal logic II: Non-normal propositional calculi. In: J.W. Addison, L. Henkin, A. Tarski (eds.): The Theory of Models. Amsterdam: North-Holland, 206–220

Kripke, S. A. (1965): Semantical analysis of intuitionistic logic. In: J.N.Crossley, M.A.E.Dummett (eds.): Formal Systems and Recursive Functions. Amsterdam: North-Holland, 92–130

Lee, R. C. T. (1967): A completeness theorem and a computer program for finding theorems derivable from axioms. Ph.D. Thesis, Dept. of Elec. Eng. and Comput. Sci., U. of Calif., Berkeley

Lee, R. C. T., Chang, C. L. (1971): Some properties of fuzzy logic. Information and Control, 19, 417–431

Lee, R. C. T. (1972): Fuzzy logic and the resolution principle. J. of the Assoc. for Computing Machinery, 19, 109–119

Levi, I. (1959): Putnam's three truth-values. Philosophical Studies, 10, 65–69

Lewis, C. I. (1933): Note concerning many-valued logical systems. Journal of Philosophy, 30, 364

Lifschitz, V. (1986): Pointwise circumscription: preliminary report. In Proceedings of the Fifth National Conference on Artificial Intelligence, 406–410

Lindenbaum, A. (1930): Remarques sur une question de la methode axiomatique. Fundamenta Mathematicae, 15, 313–321

Łoś, J. (1948): Many-valued logics and the formalization of intensional functions. Kwartalnik Filozoficzny, 17, 59–78 (In Polish)

Łukasiewicz, J. (1920): On 3-valued logic. Ruch Filozoficzny, 5, 169–171 (In Polish)

Łukasiewicz, J. (1921): Two-valued logic. Przegląd Filozoficzny, 23, 189–205 (In Polish)

Łukasiewicz, J. (1923): On determinism. In: J.Słupecki (ed.), Jan Łukasiewicz, Z Zagadnień Logiki i Filozofii, Warsaw, 1961; tr. in S.McCall (ed.), Polish Logic: 1920–1939, Oxford (1967), 19–39

Łukasiewicz, J. (1929): Elements of mathematical logic. Warsaw, Państwowe Wydawnictwo Naukowe, 2d ed., (1958) (In Polish)

Łukasiewicz, J. (1930): Philosophische Bemerkungen zu mehrwertigen Systemem des Aussagenkalküls. Comptes rendus des séeances de la Société des Sciences et des Lettres de Varsovie, Classe III, 23, 51–77. Tr. in S.McCall (ed.), Polish Logic: 1920–1939 (Oxford, 1967), 40–65. Polish tr. in Łukasiewicz (1961)

Łukasiewicz J., Tarski, A. (1930): Untersuchungen über den Aussagenkalkül. Comptes rendus des séances de la Société des Sciences et des Lettres de Varsovie, Classe III, 23, 1–21, 30–50. English tr. in J.H.Woodger (tr.), Alfred Tarski: Logic, Semantics, Metamathematics (Oxford, 1956), 38–59. Polish tr. in Łukasiewicz (1961)

Łukasiewicz, J. (1934): From the history of the logic of propositions. Przegląd Filozoficzny, 37, 417–437. Tr. in S.McCall (ed.), Polish Logic: 1920–1939 (Oxford, 1967), 66–87

Łukasiewicz, J. (1935–36): Zur vollen dreiwertigen Aussagenlogik, Erkenntnis, 5, 176

Łukasiewicz, J. (1936): Bedeutung der logischen Analyse für die Erkenntnis. Actes du Huitieme Congres International de Philosophie, Prague, 75–84

Łukasiewicz, J. (1938): Die logik und das Grundlagenproblem. In: F. Gonseth (ed.): Les Entretiens de Zurich sur les fondements et la methode des sciences mathematiques. Zurich: S.A.Leemann Freres & Cie, 1941. 82–100; Discussion, 100–108

Łukasiewicz, J. (1952): On the Intuitionistic Theory of Deduction. Indagationes Mathematicae, 14, 202–212

Łukasiewicz, J. (1970): Selected Works, ed. by L.Borkowski, Amsterdam: North-Holland

Malinowski, G. (1977): Classical characterization of n-valued Łukasiewicz calculi. Reports on Mathematical Logic, 9, 41–45

Malisoff, W. M. (1936): Meanings in multi-valued logic (Toward a general semantics). Erkenntnis, 6, 133–136

Malisoff, W. M. (1941): Meanings in multi-valued logics. Philosophy of Science, 8, 271–274

Margaris, A. (1958): A problem of Rosser and Turquette. Journal of Symbolic Logic, 23, 271–279

Margenau, H. (1934): On the application of many-valued systems of logic to physics. Philosophy of Science, 1, 118–121

Margenau, H. (1939): Probability, many-valued logics, and physics. Philosophy of Science, 6, 65–87

Martin, N. M. (1950): Some analogues of the Sheffer stroke function in n-valued logic. Koninklijke Nederlandse Akademie van Wetenschappen, Proceedings of the section of sciences, 53, 1100–1107; also Indagationes Mathematicae, 12, 393–400

Martin, N. M. (1951): A note on Sheffer functions in n-valued logic. Methodos, 3, 240–242

Martin, N, M. (1952): Sheffer functions and axiom sets in m-valued propositional logic, Ph.D. thesis Los Angeles: University of California

Martin, N. M. (1954): The Sheffer functions of three-valued logic. Journal of Symbolic Logic, 19, 45–51

Martynjuk, V. V. (1960): Investigation of certain classes of functions in many-valued logics. Problemy Kibernetyki, 3, 49–60

McCarthy, J. (1980): Circumscription — a form of non-monotonic reasoning. Artificial Intelligence, 13, 27–39

McCarthy, J. (1977): Epistemological problems of artificial intelligence. In Proceedings of the Fifth International Joint Conference on Artificial Intelligence, Cambridge, MA, 1038–1044

McCarthy, J. (1986): Applications of circumscription to formalizing common sense knowledge. Artificial Intelligence, 28, 89–116

McDermott D., Doyle, J. (1980): Non-monotonic logic I. Artificial Intelligence, 13, 41–72

McNaughton, R. (1951): A theorem about infinite-valued sentential logic. Journal of Symbolic Logic, 16, 1–13

Menu, J., Pavelka, J. (1976): A note on tensor products on the unit interval. Comment. Math. Univ. Carolinae, 17, 71–83

Meredith, C. A. (1958): The dependence of an axiom of Łukasiewicz. Transactions of the American Mathematical Society, 87, 54

Mlëziva, M. (1961): On the axiomatization of three-valued propositional logic. Casopis pro Pestovani Matematiky, Prague, 86, 392–403

Moch, F. (1956): Des antinomies classiques a la Logique de Mme. Destouches-Février, Comptes rendus hebdomadaires des séances de l'Académie des Sciences, 242, 1562–1563

Moisil, G. C. (1940): Recherches sur les logiques non-chrysippiennes. Annales scientifiques de l'Universite de Jassy, 26, 431–466

Moisil, G. C. (1941): Sur les anneaux de caracteristique 2 ou 3 et leurs applications. Bulletin de l'Ecole Polytechnique de Bucarest, 12, 66–90

Moisil, G. C. (1959): Rapport sur le développement dans la R.P.R. de la théorie algébrique des mécanismes automatiques. Acta Logica, Bucharest, 2, 145–199

Moisil, G. C. (1960): On the logic of Bochvar. Academia Republicii Populare Romine; Analele Romino-Sovietice Seria Matematica-Fizica, Bucharest, 14, 19–25

Moisil, G. C. (1960): Mathematical logic and modern technology: many-valued logic and relay-contact circuits. Probleme Filosofice ale steintelor naturii, Bucharest: Académie de la République populaire roumaine

Moisil, G. C. (1961): The predicate calculus in three-valued logic. Analele Universitatii Bucuresti Seria Acta Logica, Bucharest, 4, 103–112

Moisil, G. C. (1961): Sur la logique a trois valeurs de Łukasiewicz. Analele Universitatii Bucuresti Seria Acta Logica, 5, 103–117

Moisil, G. C. (1962): Les Logiques non-chrysippiennes et leurs applications. Proceedings of a Colloquium on Modal and Many-Valued Logics, Helsinki, 137–149

Moisil, G. C. (1972): Essais sur les logiques non-chrysippiennes. Editions de l'Academie de la Republique Socialiste de Roumanie, Bucarest

Moore, R. C. (1985): Semantical considerations on non-monotonic logic. Artificial Intelligence, 25, 75–94

Mostowski, A. (1985): Mathematical logic. Monografie matematyczne, 18, Varsovie et Wroclaw

Mostowski A., and collaborators, (1955): The Present State of Investigations on the Foundations of Mathematics, Warsaw, Państwowe Wydawnictwo Naukowe

Mostowski, A. (1957): L'oeuvre scientifique de Jan Łukasiewicz dans le domaine de la logique mathématique. Fundamenta Mathematicae, 44, 1–11

Mostowski, A. (1961): Axiomatizability of some many-valued predicate calculi. Fundamenta Mathematicae, 50, 165–190

Mostowski, A. (1961): An example of a non-axiomatizable many-valued logic. Zeitschrift für mathematische Logik und Grundlagen der Mathematik, 7, 72–76

Nagel, E. (1946): Professor Reichenbach on quantum mechanics: a rejoinder. Journal of Philosophy, 43, 247–250

Nakamura, A. (1962): On the infinitely many-valued threshold logics and von Wright's system M. Zeitschrift für mathematische Logik und Grundlagen der Mathematik, 8, 147–164

Nakamura, A. (1962): On an axiomatic system of the infinitely many-valued threshold logics. Zeitschrift für mathematische Logik und Grundlagen der Mathematik, 8, 71–76

Nelson, E. (1959): Regular probability measures on function space. Ann. Math., 69, 630–643

von Neumann, J. (1927): Zur Hilbertschen Beweistheorie. Mathematische Zeitschrift, 26, 1–46

von Neumann, J. (1962): Quantum logics, unpublished, reviewed by A.H.Taub in John von Neumann. Collected Works, New York, 4, 195–197

Nilsson, N. J. (1965): Learning Machines. New York: McGraw-Hill

Nilsson, N. J. (1971): Problem Solving Methods in Artificial Intelligence. New York: McGraw-Hill

Nilsson, N. J. (1986): Probabilistic logic. Artificial Intelligence, 28, 71–87

Novák, V. (1985): Fuzzy Sets and their Applications, SNTL, Prague (In Czech.)

Novák, V. (1987): First order fuzzy logic. Studia Logica 46(1), Ossolineum — North-Holland Pn.Co., Wrocław 1982, 87–109

Orci, I.P. (1985): Programming in fuzzy logic for expert systems design. In: Proc. 5th Journees Intern. Les Systemes Experts et leurs Applications, Avignon, France, 1179–1182

Orłowska, E. (1985): Mechanical proof methods for Post logics. Logique et Analyse, 110–111, 173–192

Orłowska, E., Pawlak, Z. (1984): Expressive power of knowledge representation systems. International Journal of Man-Machine Studies, 20, 485–500

Orłowska, E., Pawlak, Z. (1984): Representation of nondeterministic information. Theoretical Computer Science, 29

Pavelka, J. (1979): On fuzzy logic. Zeitschrift für mathematik Logik und Grundlagen der Mathematic, 25, 45–52, 119–134, 447–464

Pavelka, J. (1976): A note on closed categories. Comment. Math. Univ. Carolinae, 17, 261–272

Pavelka, J. (1976): On *L*-fuzzy semantics. Weiterbildungszentrum für Mathematische Kybernetik und Rechentechnik, Technische Universität, Dresden

Pawlak, Z. (1981): Rough relations. ICS PAS Reports 435

Pawlak, Z. (1982): Rough sets. International Journal of Information and Computer Science, 11(5), 341–356

Pawlak, Z. (1984): Rough classification. International Journal of Man-Machine Studies, 20, 469–483

Pearl, J. (1982): Reverend Bayes on inference engines: A distributed hierarchical approach. Proc. of the National Conference on A.I., Pittsburgh, 133–136

Peirce, C. S. (1914): Collected Papers of Charles Sanders Peirce, 3-4, ed. by C.Hartshorne and P.Weiss, Cambridge, MA: Harvard University Press, 1933

Peirce, C. S. (1885): On the algebra of logic. American Journal of Mathematics, 7

Picard, S. (1935): Sur les fonctions definies dans les ensembles finis quelconques. Fundamenta Mathematicae, 24, 298–301

Piróg-Rzepecka, K. (1977): Systemy Nonsense-Logics, PWN, Warszawa — Wrocław

Post, E. L. (1920): Introduction to a general theory of elementary propositions. Bulletin of the American Mathematical Society, 26, 437

Post, E. L. (1921): Introduction to a general theory of elementary propositions. American Journal of Mathematics, 43, 163–185. Reprinted in J. van Heijenoort (ed.), From Frege to Gödel: A Source Book in Mathematical Logic, 1879–1931, Cambridge, Mass., 1967, 265–283

Prade, H., Dubois, D. (1987): Necessity measures and the resolution principle. IEEE Transactions on Systems, Man and cybernetics, SMC–17, 3, 474–478

Price, H. H. (1973): Belief and evidence. In R.M.Chisholm and R.J.Swartz (eds.): Empirical Knowledge. Englewood Cliffs NJ: Prentice Hall, 95–126

Prior, A. N. (1952): In what sense is modal logic many-valued?. Analysis, 12, 138–143

Prior, A. N. (1953): Three-valued logic and future contingents. Philosophical Quarterly, 3, 317–326

Prior, A. N. (1955): Many-valued and modal systems: An intuitive approach. Philosophical Review, 64, 626–630

Prior, A. N. (1955): Curry's paradox and 3-valued logic. Australasian Journal of Philosophy, 33, 177–182

Prior, A. N. (1955): Formal Logic. Oxford: Clarendon Press, 1955; 2d ed., 1962. See especially Chap. II of Pt. III, Three-Valued and Intuitionistic Logic

Prior, A. N. (1957): Time and Modality. London: Oxford University Press

Prior, A. N. (1957): The necessary and the possible: The first of three talks on 'The logic game'. The Listener, 57, 627–628

Prior, A. N. (1957): Symbolism and analogy: The second of three talks on 'The logic game'. The Listener, 57, 675–678

Prior, A. N. (1957): Many-valued logics: The last of three talks on 'The logic game'. The Listener, 57, 717–719

Putnam, H. (1957): Three-valued logic. Philosophical Studies, 8, 73–80

Rasiowa, H., Sikorski, R. (1950): A proof of the completeness theorem of Gödel. Fundamenta Mathematicae, 37, 193–200

Rasiowa, H., Sikorski, R. (1970): The mathematics of metamathematics, Warsaw PWN

Rasiowa, H. (1977): Many-valued algorithmic logic as a tool to investigate programs. In: Dunn, Epstein (ed.): Modern Uses of Multiple-Valued Logic, 79–102

Rasiowa, H., Skowron, A. (1984): Rough concepts logic. Proc. of Computation Theory Fifth Symposium, Zaborów, Poland, 288–297

Rasiowa, H., Skowron, A. (1985): Approximation logic. Proc. of the International Spring School, Mathematical Methods of Specyfication and Synthesis of Software Systems'85

Rasiowa, H. (1973): On generalized Post algebras of order ω^+ and ω^+-valued predicate calculi. Bull. Ac. Pol. Sci., Ser. Math. Astron. Phys., 21, 209–219

Rasiowa, H. (1974): An algebraic approach to non-classical logics. Studies in Logic and the Foundations of Mathematics, 78. Amsterdam: North-Holland

Rasiowa, H. (1986): Rough concepts and multiple-valued logic. Proc. ISMVL'86, May, Blacksburg, VA, IEEE Computer Society Press

Rasiowa, H. (1987): Logic of approximation reasoning. 1st Workshop on Computer Science Logic (CSL'87), Karlsruhe, FRG, October 1987, E.Börger, H.Kleine Büning, M.M.Richter (eds.). Lect. Notes in Comput. Sci. 329. Berlin: Springer 1988, 188–210

Rasiowa, H. (1989): Logics of approximation reasoning semantically based on partially ordered sets, prepared for ASL Logic Colloquium'89, Berlin, FRG

Rasiowa, H., Epstein, G. (1987): Approximation reasoning and Scott's information systems. Proc. of the 2nd Int. Symp. on Methodologies for Intelligent Systems, North-Holland, 33–42

Rasiowa, H. (1991): On approximation logics: A survey. Institute of Mathematics University of Warsaw, Poland 1991

Reichbach, J. (1962): On the connection of the first-order functional calculus with many-valued propositional calculi. Notre Dame Journal of Formal Logic, 3, 102–107

Reichbach, J. (1963): About connection of the first-order functional calculus with many-valued propositional calculi. Zeitschrift für mathematische Logik und Grundlagen der Mathematik, 9, 117–124

Reichbach, J. (1964): A note about connection of the first-order functional calculus with many-valued propositional calculi. Notre Dame Journal of Formal Logic, 5, 158–160

Reichenbach, H. (1937): Les fondements logiques du calcul des probabilites. Annales de l'Institut Henri Poincare, 7, 267–348

Reichenbach, H. (1944): Philosophical Foundations of Quantum Mechanics. Berkeley, Los Angeles, University of California Press

Reichenbach, H. (1946): Reply to Ernest Nagel's criticism of my views on quantum mechanics. Journal of Philosophy, 43, 239–247

Reichenbach, H. (1949): Philosophische Grunglagen der Quanten-mechanic. Basel, Verlag Birkhäuser

Reichenbach, H. (1951): Über die erkenntnistheoretische Problemlage und den Gebrauch einer dreiwertigen Logik in der Quantenmechanik. Zeitschrift für Naturforschung, 6a, 569–575

Reinfrank M., Freitag, H. (1988): An integrated non-monotonic deduction and reason maintenance system. In: H.Stoyan (ed.): Begründungsverwaltung. Informatik-Fachberichte 162. Berlin: Springer

Reiser, O. L. (1952): Physics, probability, and multi-valued logic. Philosophical Review, 61, 147–159

Reiter, R. (1980): A logic for default reasoning. Artificial Intelligence, 13, 81–132

Rescher, N. (1955): Some comments on two-valued logic. Philosophical Studies, 6, 54–58

Rescher, N. (1962): Quasi-truth-functional systems of propositional logic. Journal of Symbolic Logic, 27, 1–10

Rescher, N. (1963): A probabilistic approach to modal logic. Acta Philosophica Fennica, 16, 215–226

Rescher, N. (1964): Quantifiers in many-valued logic. Logique et Analyse, 7, 181–184

Rescher, N. (1969): Many-Valued Logic. New York: McGraw-Hill

Ridder, J. (1948): Über mehrwertige Aussagenkalküle und mehrwertige engere Prädikatenkalküle I–III. Koninklijke Nederlandsche Akademie van Wetenschappen, Proceedings of the section of sciences, 51, 670–680, 836–845, 991–995; also Indagationes mathematicae, 10, 221–231, 264–273, 324–328

Ridder, J. (1949): Sur quelques logiques multivalentes. Actes du Xieme Congres International de Philosophie, Amsterdam, 11–18 aout, 1948 — Proceedings of the Tenth International Congress of Philosophy, Amsterdam: North-Holland, 728–730

Rine, D. C.(ed.) (1977): Computer Science and Multiple-Valued Logic. Theory and Applications. Amsterdam: North-Holland

Robinson, J. A. (1965): A machine-oriented logic based on the resolution principle. J.ACM, 12, 23–41

Rollinger, C. R. (1983): How to represent evidence — aspects of uncertainty reasoning. Proc. of the 8th Intern. Joint Conference on A.I., Karlsruhe, 358–361

Rose, A. (1950): A lattice-theoretic characterization of three-valued logic. Journal of the London Mathematical Society, 25, 255–259

Rose, A. (1950): Completeness of Łukasiewicz-Tarski propositional calculi. Mathematische Annalen, 122, 296–298

Rose, A. (1951): Conditional disjunction as a primitive connective for the m-valued propositional calculus. Mathematische Annalen, 123, 76–78

Rose, A. (1951): Axiom systems for three-valued logic. Journal of the London Mathematical Society, 26, 50–58

Rose, A. (1951): Systems of logic whose truth-values form lattices. Mathematische Annalen, 123, 152–165

Rose, A. (1951): A lattice-theoretic characterization of the α_0-valued propositional calculus. Mathematische Annalen, 123, 285–287

Rose, A. (1951): The degree of completeness of some Łukasiewicz-Tarski propositional calculi. Journal of the London Mathematical Society, 26, 47–49

Rose, A. (1951): An axiom system for three-valued logic. Methodos, 3, 233–239

Rose, A. (1952): Some generalized Sheffer functions. Proceedings of the Cambridge Philosophical Society, 48, 369–373

Rose, A. (1952): The degree of completeness of the m-valued Łukasiewicz propositional calculus. Journal of the London Mathematical Society, 27, 92–102

Rose, A. (1952): A formalisation of Post's m-valued propositional calculus, Mathematische Zeitschrift, 56, 94–104

Rose, A. (1952): Le degre de saturation du calcul propositionnel implicatif a trois valeurs de Sobociński. Comptes rendus hebdomadaires des séances de l'Académie des Sciences, 235(1953), 1000–1002

Rose, A. (1953): The degree of completeness of the α_0-valued Łukasiewicz propositional calculus. Journal of the London Mathematical Society, 28, 176–184

Rose, A. (1953): Conditional disjunction as a primitive connective for the Erweiterter Aussagenkalkul, Journal of Symbolic Logic, 18, 63–65

Rose, A. (1953): The m-valued calculus of non-contradiction. Journal of Symbolic Logic, 18, 237–241

Rose, A. (1953): Fragments of the m-valued propositional calculus. Mathematische Zeitschrift, 59, 206–210

Rose, A. (1953): A formalization of Sobociński's three-valued implicational propositional calculus. Journal of Computing Systems, 1, 165–168

Rose, A. (1953): A formalization of an α_0-valued propositional calculus. Proceedings of the Cambridge Philosophical Society, 49, 367–376

Rose, A. (1953): Some self-dual primitive functions for propositional calculi. Mathematische Annalen, 126, 144–148

Rose, G. F. (1953): Propositional calculus and realizability. Transactions of the American Mathematical Society, 75, 1–19

Rose, A. (1954): Sur les fonctions definissables dans une logique a un nombre infini de valeurs. Comptes rendus hebdomadaires des séances de l'Académie des Sciences, 238, 1462–1463

Rose, A. (1955): Le degré de saturation du calcul propositionel implicatif a m valeurs de Łukasiewicz. Comptes rendus hebdomadaires des séances de l'Académie des Sciences, 240, 2280–2281

Rose, A. (1955): A Gödel theorem for an infinite-valued Erweiterter Aussagenkalkül. Zeitschrift für mathematische Logik und Grundlagen der Mathematik, 1, 89–90

Rose, A. (1956): An alternative formalisation of Sobociński's three-valued implicational propositional calculus. Zeitschrift für mathematische Logik und Grundlagen der Mathematik, 2, 166–172

Rose, A. (1956): Formalisation du calcul propositionnel implicatif a α_0 valeurs de Łukasiewicz. Comptes rendus hebdomadaires des séances de l'Académie des Sciences, 243, 1183–1185, 1263–1264

Rose, A. (1956): Some formalisations of α_0-valued propositional calculi. Zeitschrift für mathematische Logik und Grundlagen der Mathematik, 2, 204–209

Rose, A. (1958): Many-valued logical machines. Proceedings of the Cambridge Philosophical Society, 54, 307–321

Rose, A. (1958): Sur les definitions de l'implication et de la négation dans certains systémes de logique dont les valeurs forment des treillis. Comptes rendus hebdomadaires des séances de l'Académie des Sciences, 246, 2091–2094

Rose, A. (1958): Applications of logical computers to the construction of electrical control tables for signalling frames. Zeitschrift für mathematische Logik und Grundlagen der Mathematik, 4, 222–243

Rose, A. (1960): An extension of a theorem of matgaris. Journal of Symbolic Logic, 25, 209–211

Rose, A. (1960): Sur les schémas d'axiomes pour les calculs propositionnels a m valeurs ayant des valeurs sur désignées. Comptes rendus hebdomadaires des séances de l'Académie des Sciences, 250, 790–792

Rose, A. (1961): Self-dual binary and ternary connectives for m-valued propositional calculi. Mathematische Annalen, 143, 448–462

Rose, A. (1961): Sur certains calculus propositionnels a m valeurs ayant un seul foncteur primitif lequel constitue son propre dual. Comptes rendus hebdomadaires des séances de l'Académie des Sciences, 252, 3176–3178, 3375–3376

Rose, A. (1962): An alternative generalisation of the concept of duality. Mathematische Annalen, 147, 318–327

Rose, A. (1962): Extension of some theorems of Anderson and Belnap. Journal of Symbolic Logic, 27, 423–425

Rose, A. (1962): A simplified self m-al set of primitive functors for the m-valued propositional calculus. Zeitschrift für mathematische Logik und Grundlagen der Mathematik, 8, 257–266

Rose, A. (1962): Sur un ensemble complet de foncteurs primitifs independants pour le calcul propositionnel trivalent lequel constitue son propre trial. Comptes rendus hebdomadaires des séances de l'Académie des Sciences, 254, 2111

Rose, A. (1962): Sur les applications de la logique polyvalente a la construction des machines Turing. Comptes rendus hebdomadaires des séances de l'Académie des Sciences, 255, 1836–1838

Rose, A. (1962): Sur un ensemble de foncteurs primitifs pour le calcul propositionnel a m valeurs lequel constitue son propre m-al. Comptes rendus hebdomadaires des séances de l'Académie des Sciences, 254, 1897–1899

Rosenberg, I. (1973): The number of maximal closed classes in the set of functions over a finite domain. Journal of Combinatorial Theory Series A, 14, 1–7

Rosenbloom, P. C. (1942): Post algebras. I. Postulates and general theory. American Journal of Mathematics, 64, 167–188

Rosser, J. B. (1939): The introduction of quantification into a three-valued logic. Reprinted for the members of the Fifth International Congress for the Unity of Science, (Cambridge, 1939) from Journal of Unified Science (Erkenntnis), 9 (never published). Abstracted in Journal of Symbolic Logic, 4, 170–171

Rosser, J. B. (1941): Many-valued logic. School Science and Mathematics, 41, 99–100

Rosser, J. B. (1941): On the many-valued logics. American Journal of Physics, 9, 207–212; reprinted in Papers from the Second American Congress on General Semantics, University of Denver, August, (Chicago, 1943), 79–86

Rosser, J. B. (1943): On the many-valued logics. Papers from the Second American Congress on General Semantics, University of Denver, August, 1941 (Chicago, Institute of General Semantics, 1943), 79–86

Rosser, J. B. (1953): Logic for Mathematicians. New York, McGraw-Hill

Rosser, J. B. (1960): Axiomatization of infinite-valued logics. Logique et Analyse, 3, 137–153

Rosser, J. B., Turquette, A. R. (1945): Axiom schemes for m-valued propositional calculi. Journal of Symbolic Logic, 10, 61–82

Rosser, J. B., Turquette, A. R. (1948): Axiom schemes for m-valued functional calculi of first order. Part I. Definition of axiom schemes and proof of plausibility, Journal of Symbolic Logic, 13, 177–192

Rosser, J. B., Turquette, A. R. (1949): A note on the deductive completeness of m-valued propositional logic. Journal of Symbolic Logic, 14, 219–225

Rosser, J. B., Turquette, A. R. (1951): Axiom schemes for m-valued functional calculi of first order. Part II. Deductive completeness. Journal of Symbolic Logic, 16, 22–34

Rosser, J. B., Turquette, A. R. (1952): Many-valued logics. Amsterdam: North-Holland

Rougier, L. (1939): La relativite de la logique. Journal of Unified Science, 8, 193–217

Rousseau, G. (1969): Logical systems with finitcly many truth-values. Bulletin de l'Academie Polonaise des Sciences, Serie des sciences mathematiques, astronomiques et physiques, 17, 189–194

Rousseau, G. (1970): Post algebras and pseudo-Post algebras. Fund. Math., 67, 133–145

Rousseau, G. (1970): Sequents in many-valued logics II. Fundamenta Mathematicae, 67, 125–131

Rutledge, J. D. (1959): A preliminary investigation of the infinitely many-valued predicate calculus, Ph.D.thesis, Cornell

Rutledge, J. D. (1960): On the definition of an infinitely many-valued predicate calculus. Journal of Symbolic Logic, 25, 212–216

Salomaa, A. (1959): On many-valued systems of logic. Ajatus, 22, 115–159

Salomaa, A. (1963): Some analogues of Sheffer functions in infinite-valued logics. Acta Philosophica Fennica, 16, 227–235

Saloni, Z. (1972): Gentzen rules for the m-valued logic. Bulletin P.A.Sc. Ser.Math., 20, 819–826

Salwicki, A. (1970): Formalized algorithmic languages. Bulletin de l'Académie Polonaise des Sciences, Série des sciences mathematiques, astronomiques et physiques, 18, 227–232

Sandewall, E. (1985): A functional approach to non-monotonic logic. In Proceedings of the Ninth International Joint Conference on Artificial Intelligence, 100–106

Scarpellini, B. (1962): Die Nichtaxiomatisierbarkeit des unendlichwertigen Prädikatenkalküls von Łukasiewicz. Journal of Symbolic Logic, 27, 159–170

Schiller, F. C. S. (1935): Multi-valued logics — and others. Mind, 44, 467–483

Schmierer, Z. (1936): On characteristic functions in many-valued systems of logic. Przegląd filozoficzny, 39, 437 (In Polish)

Scholz, H. (1957): In memoriam Jan Łukasiewicz. Archiv für mathematische Logik und Grundlagenforschung, 3, 1–18

Schröter, K. (1955): Methoden zur Axiomatisierung beliebiger Aussagen — und Prädikatenkalküle. Zeitschrift für mathematische Logik und Grundlagen der Mathematik, 1, 241–251

Schwartz, D. G. (1985): The case for an interval based representation of linguistic truths. Fuzzy sets and systems, 17, 153–165

Scott, D. (1973): Background to formalisation. In: H. Leblanc (ed.): Truth, Syntax and Modality. Amsterdam: North-Holland, 244–273

Scott, D. (1974): Completeness and axiomatizability in many valued logics. In Proceedings of the Tarski Symposium, American Mathematical Society, Providence, Rhode Island, 411–436

Scott, D. (1982): Domains for denotational semantics. A corrected and expanded version of a paper prepared for ICALP'82, Aarhus, Denmark, Lecture Notes in Computer Science, 140. Berlin: Springer

Scott, D. S. (1982): Some ordered sets in computer science. In I.Rival, (ed.), Ordered Sets, 677–718, Boston: Reidel

Scott, D., Krauss, P. (1966): Assigning probabilities to logical formulas. In J.Hintikka and P.Suppes (eds.), Aspects of Inductive Logic. Amsterdam: North-Holland, 219–264

Scott, D., Solovay, R. (1969): Boolean valued models for set theory. Proceedings of the American Mathematical Society Summer Inst. Axiomatic Set Theory, 1967, University of California, Los Angeles, Proceedings of the Symposia in Pure Mathematics, 13

Segerberg, K. (1965): A contribution to nonsense-logic. Theoria 31, 199–217

Shannon, C. E. (1938): A symbolic analysis of relay and switching circuits. Transactions of the American Institute of Electrical Engineers, 57, 713–723

Shaw-kwei, M. (1954): Logical paradoxes for many-valued systems. Journal of Symbolic Logic, 19, 37–40

Shoham, Y. (1986): Chronological ignorance. In Proceedings of the Fifth National Conference on Artificial Intelligence, 389–393

Sierpiński, W. (1960–61): Sur un problème de la logique à n valeurs. Fundamenta Mathematicae, 49, 167–170

Sikorski, R. (1964): Boolean algebras, second ed. Berlin: Springer

Skolem, T. (1929): Über einige Grundlagenfragen der Mathematik. Skrifter utgitt av Det Norske Videnskaps-Akademi i Oslo, 4, 1–49

Skolem, T. (1957): Mengenlehre gegründet auf einer Logik mit unendlich vielen Wahrheitswerten. Sitzungsberichte der Berliner Mathematischen Gesellschaft, (1957–58), 41–56

Skolem, T. (1957): Bemerkungen zum Komprehensionsaxiom. Zeitschrift für mathematische Logik und Grundlagen der Mathematik, 3, 1–17

Skolem, T. (1960): A set theory based on a certain 3-valued logic. Mathematica Scandinavica, 8, 127–136

Skowron, A., Stepaniuk, J. (1992): Towards on approximation theory of discrete problems, Part I. Fundamenta Informaticae (to appear)

Skowron, A., Stepaniuk, J. (1991): Searching for classifiers. The First World Conference on Foundations of AI, Paris

Slagle, J. R. (1970): Interpolation theorems for resolution in lower predicate calculus. J.ACM, 17, 535–542

Slagle, J. R. (1971): Artificial Intelligence, the Heuristic Programming Approach. New York: McGraw-Hill

Slage, J. R., Chang, C. L., Lee, R. C. T. (1969): Completeness theorems for semantic resolution in consequence finding. Proc. of the First International Joint Artificial Intelligence Conf., Washington, D.C.

Słupecki, J. (1936): Der volle dreiwertige Aussagenkalkül. Comptes rendus des séances de la Société des Sciences et des Lettres de Varsovie, Classe III, 29, 9–11

Słupecki, J. (1939): Proof of the axiomatizability of full many-valued systems of propositional calculus. Comptes rendus des séances de la Société des Sciences et des Lettres de Varsovie, Classe III, 32, 110–128 (In Polish)

Słupecki, J. (1939): A criterion of completeness of many-valued systems of propositional logic. Comptes rendus des seances de la Societe des Sciences et des Lettres de Varsovie, Classe III, 32, 102–110
Słupecki, J. (1946): The complete three-valued propositional calculus. Annales Universitatis Mariae Curie-Sklodowska, 1, 193–209
Smiley, T. J. (1962): Analytic implication and 3-valued logic. Journal of Symbolic Logic, 27, 378
Smullyan, R. M. (1968): First order logic. Berlin: Springer
Sobociński, B. (1936): Axiomatization of certain many-valued systems of the theory of deduction. Roczniki prac naukowych zrzeszenia asystentów Uniwersytetu Józefa Piłsudskiego w Warszawie, 1, 399–419
Sobociński, B. (1952): Axiomatization of a partial system of three-valued calculus of propositions. Journal of Computing Systems, 1, 23–55
Sobociński, B. (1956): In memoriam Jan Łukasiewicz. Philosophical Studies (Maynooth), 6, 3–49
Sobociński, B. (1957): La génesis de la Escuela Polaca de Lógica. Oriente europeo, 7, 83–95 (In Spanish)
Sobociński, B. (1957): J.Łukasiewicz (1878–1956). Rocznik Polskiego Towarzystwa Naukowego na Obczyźnie, London, 1957, 3–21
Sobociński, B. (1961): A note concerning the many-valued propositional calculi. Notre Dame Journal of Formal Logic, 2, 127–128
Storer, T. (1946): The logic of value imperatives. Philosophy of Science, 13, 25–40
Storer, T. (1954): The notion of tautology. Philosophical Studies, 5, 75–78
Suchoń, W. (1974): Définition des foncteurs modaux de Moisil dans le calcul n-valent des propositions de Łukasiewicz avec implication et négation. Reports on Mathematical Logic, 2, 43–47
Sueki, T. (1952-54): The formalization of two-valued and n-valued systems, I, II, III. Reports of the University of Electro-Communications, Tokyo, 4 (1952), 1–24; 5 (1953), 1–18; 6 (1954), 1–19
Sugihara, T. (1950): Many-valued logic. Tetsugaku kenkyu, 33, 684–703
Sugihara, T. (1951): Many-valued logical characteristics of Brouwerian logic. Kagaku, 21, 294–295
Sugihara, T. (1952): Negation in many-valued logic. Memoirs of Liberal Arts College, Fukui University, 1, 1–5
Sugihara, T. (1956): Four-valued propositional calculus with one designated truth-value. Memoirs of the Liberal Arts College, Fukui University, 5, 41–48
Sugihara, T. (1958): A three-valued logic with meaning-operator. The Memoirs of Fukui University, Liberal Arts Department, 8, 59–60
Summersbee, S., Walters, A. (1963): Programming the functions of formal logic. II. Multi-valued logics. Notre Dame Journal of Formal Logic, 4, 293–305

Surma, S. J. (1971): Jaśkowski's matrix criterion for the intuitionistic propositional calculus. Prace z logiki, 6, 21–54

Surma, S. J. (1973): A historical survey of the significant methods of proving Post's theorem about the completeness of the classical propositional calculus. In Studies in the History of Mathematical Logic, S.J.Surma (ed.), Ossolineum, Wrocław, 19–32

Surma, S. J. (1977): An algorithm for axiomatizing very finite logic. In: D. C. Rine (ed.): Computer science and multiple-valued logic: theory and applications. Amsterdam: North-Holland, 137–143

Suszko, R. (1957): A formal theory of the logical values, I. Studia Logica, 6, 145–237

Suszko, R. (1975): Remarks on Łukasiewicz's three-valued logic. Bulletin of the Section of Logic, 4, (3), 87–90

Swift, J. D. (1952): Algebraic properties of n-valued propositional calculi. American Mathematical Monthly, 59, 612–621

Szczerba, L. W. (1987): Rough quantifiers. Bulletin of the Polish Academy of Sciences — serie Mathematics, 35, 251–254

Takekuma, R. (1954): On a nine-valued propositional calculus. Journal of Computing Systems, 1, 225–228

Tarski, A. (1956): Logic, Semantics, Metamathematics: Papers from 1923 to 1938. Translated by J.H.Woodger, Oxford: Clarendon Press

Teh, H. H. (1963): On 3-valued sentential calculus. An axiomatic approach. Bulletin of Mathematical Society, Singapore, Nanyang University, 1–37

Thiele, H. (1958): Theorie der endlich-wertigen Łukasiewiczschen Prädikatenkalküle der ersten Stufe. Zeitschrift für mathematische Logik und Grundlagen der Mathematik, 4, 108–142

Tokarz, M. (1974): A method of axiomatization of Łukasiewicz logics. Bulletin of the Section of Logic, 3, (2), 21–24

Traczyk, T. (1962): On axioms and some properties of Post algebras. Bulletin de l'Académie Polonaise des Sciences: Série des Sciences Mathématiques, Astronomiques et Physiques, Warsaw, 10, 509–512

Traczyk, T. (1964): An equational definition of a class of Post algebras. Bulletin de l'Academie Polonaise des Sciences, Cl.III, 12, 147–149

Traczyk, T. (1967): On Post algebras with uncountable chain of constants, algebras and homomorphisms. Bull. Ac. Pol. Sci., Ser. Math. Astron. Phys., 15, 673–680

Turner, R. (1984): Logics for Artificial Intelligence. Chichester, UK: Ellis Horwood

Turowicz, A. B. (1960): Sur une méthode algébrique de vérification des théoremes de la logique des énoncés. Studia Logica, 9, 27–36

Turquette, A. R. (1944): A study and extension of m-valued symbolic logics. Cornell University, Abstracts of Theses Accepted in Partial Satisfaction of the Requirements for the Doctor's Degree, 1943, Ithaca, NY: Cornell University Press, 49–51

Turquette, A. R. (1945): Review of Reichenbach's philosophical foundations of quantum mechanics. Philosophical Review, 54, 513–516

Turquette, A. R. (1954): Many-valued logics and systems of strict implication. Philosophical Review, 63, 365–379

Turquette, A. R. (1958): Simplified axioms for many-valued quantification theory. Journal of Symbolic Logic, 23, 139–148

Turquette, A. R. (1961): Solution to a problem of Rose and Rosser. Proceedings of the American Mathematical Society, 12, 253–255

Turquette, A. R. (1963): Modality, minimality, and many-valuedness. Acta Philosophica Fennica, 16, 261–276

Turquette, A. R. (1963): Independent axioms for infinite-valued logic. Journal of Symbolic Logic, 28, 217–221

Umezawa, T. (1959): On intermediate many-valued logics. Journal of the Mathematical Society of Japan, Rd.11, 2, 116–128

Urquhart, A. (1973): An interpretation of many-valued logic. Zeitschrift für Mathematische Logik und Grundlagen der Mathematik, 19, 111–114

Urquhart, A. (1986): Many-valued logic. In: D.Gabbey and F.Guenther (eds.), Handbook of Philosophical Logic, 3, 71–116 Dordrecht: Reidel

Ushenko, A. P. (1936): The many-valued logics. Philosophical Review, 45, 611–615

Vaccarino, G. (1953): Le logiche polivalenti e non aristoteliche. Archimede, 5, 226–231

Vaidyanathaswamy, R. (1938): Quasi-Boolean algebras and many-valued logics. Proceedings of the Indian Academy of Sciences, 8, 165–170

Vasil'ev, N. A. (1924): Imaginary (Non-Aristotelian) logic. Atti del V Congresso Internazionale di Filosofia, Napoli, 1924, (Naples,1925), 107–109

Wahlster, W. (1977): Die Repräsentation von vagem Wissen in natürlichsprachlichen Systemen der Künstlichen Intelligenz. Universität Hamburg IfI–HH–B–38/77

Waismann, F. (1946): Are there alternative logics. Proceedings of the Aristotelian Society, 46, 77–104

Wajsberg, M. (1931): Axiomatization of the 3-valued propositional calculus. Comptes rendus des séances de la Société des Sciences et des Lettres de Varsovie, Classe III, 24, 126–148. Translated in S.McCall (ed.), Polish Logic: 1920–1939, (Oxford, 1967), 264–284

Wajsberg, M. (1933): Ein erweterter Klassenkalkül. Monatshefte für Mathematik und Physik, 40, 113–126

Wajsberg, M. (1935): Beiträge zum Metaaussagenkalkül I. Monatshefte für Mathematik und Physik, 42, 221-242

Wajsberg, M. (1935): Axiomatization of the three-valued sentential calculus. Comptes Rendus des seances de la Societe des Sciences et des Lettres de Varsovie, 24, 126–148

Wajsberg, M. (1937): Metalogische Beitraḡe. Wiadomości Matematyczne, 43, 131–168. English translation in S.McCall (ed.), Polish Logic: 1920–1939 (Oxford, 1967), 285–318

Wang, H. (1961): The calculus of partial predicates and its extension to set theory.I. Zeitschrift für mathematische Logik und Grundlagen der Mathematik, 7, 283-288

Webb, D. L. (1935): Generation of any n-valued logic by one binary operation. Proceedings of the National Academy of Sciences, 21, 252–254

Webb, D. L. (1936): The algebra of n-valued logic. Comptes rendus des séances de la Société des Sciences et des Lettres de Varsovie, Classe III, 29, 153–168

Webb, D. L. (1936): Definition of Post's generalized negative and maximum in terms of one binary operation. Bulletin of the American Mathematical Society, 58, 193–194

Whitehead, A. N., Russell, B. (1910): Principia Mathematica, Vol. I, Cambridge University Press

Wilhelmy, A. (1962): Bemerkungen zur Semantik quantifizierter mehrwertiger logistischer Systeme.In Max Käsbauer (ed.), Logik und Logikkalkül, Freiburg, München, K.Alber, 179–188

Wittgenstein, L. (1922): Tractatus logico-philosophicus. London: Routledge & Kegan Paul

Wolf, R. G. (1977): A Survey of many-valued logic (1966–1974). Appendix II, in Dunn, Epstein (ed.), 167–324

Wójcicki, R. (1970): Some remarks on the consequence operation in sentential logics. Fundamenta Mathematicae, 68, 269–279

Wójcicki, R. (1977): Strongly finite sentential calculi. Selected papers on Łukasiewicz sentential calculi, R.Wójcicki, G.Malinowski (ed.), Ossolineum, Wrocław, 53–77

Wójcicki, R. (1984): Lectures on propositional calculi. Ossolineum, Wrocław

Yablonskii, S. V. (1958): Functional constructions in a k-valued logic. Trudy matematičeskogo instituta imeni V.A.Steklova, 51, 5–142

Yager, R. R. (1979): Finite linearly ordered fuzzy sets with applications to decisions. Int. J. Man-Machine Studies, 12, 299–322

Yager, R. R. (1982): Some procedures for selecting fuzzy set-theoretic operators. Int. J. General Systems, 8, 115–124

Yasuura, K. (1955): On the representation of many-valued propositional logics by relay circuits. Kyusyu Daigaku kogaku syuho (Technology Reports of the Kyushu University), 28, 94–96 (In Japanese)

Yonemitsu, N. (1954): Note on completeness of m-valued propositional calculi. Mathematica Japonicae, 3, 57–61

Zadeh, L. A. (1965): Fuzzy sets. Information and Control, 8, 338–353

Zadeh, L. A. (1972): A fuzzy-set-theoretic interpretation of linguistic hedges. J. of Cybernetics, 2, 4–34

Zadeh, L. A. (1975): Fuzzy logic and approximate reasoning (In memory of Grogor Moisil). Synthese, 30, 407–428

Zadeh, L. A. (1976): A fuzzy-algorithmic approach to the definition of complex or imprecise concepts. International Journal of Man-machine Studies, 8, 249–291

Zadeh, L. A. (1978): Fuzzy sets as a basis for a theory of possibility. Fuzzy Sets and Systems, 1, 3–28

Zadeh, L. A. (1979): Fuzzy sets and information granularity. In M.M.Gupta, R.K.Ragade and R.R.Yager (eds.), Advances in Fuzzy Set Theory and Applications. New York: North-Holland 3–18

Zawirski, Z. (1931): Jan Łukasiewicz' 3-valued logic. On the logic of L.E.J. Brouwer. Attempts at applications of many-valued logic to contemporary natural science. Sprawozdania Poznańskiego Towarzystwa PrzyjaciółNauk, 2, 2–4 (In Polish)

Zawirski, Z. (1934): Significance of many-valued logic for cognition and its connection with the calculus of probability. Przegląd Filozoficzny, 37, 393–398 (In Polish)

Zawirski, Z. (1934): Le rapport de la logique a plusieurs valeurs au calcul des probabilités). Prace Komisji Filozoficznej Poznańskiego Towarzystwa PrzyjaciółNauk, 4, 155–240 (In Polish)

Zawirski, Z. (1935): Über das Verhältnis mehrwertiger Logic zur Wahrscheinlichkeitsrechnung. Studia Philosophia, 1, 407–442

Zawirski, Z. (1936): Les rapports de la logique polyvalente avec le calcul des probabilites. Actes du Congres International de Philosophie Scientifique, Paris, 4, 40–45

Zawirski, Z. (1946): Genesis and development of intuitionistic logic. Kwartalnik filozoficzny, 16, 165–222 (In Polish)

Zich, O. V. (1938): Sentential calculus with complex values. Ćeska Mysl, 34, 189–196 (In Czeh.)

Zinov'ev, A. A. (1959): The problem of truth-values in many-valued logic. Voprosy Filosofii, 3, 131–136 (In Russian)

Zinov'ev, A. A. (1960): Philosophical problems of many-valued logic. Moscow, Institut Filosofii, Izdatél'stvo Akademii Nauk SSSR, 1960 (in Russian) translation by G.Küng and D.D.Comey, see Zinov'ev, 1963

Zinov'ev, A. A. (1961): A method of describing the truth-functions of the n-valued propositional calculus. Studia Logica, 11, 217–222

Zinov'ev, A. A. (1962): Two-valued and many-valued logic. Filosofskie Voprosy Sovremennoj Formal'noj Logiki, Moscow, Institut Filosofii, Izdatel'stvo Akademii Nauk SSSR

Zinov'ev, A. A. (1963): Philosophical problems of many-valued logic. Ed. and translation by G. Küng and D.D.Comey, Dordrecht: Reidel

Zinov'ev, A. A. (1963–64): Two-valued and many-valued logic. Soviet Studies in Philosophy, 2, 69–84

Zwicky, F. (1933): On a new type of reasoning and some of its possible consequences. Physical Review, 43, 1031–1033.

Index of Symbols

$\alpha \in A$, 1
$a \notin A$, 1
$\sim (\alpha \in A)$, 1
$A \subseteq B$, 1
$A = B$, 1
$\{x : P(x)\}$, 1
$\{x \in B : P(x)\}$, 1
$\{x, y\}, \{x, y, z, \ldots, v\}$, 1
$\{x\}$, 1
$(x, y) = \{\{x\}, \{x, y\}\}$, 1
(x, y, z), 2
$A \cup B$, 2
$A \cap B$, 2
$A - B$, 2
$P(A)$, 2
$A \times B$, 2
$A \times B \times \ldots \times C$,2
A^n, 2
$D(r)$, 2
$D_{-1}(r)$, 2
r_{-1}, 2
$s \circ r$, 3
i_A, 3
A^2, 3
$f : A - \circ \rightarrow B$, 4
$f : A \rightarrow B$, 4
$f(X)$, 4
$f^{-1}(Y)$, 4
$(A_i)_{i \in I}$, 5
$\{A_i : i \in I\}$, 5
$\bigcup (A_i)_{i \in I}$, 5
$\bigcup_{i \in I} A_i$,
$\bigcap (A_i)_{i \in I}$, 5
$\bigcap_{i \in I} A_i$, 5
$\prod_{i \in I} A_i$, 5
(n), 5
$\bar{n}$, 5
$[x]_r$, 6
A/r, 6
$h_r(x)$, 6
r^+, 6
r^*, 6
$x \leq y$, 7
$x < y$, 8
$(x, y) \leq^* (z, w)$, 8
$f \subseteq^* g$, 8
A^*, 9
e_A, 9
$d(u)$, 9
X^+, 11
X_+, 11
$(U, \{s_i : i \in I\})$, 12
$(U, \{s_1, \ldots, s_k\})$, 12
$(U, s_1, \ldots, s_k)$, 12
$\approx$, 15
$U/ \approx$, 15
$Q(X)$, 17
$E(\mathfrak{M})$, 20
$V(\mathfrak{M})$, 20
$N(\mathfrak{M})$, 21
(V, S), 25
Pr, 26
$i(\Omega_0)$, 27
$i(\Omega_1)$, 27
$i(\Omega_2)$, 27
$| \cdot |$, 28
L_n, 28
$J_k(x)$, 29

$h_m(x)$, 29
$I_k(x)$, 29
$i_X(\Omega_j)$, 30
Ω_0^X, 31
s_1^n, 34
s_n^n, 34
s_i^n, 34
$\bigwedge_{j=i}^{n-1} b_j$, 34
P_n, 38, 47
∇, 49
Cn, 52
P_{nr}, 54
(Ł$_3$), 63
$\mathfrak{M}$Ł$_3$, 64
Tx, 64
$\mathfrak{M}B_3^{I_1}$, 65
$\mathfrak{M}B_3^{I_2}$, 65
$\mathfrak{M}B_3^E$, 65
$\mathfrak{M}C_2$, 66
$\mathfrak{M}F_3$, 66
$\mathfrak{M}H$, 68
$\mathfrak{M}A$, 69
$\mathfrak{M}S_1$, 70
$\mathfrak{M}S_2$, 71
$\mathfrak{M}S_3$, 71
$\mathfrak{M}PR$, 71
$\mathfrak{M}He$, 73
$\mathfrak{M}Kl_1$, 74
$\mathfrak{M}Kl_2$, 74
$\mathfrak{M}R$, 75
$\mathfrak{M}Sl$, 76
$\mathfrak{M}Sb$, 77
$\mathfrak{M}Slnr$, 79
$\mathfrak{M}Sbn$, 82
$\mathfrak{M}Gn$, 84
$\mathfrak{M}Cnr$, 85
$LCnr$, 86
$E(\mathfrak{M}Cnr)$, 89
Cn_I, 96
$X \vdash \alpha$, 98
$(G, r, \models)$, 102
F_i, 105
P_i, 105
$\forall$, 106
$\exists$, 106
$zw(\alpha)$, 107
$zw(t)$, 107
$ct(t)$, 107
$ct(\alpha)$, 107
$(t)x_k/t_m$, 108
$(\alpha)x_k/t_m$, 108
$(x_i)x_k/t_m$, 108
$(c_k)x_k/t_m$, 108
$(\forall x, \alpha)x_k/t_m$, 108
$(\exists x, \alpha)x_k/t_m$, 108
$w(t, b)$, 110
$w(\alpha, s)$, 110, 111
$\bigvee_{j=1}^{s} \alpha_j$, 36, 126
S_0, 123
S_{k+1}, 123

Author Index

Adams, E.W., 231, 241, 250, 253
Ackermann, W., 105,
Åqvist, L., 69, 70
Bernays, P., 105
Beth, E.W., 123, 98
Birkhoff, G., 24, 158
Bochvar, D.A., 65, 66, XII
Boole, G., 23
Brouwer, L.E.J., 73, 158
Carnielli, W., 123, 130, 131, 133, 134, 138, 140, 141
Cat Hao, Ng., 46
Chang, C.C., XII
Cignoli, R., 32, 40, 42, 45
Craig, W.
Destouches-Fevrier, P., 24
Dwinger, P., 46, 51
Epstein, G., 46
Evans, T., XII
Fagin, R., 231
Finn, W.K., 66
Fitting M., 103
Foxley, E., XII
Frege, G., 105
Gentzen, G.
Glivenko, W., 97
Gougen, J.A., 144, 146, 155
Gödel, K., 84, 98, XII
Hallden, S., 68, 69, 70
Halpern, J.Y., 231
Hardy, L., XII
Heyting, A., 32, 34, 36, 45, 46, 73, 158, 222, XII
Hilbert, D., 105, 171
Hintikka, J., 123, 129, 131, 132, 138, 139, 140, 141
Horn, A., 46
Jaśkowski, S.
Kalmar, L., 85, 89
Kirkerud, B., 24
Kleene, S.C., 24, 74, XII
Krauss, P., 231, 235, 236, 243
Kripke, S.A., 98, 102, 103
Lindenbaum, A., 21, 233
Łukasiewicz, J., 23, 27, 28, 30, 32, 34, 35, 36, 37, 38, 39, 40, 41, 42, 45, 46, 55, 56, 63, 64, 73, 79, 85, 126, 158, 159, 161, 170, 175, 190, 231, 232
Martin, N.M., XII
McNaughton, R., 41
Moisil, G.C.
Mostowski, A., 105
von Neumann, J.
Novak, V., 192, 193, 199, 202, 204, 207
Orłowska, E., 46
Pavelka, J., 143, 146, 148, 155, 161, 162, 165, 167, 170, 171, 175, 177, 192, 193, 194, 202, 204
Pawlak, Z., 209
Peirce Ch.S., 23, 105
Picard, S., 80
Piróg-Rzepecka, K., 24, 63, 70, 71
Post, E.L., 23, 32, 46, 47, 48, 50, 51, 52, 56, 63, 64, 67, 71, 79, 85, 113
Price, H.H.
Rasiowa, H., 24, 30, 32, 46, 48, 49, 50, 51, 52, 53, 101, 113, 122, 209, 213, 218, 219, 220, 230
Reichenbach, H., 24, 75, 231
Rescher, N., 63
Rose, A., XII
Rosenbloom, P.C., 46
Rosser, J.B., 40
Rousseau, G., 46

Russell, B., 105
Saloni Z., 46
Schroeder, E., 105
Scott, D., 231, 235, 236, 243
Segerberg, K., 70, 71
Sheffer, E.M., 61
Sikorski, R., 101, 241
Skolem, T., 140
Skowron, A., 209
Słupecki, J., 24, 64, 76, 79, 80, 81, XII
Sobociński, B., 24, 77, 82, 83, XII
Suchoń, W., 41, 45, 123
Surma, S.J., 85, 123, 141
Tarski, A., 145, 233, XII
Traczyk, T., 46, 50, 51
Turquette, A.R., 40
Vasil'ev, N.A., 23
Wajsberg, M., 64, XII
Webb, D.L., 24, 61
Whitehead, A.N., 105
Zinov'ev A., 29

Subject Index

C-formula, 220
D-filter, 49
I-rough set, 210
I_j-definable, 212
I_j-definable relations in U, 212
j-argument propositional connectives, 196
k-th degree of complexity, 124
m-argument operation, 12
m-ary operation, 12
n-argument function, 5
n-string, 8
n-th Cartesian power, 2
n-tuples, 2, 8
n-valued calculus of Łukasiewicz, 27, 28
n-valued Post algebra, 46
n-valued Post space, 50
n-valued Postian calculus, 51
n-valued Łukasiewicz algebra, 34
P-ordering, 248
psP-functions of type T, 221
QU-Hintikka set, 138
R-free, 15
α is satisfiable, 20
α is tautologous, 20
(generalized) Cartesian power, 5
(generalized) intersection (meet), 5
(generalized) union (join), 5
abstract algebra, 12
algebra, 12
analytic consistency property, 139
antisymmetric, 3
approximately definable, 211
approximately I_j-definable, 213
approximating translation, 220
approximation logic, 227
approximation space, 211
associative, 3
auxiliary symbols, 106

basic plain semi-Post algebra (psP-algebra) of type T, 221
bijection, 4
bijective, 4
binary connectives, 106
binary relation, 2
Boolean algebra,
boundary, 210
bound variables, 107

canonical map, 6
cardinality, 5
Cartesian product, 2, 17
chain, 7
characteristic function, 8
closed atomic U-formula, 136
complement of B relative to A, 2
complete psP-algebra, 223
composition of r and s, 3
concatenation, 9
conditional formulas, 244
congruence modulo f, 7
conjunction, 63
connected, 3
consequence operation, 26
coset, 6
coset, of x modulo r, 6
countable, 5
countably infinite, 5

definable relation 211
degenerated algebra, 12
degree k, 9, 124
designated formulas, 124
difference, 2
disjunction, 63
dissatisfied, 112
domain, 2

element, 1
elementary, 211
elementary n-ary relation, 211
elementary n-argument relations in U, 211
empty set, 2
empty sequence, 9
endomorphism, 13
epimorphism,
equivalence, 63
equivalence class, 6
equivalence relation, 6
existential, 106
extension of f, 8

falsity, 63
finite path, 10
first-order predicate calculus, 114
formulas, 25, 106
free generators, 15
free in class R, 15
free individual variables, 107
full relation, 3
function, 4
function symbols, 105

generalized abstract algebra, 17
generalized algebra, 17
generalized operation, 17
generators, 13
greatest lower bound, 7

Hintikka set, 129
homomorphism, 13

identity realtion, 3
image, 4
immediate reduction, 249
immediate successor, 10
implication, 63
independent unions, 237
indexed family of sets, 5
indiscernibility, 211
individual constants, 107, 195
individual variables, 105, 195
inductive closure, 11
infallible, 20
initial segments, 5
injection, 4
injective, 4
inner-$I - j$-undefinable, 213
interpretation function, 25
interpretation of a language, 109
interpretation universe, 109
intersection, 2
intuitionist model, 102
intuitionistic calculus, 95
indiscernibility, 212
inner-undefinable, 211
inverse, 2
inverse relation, 2
irreducible D-filter, 49
isomorphism, 13

join, 2

Kelley property, 241

label, 9
language of the Łukasiewicz calculus, 28
Law of the Excluded Middle, 23
least equivalence relation, 7

least upper bound, 7
level k, 124
lexicographic order, 8, 10
Lindenbaum-Tarski formula algebra, 233
Lindenbaum-Tarski quotient algebras, 233
linearly ordered, 7
lower approximation, 210
lower approximation of R in U, 211
lower bound, 7

many-valued calculi of Gödel, 84
many-valued calculus Cnr, 85
many-valued calculus of Sobociński, 82
many-valued calculus of Słupecki, 79
many-valued first order predicate calculus of Post, 113
map, 4
mapping, 4
matrix, 20
maximal, 7
maximal D-filter
measure algebra, 234
meet, 2
metaoperators for designation of formulas, 124
minimal, 7
minimal adequate matrix, 25
model, for the n-valued predicate calculus, 136
monomorphism, 13

natural epimorphism, 15
node, 9

order relation, 7
ordered pair, 1
ordered set, 7
ordering, 7
ordinal partition, 249
outer-I_j-undefinable, 213
outer-undefinable, 211

plain semi-Post algebra, 221
partial function, 4
partition of A, 6
point of the tree, 9
power set, 2
prefix, 9
preimage, 4
prime D-filter, 50
probabilistic, 234
probabilistic assertion, 240
probabilistic consequence, 246
probability model, 236, 240
probability system, 234
product, 2
product of more factors, 2
proof, 26
proper n-valued Łukasiewicz algebra, 37
proper Łukasiewicz algebra, 37
propositional calculus, 25
propositional constants, 105
propositional variables, 26, 30

quantifier Hintikka set over U, 138
quantifiers, 106
quasi-Boolean, 33
quotient (set) of A modulo r, 6
quotient algebra, 15

range, 2
reasonable consequence, 245
reducing sequence, 8
reflexive, 3, 6
relation, 2
relation symbols (predicates), 105
restriction of g, 8
root, 10
rough logic, 217
rough set modulo relation, 210
rule, 26
rule of substitution, 108

satisfied, 112
saturated, 129
sequence, 8
sequence of lenght n, 8
set, 1
set of designated values, 124
set of generators, 13
similar, 13
singleton, 1
Stone filter, 36
strict consequence, 245
strict identity system, 234
strict ordering, 8
subalgebra, 13, 18
submatrix, 20
subtree, 10
suffix, 9
surjection, 4
surjective, 4
symmetric, 3
symmetric Heyting algebra, 34
symmetric probability systems, 238

tautological consequence, 245
tautology
tautology of the n-valued first-order predicate calculus, 112
terms, 106
three-valued calculus of Åqvist, 69
three-valued calculus of Bochvar, 65
three-valued calculus of Finn, 66
three-valued calculus of Hallden, 68
three-valued calculus of Heyting, 73
three-valued calculus of Kleene, 74
three-valued calculus of Piróg-Rzepecka, 71
three-valued calculus of Reichenbach, 75
three-valued calculus of Segerberg, 70
three-valued calculus of Sobociński, 77
three-valued calculus of Słupecki, 76
three-valued calculus of Łukasiewicz-Słupecki, 64
three-valued Łukasiewicz calculus, 63
totally I_j-undefinable, 213
totally-undefinable, 211
transformation, 4
transitive, 3
transitive closure, 6
transitive reflexive closure of r, 6
tree, 9
tree domain, 9
triples, 2
truth, 63
truth-functional, 244
two-valued Post algebra, 47

ultraproducts, 238
unary connectives, 106
underlying space
uniform, 246
union, 2
univalent, 3
universal, 106
universe, 12
unordered pair, 1
unreasonable inference, 253
upper approximation, 210
upper approximation of R in U, 211
upper bound, 7
upper approximation, 210

valuation of propositional variables, 26, 30
value of α on, 110
value of a term, 110

well-formed formulas, 25
well-founded, 8
well-ordered, 8